Low-Dimensional Topology and Quantum Field Theory

NATO ASI Series

Advanced Science Institutes Series

A series presenting the results of activities sponsored by the NATO Science Committee, which aims at the dissemination of advanced scientific and technological knowledge, with a view to strengthening links between scientific communities.

The series is published by an international board of publishers in conjunction with the NATO Scientific Affairs Division

A	Life Sciences	Plenum Publishing Corporation
B	Physics	New York and London
C	Mathematical and Physical Sciences	Kluwer Academic Publishers
D	Behavioral and Social Sciences	Dordrecht, Boston, and London
E	Applied Sciences	
F	Computer and Systems Sciences	Springer-Verlag
G	Ecological Sciences	Berlin, Heidelberg, New York, London,
H	Cell Biology	Paris, Tokyo, Hong Kong, and Barcelona
I	Global Environmental Change	

Recent Volumes in this Series

Volume 310 —Integrable Quantum Field Theories
　　　　　　edited by L. Bonora, G. Mussardo, A. Schwimmer, L. Girardello, and
　　　　　　M. Martellini

Volume 311 —Quantitative Particle Physics: *Cargèse 1992*
　　　　　　edited by Maurice Lévy, Jean-Louis Basdevant, Maurice Jacob,
　　　　　　Jean Iliopoulos, Raymond Gastmans, and Jean-Marc Gérard

Volume 312 —Future Directions of Nonlinear Dynamics in Physical and Biological
　　　　　　Systems
　　　　　　edited by P. L. Christiansen, J. C. Eilbeck, and R. D. Parmentier

Volume 313 —Dissociative Recombination: Theory, Experiment, and Applications
　　　　　　edited by Bertrand R. Rowe, J. Brian A. Mitchell, and André Canosa

Volume 314 —Ultrashort Processes in Condensed Matter
　　　　　　edited by Walter E. Bron

Volume 315 —Low-Dimensional Topology and Quantum Field Theory
　　　　　　edited by Hugh Osborn

Volume 316 —Super-Intense Laser–Atom Physics
　　　　　　edited by Bernard Piraux, Anne L'Huillier, and Kazimierz Rzążewski

Series B: Physics

Low-Dimensional Topology and Quantum Field Theory

Edited by
Hugh Osborn
University of Cambridge
Cambridge, United Kingdom

Plenum Press
New York and London
Published in cooperation with NATO Scientific Affairs Division

Proceedings of a NATO Advanced Research Workshop on
Low-Dimensional Topology and Quantum Field Theory,
held September 6–12, 1992,
in Cambridge, United Kingdom

NATO-PCO-DATA BASE

The electronic index to the NATO ASI Series provides full bibliographical references (with keywords and/or abstracts) to more than 30,000 contributions from international scientists published in all sections of the NATO ASI Series. Access to the NATO-PCO-DATA BASE is possible in two ways:

—via online FILE 128 (NATO-PCO-DATA BASE) hosted by ESRIN, Via Galileo Galilei, I-00044 Frascati, Italy

—via CD-ROM "NATO Science and Technology Disk" with user-friendly retrieval software in English, French, and German (©WTV GmbH and DATAWARE Technologies, Inc. 1989). The CD-ROM also contains the AGARD Aerospace Database.

The CD-ROM can be ordered through any member of the Board of Publishers or through NATO-PCO, Overijse, Belgium.

Library of Congress Cataloging-in-Publication Data

Low-dimensional topology and quantum field theory / edited by Hugh
  Osborn.
      p.   cm. -- (NATO ASI series. Series B. Physics ; v. 315)
  Includes bibliographical references and index.
  ISBN 0-306-44578-6
  1. Quantum field theory--Congresses. 2. Low-dimensional topology-
-Congresses. 3. Mathematical physics--Congresses. I. Osborn,
Hugh. II. Series.
QC174.45.A1L68 1993
530.1'43--dc20   93-28686
                                                                 CIP

ISBN 0-306-44578-6

©1993 Plenum Press, New York
A Division of Plenum Publishing Corporation
233 Spring Street, New York, N.Y. 10013

All rights reserved

No part of this book may be reproduced, stored in a retrieval system, or transmitted in any form or by any means, electronic, mechanical, photocopying, microfilming, recording, or otherwise, without written permission from the Publisher

Printed in the United States of America

Preface

The motivations, goals and general culture of theoretical physics and mathematics are different. Most practitioners of either discipline have no necessity for most of the time to keep abreast of the latest developments in the other. However on occasion newly developed mathematical concepts become relevant in theoretical physics and the less rigorous theoretical physics framework may prove valuable in understanding and suggesting new theorems and approaches in pure mathematics. Such interdisciplinary successes invariably cause much rejoicing, as over a prodigal son returned. In recent years the framework provided by quantum field theory and functional integrals, developed over half a century in theoretical physics, have proved a fertile soil for developments in low dimensional topology and especially knot theory. Given this background it was particularly pleasing that NATO was able to generously support an Advanced Research Workshop to be held in Cambridge, England from 6th to 12th September 1992 with the title Low Dimensional Topology and Quantum Field Theory. Although independently organised this overlapped as far as some speakers were concerned with a longer term programme with the same title organised by Professor M Green, Professor E Corrigan and Dr R Lickorish. The contents of this proceedings of the workshop demonstrate the breadth of topics now of interest on the interface between theoretical physics and mathematics as well as the sophistication of the mathematical tools required in current theoretical physics.

As director of the workshop I would first like to thank all the participants for their enthusiasm throughout the meeting and readiness to attend lectures from early morning to late evening. The authors of articles to this volume also deserve much thanks for their rapid response in providing their contributions and for extending my knowledge of dialects of TEX. I would also like to express my thanks to Professor E Corrigan for much assistance and smoothing over of difficulties, despite conflicting pressures, before and during the workshop. I should also mention the unfailing help and cooperation I received from Mrs L Troake, the warden of Wolfson Court, Girton College where most of the participants stayed during the meeting. My thanks are also due to Dr P Landshoff for resolving various administrative hassles in concluding the final paperwork. Finally I would like to express my unreserved gratitude to Gérard Watts and David McAvity for helping with their TEXpertise in ensuring that the articles in this volume have at least approximately a uniform style. The skills of hacking should no longer be derided.

Hugh Osborn

April 1993

Contents

Combinatorial recoupling theory and 3-manifold invariants L.H. Kauffman	1
Quantum field theory and A,B,C,D IRF model invariants O.J. Backofen	19
On combinatorial three-manifold invariants G. Felder and O. Grandjean	31
Schwinger-Dyson equation in three dimensional simplicial quantum gravity H. Ooguri	51
Observables in the Kontsevich model P. Di Francesco	73
Matrix models in statistical mechanics and quantum field theory, recent examples and problems G.M. Cicuta	85
Dilogarithms and W-algebras W. Nahm	95
Dilogarithm identities and spectra in conformal field theory A.N. Kirillov	99
Physical states in topological coset models J. Sonnenschein	109
Finite W symmetry in finite dimensional integrable systems T. Tjin	123
On the "Drinfeld-Sokolov" reduction of the Khizhnik-Zamolodchikov equation P. Furlan, A.Ch. Ganchev, R. Paunov and V.B. Petkova	131
Noncritical dimensions for critical string theory: life beyond the Calabi-Yau frontier R. Schimmrigk	143
W_∞ algebra in two-dimensional black holes T. Eguchi, H. Kanno and S.K. Yang	159

Graded Lie derivatives and short distance expansions in two dimensions 169
    R. Dick

2D Black holes and 2D gravity 177
    F. Ardalan

The structure of finite dimensional affine Hecke algebra quotients
    and their realization in 2D lattice models 183
    D. Levy

An exact renormalisation in a vertex model 193
    B.W. Westbury

New representations of the Temperley-Lieb algebra with applications 203
    H.N.V. Temperley

Order-disorder quantum symmetry in G-spin models 213
    K. Szlachányi

Quantum groups, quantum spacetime and Dirac equation 221
    A. Schirrmacher

Hamiltonian structure of equations appearing in random matrices 231
    J. Harnad, C.A. Tracy and H. Widom

On the existence of pointlike localized fields in
    conformally invariant quantum physics 247
    M. Joerss

The phase space of the Wess-Zumino-Witten model 261
    G. Papadopoulos and B. Spence

Regularization and renormalization of Chern-Simons theory 269
    G. Giavarini, C.P. Martin and F. Ruiz Ruiz

Ray-Singer torsion, topological field theories and the
    Riemann zeta function at $s=3$ 279
    C. Nash and D.J. O'Connor

Monstrous moonshine and the uniqueness of the moonshine module 289
    M.P. Tuite

Lie algebras and polynomial solutions of differential equations 297
    A. Turbiner

Torus actions, moment maps, and the symplectic geometry of
    the moduli space of flat connections on a two-manifold 307
    L.C. Jeffrey and J. Weitsman

Geometric quantization and Witten's semiclassical manifold invariants 317
    L.C. Jeffrey and J. Weitsman

Index 323

COMBINATORIAL RECOUPLING THEORY AND 3-MANIFOLD INVARIANTS

Louis H. Kauffman

Department of Mathematics, Statistics and Computer Science
University of Illinois at Chicago
Chicago, Illinois 60680

ABSTRACT

This paper discusses combinatorial recoupling theory, first in relation to the vector cross product algebra and a reformulation of the Four Colour Theorem, and secondly in relation to the Temperley-Lieb algebra, the Jones polynomial and the $SU(2)$ 3-Manifold invariants of Witten, Reshetikhin and Turaev.

1. INTRODUCTION

This paper discusses combinatorial recoupling theory, first in relation to the vector cross product algebra and a reformulation of the Four Colour Theorem, and secondly in relation to the Temperley-Lieb algebra, the Jones polynomial and the $SU(2)$ 3-Manifold invariants of Witten, Reshetikhin and Turaev.

Section 2 discusses a simple recoupling theory related to the vector cross product algebra that has implications for the colouring problem for plane maps. Section 3 discusses the combinatorial structure of the Temperley-Lieb algebra. We investigate an algebra of capforms and boundaries (the boundary logic) that underlies the structure of the Temperley-Lieb algebra. This capform algebra gives insight into the nature of the Jones-Wenzl projectors that are the basic construction for the recoupling theory for Temperley-Lieb algebra discussed in Section 4. Section 5 discusses the definition of the Witten-Reshetikhin-Turaev invariant of 3-manifolds. Section 6 explains how to translate the definition in section 5 into a partition function on a 2-cell complex by using a reformulation of the Kirillov-Reshetikhin shadow world appropriate to the recoupling theory of Section 4. The work described in 4 and 6 is joint work of the author and S. Lins and will appear in [KL92].

The foundations of the recoupling theory presented here go back to the work of Roger Penrose on spin networks in the 1960's and 1970's. On the mathematical physics side this has led to a number of interrelations with the 3-manifold invariants discussed here and theories of quantum gravity in two dimensions of space and one dimension of time. There has not been room in this paper to go into these relationships. For the record, the reference list includes papers by Penrose and also more recent authors on this topic (Hasslacher and Perry, Crane, Williams and Archer, Ooguri). It is the author's belief that the approach to the recoupling theory discussed herein will illuminate questions about these matters of mathematical physics.

2. TREES AND FOUR COLOURS

It is amusing and instructive to begin with the subject of trees and the Four Colour Theorem. This provides a miniature area that illustrates many subtle issues in relation to recoupling theory.

Recall the main theorem of [KAUF90]. There the Four Colour Theorem is reformulated as a problem about the non-associativity of the vector cross product algebra in three dimensional space. Specifically, let $V = \{i, -i, j, -j, k, -k, 1, -1, 0\}$ closed under the vector cross product (Here the cross product of a and b is denoted in the usual fashion by $a \times b$.):

$$i \times j = k, \quad j \times i = -k,$$
$$j \times k = i, \quad k \times j = -i,$$
$$k \times i = j, \quad i \times k = -j,$$
$$i \times i = j \times j = k \times k = 0,$$
$$a \times 0 = 0 \times a = 0 \text{ for any } a$$
$$-0 = 0$$
$$a \times (-1) = -1 \times a = -a \text{ for any } a$$
$$-(-a) = a \text{ for any } a.$$

V is a multiplicatively closed subset of the usual vector cross product algebra. V is not associative, since (e.g.)

$$i \times (i \times j) = i \times k = -j \text{ while } (i \times i) \times j = 0 \times j = 0.$$

Given variables $a_1, a_2, a_3, \ldots, a_n$, let L and R denote two parenthesizations of the product $a_1 \times a_2 \ldots \times a_n$. Then we ask for solutions to the equation $L = R$ with values for the A_i taken from the set $V^* = \{i, j, k\}$ and such that the resultant values of L and R are non-zero. In [KAUF90] such a solution is called a *sharp solution* to the equation $L = R$, and it is proved that *the existence of sharp solutions for all n and all L and R is equivalent to the Four Colour Theorem.*

The graph theory behind this algebraic reformulation of the Four Colour Theorem comes from the fact that an associated product of a collection of variables corresponds to a rooted planar tree. This correspondence is illustrated below. Each binary product corresponds to a binary branching of the corresponding tree.

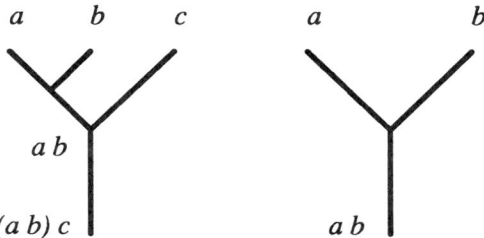

We tie the tree for $L, T(L)$, to the mirror image tree, $T(R)^*$, of the tree for R to form a planar graph $T(L)\#T(R)^*$. It is the region colouring of this graph that corresponds to a sharp solution to the equation $L = R$. (A region colouring of the graph corresponds to an edge colouring with three colours (i, j, k) so that each vertex is incident to three distinct colours, see [KAUF90] and [KAU92].)

There is in this formulation of the colouring problem an analogue to the sort of recoupling theory for networks that is common in theories of angular momentum. In particular, we are interested in the difference between the following two branching situations.

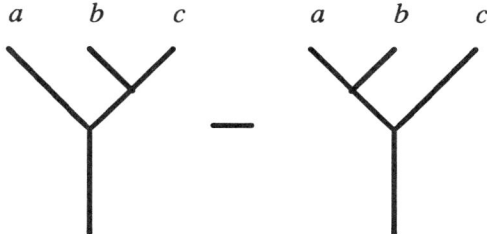

In terms of multiplication, this asks the question about the difference generated by one local shift of associated variables:

$$a \times (b \times c) - (a \times b) \times c = ?$$

Since our algebra is the vector cross product algebra, we are well acquainted with the answer to this question. This answer is

$$a \times (b \times c) - (a \times b) \times c = a(b.c) - (a.b)c$$

where a.b denotes the dot product of vectors in three space. (Thus $i.i = j.j = k.k = 1$, $i.j = i.k. = j.k = 0$ and $a.b = b.a$).

(This formula is easily proved by using the fact that quaternion multiplication is associative, and that the formula for the product of two "pure" (i.e. of the form $ai + bj + ck$ for a, b, c real) quaternions u and v is given by the equation $uv = -u.v + u \times v$.)

We can diagram this basic recoupling formula as shown below.

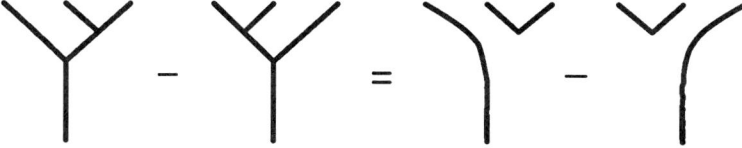

where it is understood that a concurrence of two lines represents a dot product of the corresponding labels:

We can use this formulation to investigate the behaviour of tree evaluations under changes of parenthesization. It also provides a way to investigate purely algebraically the existence of sharp solutions, and hence the existence of map colourations.

Example Find sharp solutions to the equation $a \times (b \times c) = (a \times b) \times c$.

Since $a \times (b \times c) - (a \times b) \times c = a(b.c) - (a.b)c$, we see that the difference can be zero if $b.c = 0$ and $a.b = 0$. Thus we can try $a = i$, $b = j$ and $c = i$ or k. In this case both of these work.

Example: Find sharp solutions to the equation

$$a \times (b \times (c \times d)) = ((a \times b) \times c) \times d.$$

Three applications of these recoupling formulas yield the equation

$$a \times (b \times (c \times d)) - ((a \times b) \times c) \times d =$$
$$- (a.b) c \times d + (c.d) a \times b - (a.(b \times c)) d + ((b \times c).d) a.$$

From this we are led to try $a = j$, $b = i$, $c = i$, $d = k$, and this indeed works.

Remark. We hope to find a way to use the recoupling algebra associated with vector cross product to illuminate the original colouring problem. These are basically simple structures, but there arise gaps in language from one context to another. Thus the same problem as in the last example is very quickly and confidently solved by colouring the map shown below. This map represents the tied trees $T(L)\#T(R)^*$, and we colour the regions of the map with colours W, I, J, K. The region colouring gives rise to an edge colouring by colouring an edge by the product of the colours for its adjacent regions with $IJ = k$, $JK = i$, $KI = j$, $WA = AW = a$ for all A, $AA = w$ for all A. (We use lower case letters to designate the colours of the edges.) Note that if the map is coloured so that adjacent regions receive different colours, then the edges receive only the colours i, j, k.

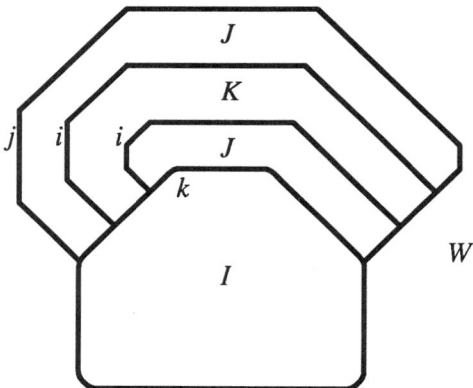

The connection between the colouring approach and the recoupling theory of the vector cross product depends crucially upon our addition of signs to the products of colours (via the vector cross product algebra). That sharp solutions do derive from colouring *including the sign* also involves the quaternions. For an equation $L = R$ to have a sharp solution up to sign (as it does from bare map colouring) implies the agreement of the signs. This is because both sides of the equation (being non-zero) can be viewed as products in the quaternions. Since the quaternions are associative, this implies that the two sides are equal, hence the signs are the same [KAU92]. Both sides must be non-zero in order for this argument to work.

The Kryuchkov Conjecture. It is interesting to examine the conjecture of Kryuchkov [Kry92] in the light of these diagrammatics. He is concerned with the possibility that one could find a colouring of L and R (i.e. a choice of values i, j, k for the variables) forming an *amicable solution*. By an amicable solution I mean a choice of values so that it is possible to go from L to R by a series of elementary recombinations (such as $(xy)z \to x(yz)$) maintaining sharp solutions throughout the procedure. Kryuchkov conjectures that there is an amicable solution for every choice of L and R.

The notation of formations (see [KAUF90]) allows a diagrammatic view of amicability. In this form we represent one colour (red) by a solid line, one colour (blue) by a dotted line, and the third colour (purple) by a superposition of a dotted line and a solid line. The elementary forms of amicability are then as shown below.

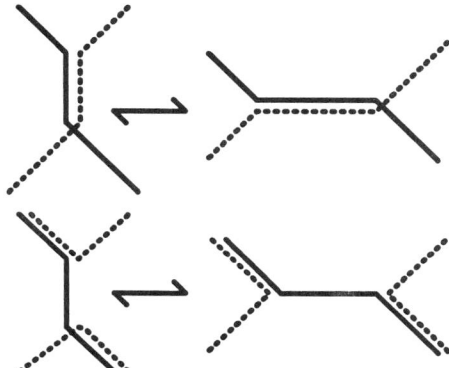

Clearly, more work remains to be done in this version of the Four Colour Theorem and its generalizations. We have begun this paper with a short review of this arena both for its intrinsic interest, and for the sake of the possible analogies with other aspects of mathematics and mathematical physics.

3. THE TEMPERLEY-LIEB ALGEBRA

We now turn to the combinatorial underpinnings of the Temperley-Lieb algebra.

First recall the tangle-theoretic interpretation of the Temperley-Lieb Algebra [KA87]. In this interpretation, the additive generators of the algebra are *flat tangles* with equal numbers of inputs and outputs. We denote by T_n the Temperley-Lieb algebra generated by tangles with n inputs and n outputs. A flat n-tangle is an *embedding* of disjoint curves and line segments into the plane so that the free ends of the segments are in one-to-one correspondence with the input and output lines of a rectangle in the plane that is denoted the *tangle box*. Except for these inputs and outputs, the disjoint curves and line segments are embedded to the interior of the rectangle.

Two such tangles are *equivalent* if there is a regular isotopy carrying one to the other occurring within the rectangle and keeping the endpoints fixed. Regular isotopy is generated by the Reidemeister Moves of type II and type III for link diagrams. (See [KA87]). The reason we adhere to regular isotopy at this point is that it is necessary to be able to freely move closed curves in such a tangle. Thus the two tangles illustrated below are equivalent via a regular isotopy in the tangle box that has intermediate stages that are out of the category of flat tangles.

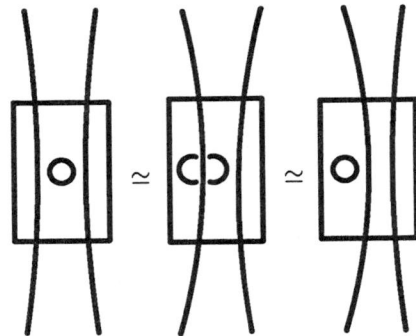

The Temperley-Lieb algebra T_n is, for our purposes freely additively generated by the flat n-tangles, over the ring $C[A, A^{-1}]$ where C denotes the complex numbers. A closed loop in a tangle is identified with an element d in this algebra to be specified later. The familiar multiplicative generators of the Temperley-Lieb algebra then appear as the following special flat tangles $U_1, U_2, \ldots, U_{n-1}$ in T_n.

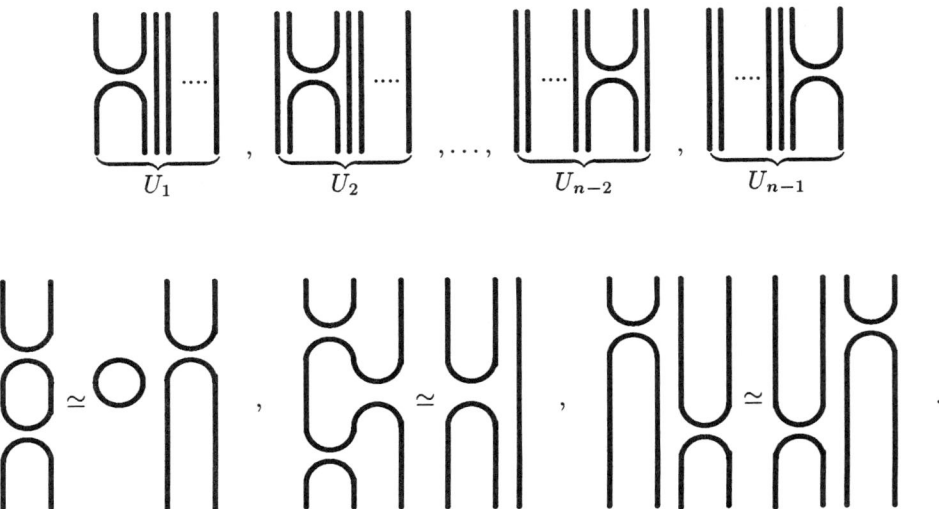

These generators enjoy the relations

$$U_i U_{i+1} U_i = U_i,$$
$$U_i U_{i-1} U_i = U_i,$$
$$U_i U_i = d U_i,$$
$$U_i U_j = U_j U_i \text{ for } |i - j| > 1,$$

and these relations generate equivalence of flat tangles [KA90].

The purpose of this section is to point out a combinatorial algebra on parenthesis structures - *the boundary logic* - that forms a foundation for the Temperley-Lieb algebra. First note that we can convert flat n-tangles to forms of parenthesization by bending the upper ends downwards and to the left as illustrated below.

Call such a structure a *capform*.

In this way we convert the tangles to capforms with $2n$ strands restricted to the bottom of the form. These are capforms with n caps. The tangle multiplication then takes the form of tieing the rightmost n strands of the left capform to the leftmost n strands of the right capform.

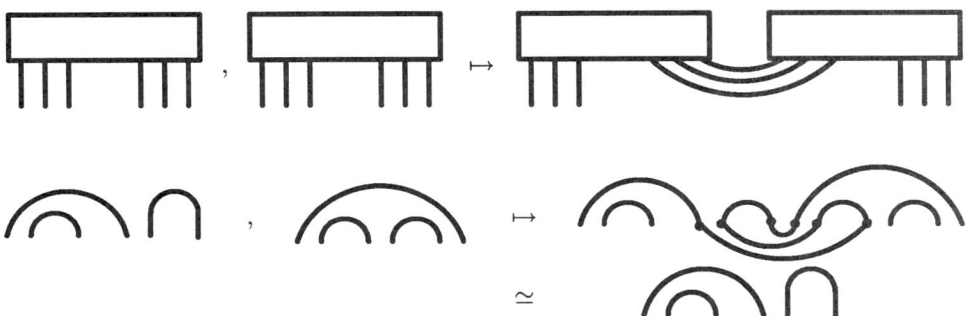

It is this operation that we shall convert into a series of more elementary operations: Regard the capform multiplication as done one pair of strands at a time. Then for a single pair of strands the pattern is to join them as shown below:

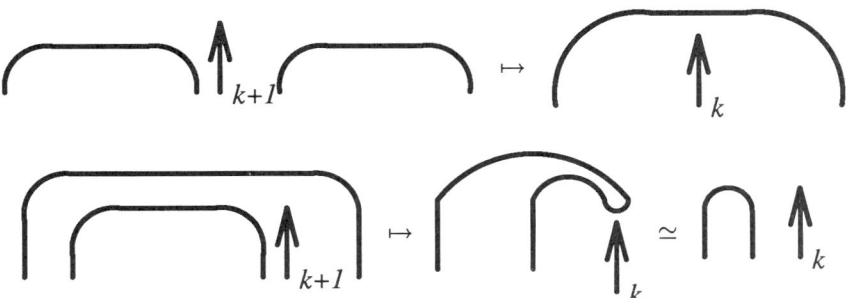

We denote this joining operation by a vertical arrow between adjacent strands. The resultant of the operation inherits an arrow in the place where the next joining can occur. An arrow with a subscript $k+1$ has an arrow with subscript k as its immediate descendant. An arrow with the subscript 0 (zero) is equal to the empty arrow. In this language the multiplication in T_n becomes, for the corresponding capforms: $X, Y \longrightarrow X \uparrow_n Y$.

It is sometimes convenient to omit the subscript on the arrow, in writing identities and also in specific calculations where the count of operations is being performed separately. We shall accordingly omit the subscripts in the text that follows.

The boundary logic of the Temperley-Lieb algebra is based on the joining and breaking of adjacent boundaries in capforms. Note the following basic equations in this boundary logic:

$$\lceil A \rceil \uparrow \lceil B \rceil = \lceil A \uparrow B \rceil$$

$$\lceil A \lceil B \rceil \uparrow C \rceil = \lceil A \lceil B \rceil \uparrow C$$

$$C \lceil \uparrow \lceil B \rceil A \rceil = C \uparrow \lceil B \rceil A \rceil$$

If we wish, we can re-express this formal structure in terms of ordinary brackets rather than the arches that have been given us by the capforms. In this form the basic equations look like

$$\langle A \rangle | \langle B \rangle = \langle A | B \rangle$$

$$\langle A \langle B \rangle | \rangle C = \langle A \rangle B | C$$

$$C \langle | \langle A \rangle B \rangle = C | A \langle B \rangle$$

Here the vertical arrow indicating the join operation has been replaced by a bold face vertical line segment.

Boundary logic encapsulates the Temperley-Lieb Algebra in a specific symbolic formalism that is suitable for machine computation. This formalism "knows" about the topology of Jordan curves in the plane! For example, take a Jordan curve, slice it by a line segment and regard the two halves as capforms. Then the successive joining of these two halves will compute a single component.

Compare this symbolic computation with the complexity of the drawing that results from using joining arcs in the usual topological mode.

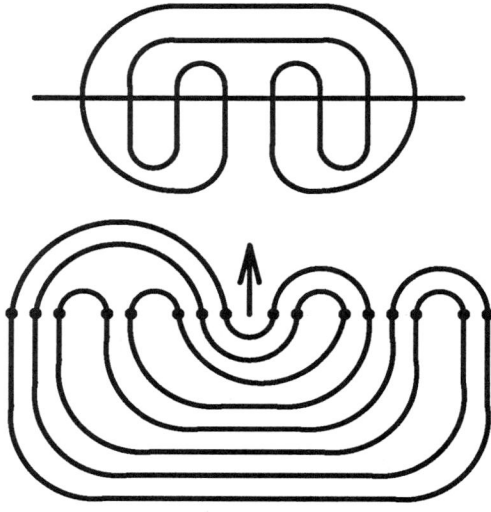

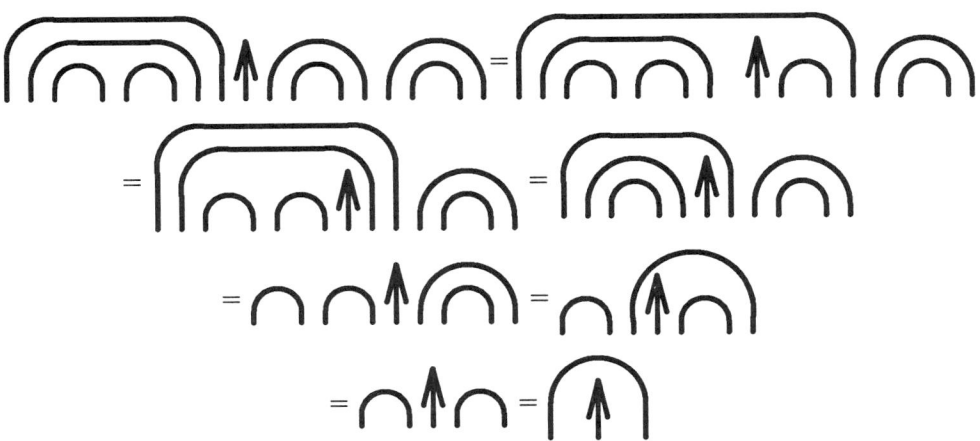

Finally, here is the same verification of connectedness performed in the boundary logic.

$$<<><>>>|<<>><<>>=$$
$$<<><>>|<>><<>>=$$
$$<<<><>|>><<>>=$$
$$<<<>>|><<>>=$$
$$<><>|<<>>=$$
$$<>|<>=$$
$$<|>$$

We have left the vertical bar in the last entry, denoting a single loop. In this form it is interesting to note that with the extra rule

$$||=|$$

the formalism contains an image of Dirac brackets:

$$P = |\rangle\langle|$$

$$PP = |\rangle\langle||\rangle\langle| \;=\; |\rangle\langle|\rangle\langle| \;=\; \langle|\rangle|\rangle\langle| = d\,P.$$

Aside from the advantages of formalization, certain structural features of the Temperley-Lieb algebra are easy to see from the point of view of the boundary logic. For example, consider the following natural map $T_n \times T_n \longrightarrow T_{n+1}$ given by the formula $A, B \to A * B = A|_{n-1}B$.

Theorem. Every element in T_{n+1} other than the identity element is of the form $A * B = A|_{n-1}B$ for some elements A and B in T_n. This decomposition is not unique.

Proof. If C in T_{n+1} is not the identity element then it is possible to draw an curve from the midpoint of the base of C to the outer region crossing less than n arcs of C. This is easily modified in a non-unique way to cross exactly $n-1$ arcs of C (possibly crossing some arcs twice). The curve so drawn then divides into the desired product.

Example

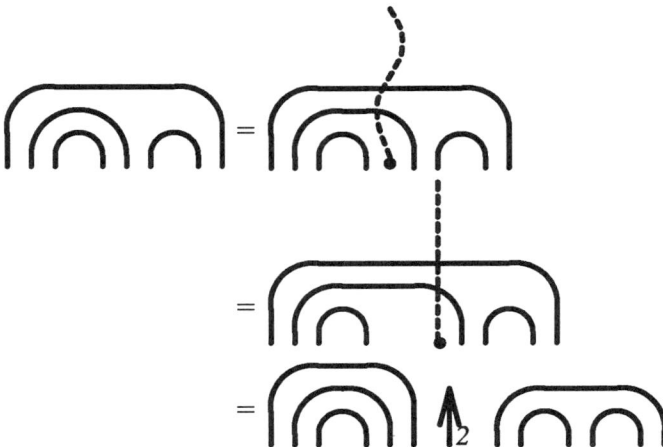

Another operation for going from T_n to T_{n+1} is $x \longrightarrow x'$ where x' is obtained by adding an innermost cap as in

The operation $x \longrightarrow x'$ takes the identity to the identity, and so together with $x \uparrow_{n-1} y$ encompasses all of T_{n+1} from T_n. This suggests combining these operations to produce inductive constructions in the Temperley-Lieb algebra. An example that fits this idea is the well known ([JO83], [KA91], [LI91]) inductive construction of the Jones-Wenzl projectors. In tangle language these projectors are constructed by the recursion

$$= \quad - (\Delta_n/\Delta_{n+1})$$

where Δ_n is a Chebyschev polynomial. These projectors are nontrivial idempotents in the Temperley-Lieb algebra, and they give zero when multiplied by the generators U_i for $i = 1, \ldots, n+1$.

Now note the capform interpretation of the terms of this summation.

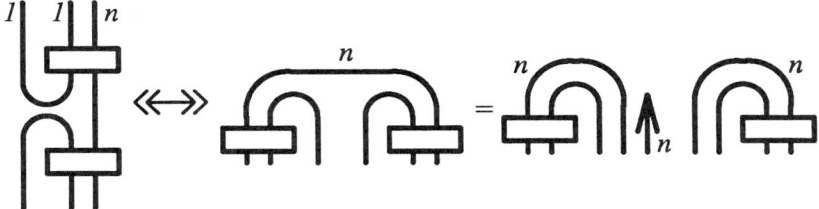

In the capform algebra the projectors are constructed via the recursion

$$g_{n+1} = g'_n - (\Delta_n/\Delta_{n+1}) g_n * g_n.$$

We shall return to these projectors in the next section.

A second use of the formalism is another reformulation of the Four Colour Theorem. There is ([KS92], [KAU92]) a completely algebraic form of the Four Colour Theorem via the Temperley-Lieb algebra. The boundary logic approach to the Temperley-Lieb algebra provides a new way to look at the combinatorics of this version of the colouring problem. This relationship will be discussed elsewhere.

4. TEMPERLEY-LIEB RECOUPLING THEORY

By using the Jones-Wenzl projectors, one builds a recoupling theory for the Temperley-Lieb algebra that is essentially a version of the recoupling theory for the $SL(2)_q$ quantum group. (See [KR88],[KAU92],[KL90],[KL92]). From the vantage of this theory it is easy to construct the Witten-Reshetikhin-Turaev invariants of 3-manifolds.

We begin by recalling the basics of the recoupling theory. The 3-vertex in this theory is built from three interconnected projectors in the pattern indicated below.

The internal lines must add up correctly and this forces the sum of the external lines to be even and it also forces the sum of any two external lines to be even and it also forces the sum of any two external line numbers to be greater than or equal to the third.

With these 3-vertices, we have a recoupling formula

$$\ = \sum_i \left\{ \begin{matrix} a & b & i \\ c & d & j \end{matrix} \right\} \ $$

Here the symbol

$$\begin{Bmatrix} a & b & i \\ c & d & j \end{Bmatrix}$$

is a generalized $6j$ symbol.

A specific formula for the evaluation of this $6j$ symbol arises as the consequence of the following identity (see [KA91], [KAU92],[KL92]):

$$b \bigcirc c = \frac{\Theta(a,b,c)}{\Delta_a} \quad \Big| \quad \delta_{aa'}$$

$$\left(\Delta_a = \ominus \, , \, \Theta(a,b,c) = \bigoplus \right)$$

From this identity it is easy to deduce that the $6j$ symbol is given by the network evaluation shown below:

$$\begin{Bmatrix} a & b & i \\ c & d & j \end{Bmatrix} = \frac{\text{Tet}}{\Theta(a,d,i)\Theta(b,c,i)} \Delta_i$$

The key ingredients are the tetrahedral and theta nets. They, in turn, can be evaluated quite specifically. (See [KL92],[LI91],[MV92]) There are a number of methods for obtaining these specific evaluations. For the general case one can induct using the recursion formula for the Jones-Wenzl projectors. In the special case where $d = -2$ there is a method to obtain the results via counting loops and colourings of loops in the networks. See[PEN79], [MOU79], [KA91],[KL92]. It should be mentioned that the case $d = -2$ corresponds to the classical theory of $SU(2)$ recoupling.

5. KNOTS AND 3-MANIFOLDS

This recoupling theory extends to trivalent graphs that are knotted and linked in three dimensional space. One way to delineate the connection is via the bracket model for the Jones polynomial [KA87]. In this model one obtains an invariant of regular isotopy of knots and links with the properties shown below.

$$\langle \times \rangle = A \langle \asymp \rangle + A^{-1} \langle)(\rangle$$

$$\langle \bigcirc \rangle = -A^{-2} - A^{-2}$$

$$\left\langle \begin{array}{c} \supset\subset \end{array} \right\rangle = \left\langle \begin{array}{c} \supset \end{array} \right\rangle \left\langle \begin{array}{c} \bigcirc \end{array} \right\rangle$$

$$\left\langle \begin{array}{c} \times \end{array} \right\rangle = \left\langle \begin{array}{c} \times \end{array} \right\rangle$$

$$\left\langle \begin{array}{c} \gamma \end{array} \right\rangle = -A^3 \left\langle \begin{array}{c} \underline{} \end{array} \right\rangle$$

$$\left\langle \begin{array}{c} \gamma \end{array} \right\rangle = -A^{-3} \left\langle \begin{array}{c} \underline{} \end{array} \right\rangle$$

Braids expand under the bracket into sums of elements in the Temperley-Lieb algebra. In this context the Jones-Wenzl projectors can be realised via sums of braids corresponding to all permutations on n-strands. (See [KA91], [KL92].) In this context the coefficient Δ_n is given by the formula $\Delta_{n-1} = (-1)^{n-1}(A^{2n} - A^{-2n})/(A^2 - A^{-2})$.

When $A = \exp(i\pi/2r)$ is a $4r$-th root of unity, then the recoupling theory goes over to this context with an extra admissibility criterion for each 3-vertex with legs a, b, c. We require that $a + b + c \leq 2r - 4$. With indices ranging over the set $\{0, 1, 2 \ldots, r - 2\}$, everything we have said about recoupling goes over. In particular, the theta net evaluations

$$\Theta(a, b, c) = \begin{array}{c} a \\ b \\ c \end{array}$$

are non-zero in this range so that the network formulas for the recoupling coefficients still hold.

It is at the roots of unity that one can define invariants of 3-manifolds. There are many ways to make this definition. A particularly neat version using the Jones-Wenzl projectors is given by Lickorish [LIC92]. We reproduce this definition here:

Let a link component, K, labelled with ω (as in $\omega * k$) denote the sum of i-cablings of this component over i belonging to the set $\{0, 1, 2 \ldots, \}$. A projector is applied to each cabling. Thus

$$\omega \bigg) = \sum \Delta_i \bigg|^i$$

and

$$\omega \bigotimes_w = \sum_{i,j} \Delta_i \Delta_j \bigotimes^{i}_{j}$$

The $\langle \omega * K \rangle$ denotes the sum of evaluations of the corresponding bracket polynomials. If K has more than one component then $\omega * K$ denotes the result of labelling each component, and taking the corresponding formal sum of products for the different cablings of individual components.

It is then quite an easy matter to prove that $\langle \omega * K \rangle$ is invariant under handle sliding in the sense shown below.

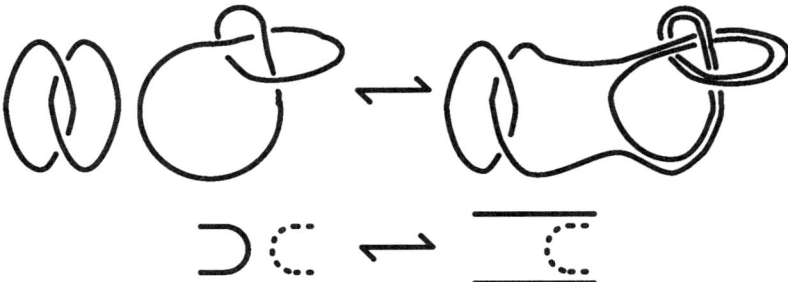

This is the basic ingredient in producing an invariant of 3-manifolds as represented by links in the blackboard framing. (A normalisation is needed for handling the fact the 3-sphere is returned after surgery on an unknot with framing plus or minus one.) A quick proof of handle sliding invariance using this definition and recoupling theory has been discovered by Justin Roberts [LIC92].

An even quicker proof using less machinery is given by Lickorish in [LIC92]. Roberts proof is based on the formula (a special case of the general recoupling formula).

$$\left)\right(\begin{smallmatrix}a&b\end{smallmatrix} = \sum_i \frac{\Delta_i}{\Theta(a,b,i)} \quad \begin{smallmatrix}a&b\\&i\\a&b\end{smallmatrix}$$

and the sequence of "events" shown below.

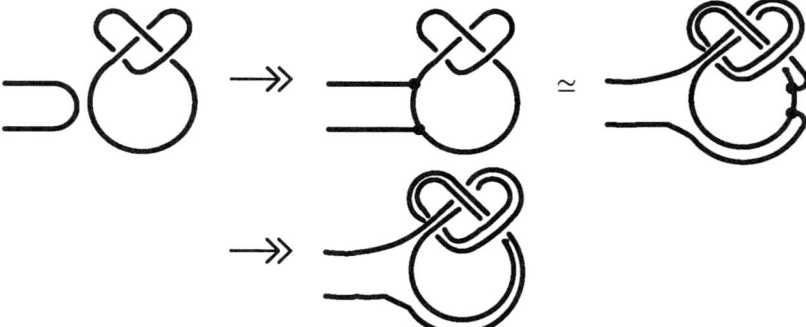

Details are omitted in the interest of brevity.

6. THE SHADOW WORLD

In this section we give a quick sketch of a reformulation of Kirillov-Reshetikhin Shadow World [KR88] from the point of view of the Temperley-Lieb Algebra recoupling theory. The payoff is an elegant expression for the Witten-Reshetikhin-Turaev Invariant as a partition function on a two-cell complex.

Shadow world formalism rewrites formulas in recoupling theory in terms of colourings of a two-cell complex. In the case of diagrams drawn in the plane this means that we allow ourselves to colour the regions of the plane as well as the lines of the diagram with indices from the set $\{0, 1, 2 \ldots r-2\}$ (working at $A = \exp(i\pi/2r)$, as in the last section). In this way a recoupling formula can be rewritten in terms of weights assigned to parts of the two-cell complex. The diagram below illustrates the process of translating between the daylight world (above the wavy line) and the shadow world (below the wavy line).

In this diagram we have indicated how if the tetrahedral evaluation is assigned to the six colours around a shadow world vertex (either 4-valent or 3-valent), the theta symbol is assigned to the three colours corresponding to an edge (the edge itself is coloured as are the regions to either side of the edge) then we can take the shadow world picture as holding the information about $6j$ symbols that are in the recoupling formula.

To complete this picture we assign Δ_i to a face labelled i, and there are phase factors corresponding to the crossings. We omit discussion of these phase factors here. The shadow world diagram is then interpreted as a sum of products of these weights over all colourings of its regions, edges and faces.

See [KL92] for a complete treatment of this topic.

The result is an expression for the handle sliding invariant $\langle \omega * K \rangle$ (of section 5) as a partition function on a two-cell complex. The extra Δ_i's coming from the assignments of ω to the components of the link can be indexed by attaching a 2-cell to each component. The result is a 2-cell complex that is a "shadow" of the 4-dimensional handlebody whose boundary is the 3-manifold constructed by surgery on the link. This marks the boundary of our constructions with the newer shadow world techniques of Turaev [TUR92], and is a good place for us to conclude this discussion.

Work for this paper was partly supported by NSF Grant Number DMS 9205277 and the Program for Mathematics and Molecular Biology, University of California at Berkeley, California. It gives the author great pleasure to thank the Newton Institute for its kind hospitality during the preparation of this work.

REFERENCES

[BL79] L.C. Biedenharn and J.D. Louck, Angular momentum in quantum physics - theory and application, in "Encyclopaedia of Mathematics and its Applications", Cambridge University Press (1979).

[CR91] L. Crane, Conformal Field Theory, Spin Geometry and Quantum Gravity, *Phys. Lett. B* **259** (1991) 243-248.

[FR79] R.A. Fenn and C.P. Rourke, On Kirby's Calculus of Links, *Topology*, **18** (1979) 1-15.

[HP81] B. Hasslacher and M.J. Perry, Spin Networks are Simplicial Quantum Gravity, *Phys. Lett. B* **103** (1981) 21-24.

[JO83] V.F.R. Jones, Index for subfactors, Invent. Math. **72** (1983) 1-25.

[JO85] V.F.R. Jones, A polynomial invariant for links via von Neumann algebras, *Bull. Amer. Math. Soc.* **129** (1985) 103-112.

[JO86] V.F.R. Jones, A New Knot Polynomial and von Neumann algebras, *Notices of AMS* **33** (1986) 219-225.

[JO87] V.F.R. Jones, Hecke algebra representations of braid groups and link polynomials, *Ann. of Math.* **126** (1987) 335-38.

[KA87] L.H. Kauffman, State Models and the Jones Polynomial, *Topology* **26** (1987) 395-407.

[KAU87] L.H. Kauffman, On Knots, *Annals of Mathematics Studies* Number 115, Princeton University Press (1987).

[KA88] L.H. Kauffman, New invariants in the theory of knots, *Amer. Math. Monthly* **95** (1988) 195-242.

[KA90] L.H. Kauffman, An invariant of regular isotopy, *Trans. Amer. Math. Soc.* **318** (1990) 417-471.

[KAU90] L.H. Kauffman, Spin Networks and Knot Polynomials, *Int. J. Mod. Phys. A* **5** (1990) 93-115.

[KAUF90] L.H. Kauffman, Map Colouring and the Vector Cross Product, *J. Comb. Theo. Ser. B.* **48** (1990) 145-154.

[KA91] L.H. Kauffman, Knots and Physics, *World Scientific* (1991).

[KA92] L.H. Kauffman, Knots, Spin Networks and 3-Manifold Invariants, *Knots 90* (edited by A. Kawauchi) (1992) pp. 271-287.

[KAU92] L.H. Kauffman, Map Colouring, q-deformed Spin Networks, and Turaev-Viro Invariants for 3-manifolds, In *The Proceedings of the Conference on Quantum Groups* Como, Italy, June 1991, Ed. M. Rasetti - World Sci. Pub., Int. J. Mod. Phys. B **6** (1992) 1765-1794.

[KL90] L.H. Kauffman and S. Lins, A 3-manifold invariant by state summation, (announcement 1990).

[KL91] L.H. Kauffman and S. Lins, Computing Turaev-Viro Invariants for 3-Manifolds. *Manuscripta Math.* **72** (1991) 81-94.

[KL92] L.H. Kauffman and S. Lins, Temperley-Lieb Recoupling Theory and Invariants of 3-Manifolds, (preprint 1993).

[KS92] L.H. Kauffman and H. Saleur, Map colouring and the Temperley-Lieb algebra, (To appear in Comm. Math. Phys).

[KL178] R. Kirby, A calculus for framed links in S^3, *Invent. Math.* **45** (1978) 35-56.

[KM91] R. Kirby and P. Melvin, On the 3-manifold invariants of Reshetikhin-Turaev for $sl(2,C)$, *Invent. Math.* **105** (1991) 473-545.

[KR88] A.N. Kirillov and N.Y. Reshetikhin, Representations of the algebra $U_q(sl2)$, q-orthogonal polynomials and invariants of links. In *Infinite Dimensional Lie Algebras and Groups* ed. by V.G. Kac. Adv. Ser. in Math. Phys. **7** (1988) 285.

[KRY92] S.I. Kryuchkov, The four-colour theorem and trees, I.V. Kurchatov Institute of Atomic Energy, Moscow (1992) - IAE - 5537/1.

[LI92] W.B.R. Lickorish, A Representation of Orientable Combinatorial 3-Manifolds, *Ann. Maths.* **76** (1992) 51.

[LIC92]   W.B.R. Lickorish, Calculations with the Temperley-Lieb algebra, *Commetarii Mathematici Helvetici* **67** (1992) 571.

[LI91]   W.B.R. Lickorish, 3-Manifolds and the Temperley-Lieb Algebra. *Math. Ann.* **290** (1991) 657.

[LI93]   W.B.R. Lickorish, Skeins and handlebodies, (preprint 1992).

[LIC93]   W.B.R. Lickorish, The Temperley-Lieb Algebra and 3-manifold invariants, (preprint 1992), (and private communication).

[LD91]   S. Lins and C. Durand, Topological classification of small graph-encoded orientable 3-manifolds, *Notas Comm. Mat.* UFPE, **177** (1991).

[MV92]   G. Masbaum and P. Vogel, 3-valent graphs and the Kauffman bracket, (preprint 1992).

[MOU79]   J.P. Moussouris, The Chromatic Evaluation of Strand Networks. *Advances in Twistor Theory* by Huston and Ward Research Notes in Mathematics, Pitman Pub. (1979), pp. 308-312.

[MOU83]   J.P. Moussouris, Quantum models of space-time based on recoupling theory, (Mathematics Thesis, Oxford University - 1983).

[OG91]   H. Ooguri and N. Sasakura, Discrete and Continuum approaches to three-dimensional quantum gravity, *Mod. Phys. Lett.* **A6** (1991) 3591

[PEN69]   R. Penrose, Angular momentum: an approach to combinatorial space-time, *in* "Quantum Theory and Beyond", T.A. Bastin ed., Cambridge Univ. Press (1969).

[PEN71]   R. Penrose, Applications of negative dimensional tensors, *in* "Combinatorial Mathematics and its Applications", D.J.A. Welsh ed., Academic Press (1971).

[PEN79]   R. Penrose, Combinatorial quantum theory and quantized directions, *in* "Advances in Twistor Theory", L.P. Hughston and R.S. Ward eds., Pitman (1979), pp. 301-307.

[PIU92]   Sergey Piunikhin, Turaev-Viro and Kauffman-Lins invariants for 3-manifolds coincide, *Journal of Knot Theory and its Ramifications*, **1** (1992) 105.

[RE87]   N.Y. Reshetikhin, Quantized universal enveloping algebras, the Yang-Baxter equation and the invariants of links, 1 and 11, LOMI reprints E-4-87 and E-17-87, Steklov Institute, Leningrad, USSR.

[RT90]   N.Y. Reshetikhin and V. Turaev, Ribbon graphs and their invariants derived from quantum groups, *Comm. Math. Phys.* **127** (1990) 1.

[RT91]   N.Y. Reshetikhin and V. Turaev, Invariants of three manifolds via link polynomials and quantum groups, *Invent. Math.* **103** (1991) 547.

[TU87]   V.G. Turaev, The Yang-Baxter equations and invariants of links, LOMI preprint E-3-87, Steklov Institute, Leningrad, USSR, *Inventiones Math.* **92** Fasc. 3, 527.

[TV90]   V.G. Turaev and O. Viro, State sum invariants of 3-manifolds and quantum $6j$ symbols, *Topology* **31** 1992 865.

[TU90]   V.G. Turaev Quantum invariants of links and 3-valent graphs in 3-manifolds, (preprint 1990).

[TU91]   V.G. Turaev, Quantum invariants of 3-manifolds and a glimpse of shadow topology, *Comptes Rendus de l'Academie des Sciences, Serie I- Math.* **313** (1991) 395.

[TU92]   V.G. Turaev, Shadow links and face models of statistical mechanics, *J. Diff. Geom.* **36** (1992) 35.

[TUR92]   V.G. Turaev, Topology of shadows, (preprint 1992).

[WA91]   K. Walker, On Witten's 3-manifold invariants, (preprint 1991).

[WI91]   R. Williams and F. Archer, The Turaev-Viro state sum model and 3-dimensional quantum gravity, *Phys. Lett.* **B 273** (1991) 438.

[WIT89]   E. Witten, Quantum field theory and the Jones polynomial, *Commun. Math. Phys.* **121** (1989) 351.

[YA92]   S. Yamada, A Topological invariant of spatial regular graphs, *in* "Knots '90", Akio Kawauchi ed., de Gruyter (1992), pp. 447-454.

QUANTUM FIELD THEORY AND A,B,C,D IRF MODEL INVARIANTS

Olaf J. Backofen

Theoretische Physik
Universität Saarbrücken
Germany

ABSTRACT

It is shown that the A,B,C,D interaction round a face model invariants can be derived exactly using Witten's Chern-Simons Theory. Using monodromy matrices of the classical Lie algebras the link invariants are obtained explicitly. This theory also allows generalizations of the previously known invariants to be incorporated.

1 INTRODUCTION

There exist several different approaches to knot theory. While historically the first link invariants were obtained by inductive means generalizing e.g. the linking number or Arf invariant (cf. introductory chapters of ref. 1) more powerful invariants could be derived using the theory of algebras, statistical mechanics or quantum field theory.

A first relationship between the new polynomial invariants of links and statistical mechanics was already implicitly contained in the pioneering paper of Jones[2], where Jones introduced his famous polynomial via a study of certain finite dimensional von Neumann algebras. Using especially the Yang-Baxter equation, state models, partition functions and several algebraic tools it was possible to specify the relationship indicated above and derive more powerful invariants [3–7].

A new powerful approach to knot theory based on the theory of exactly solvable models was established by Akutsu and Wadati[8] and Akutsu et al.[9] introducing invariants with skein relations of higher order. The consideration of models with Lie group symmetry of Cartan's type $A_{n-1}^{(1)}$, $B_n^{(1)}$, $C_n^{(1)}$, $D_n^{(1)}$ gave rise to the ABCD IRF model invariants where the $A_{n-1}^{(1)}$ case corresponds to a HOMFLY type polynomial (generalized two variable polynomials).

The consideration of state models and the graphical calculus by Kauffman[10] engendered a new valuable viewpoint of link invariants and related topics. Li and Ge et al.[11,23] examined a diagrammatic approach to the new invariants and calculated the link polynomials for the non-standard representation of the braid group.

Recently Wenzl[12,13,14] presented a general procedure to construct link invariants and subfactors for each irreducible representation of a classical Lie group defining certain traces on centralizer algebras with Markov property. This method permits also to construct the so called cable invariants related to tensor products of Lie groups.

As in most of the previous work in knot theory the evaluation of the invariants is based on two dimensional (2D) projections or algebraic approaches. Although knots are living intrinsically in three dimensions (3D) and so a 3D dimensional definition was desirable. It was E. Witten who provided the answer in his classical paper[15] by constructing knot polynomials as correlation functions of Wilson line operators in a 3D quantum field theory based on the Chern-Simons action. Moreover his theory incorporates significant generalizations of the previously known invariants.

Besides its mathematical advantages Chern-Simons theory also provides a unifying three dimensional viewpoint for 2D conformal field theory as well as 2+1 D quantum gravity[16]. Witten's derivation of link invariants considers monodromy operations of braid matrices of Cartan's type $A_{n-1}^{(1)}$. Using the Lie group $SU(n)$, the HOMFLY ($n \in N$) and Jones Polynomial ($n = 2$) are obtained respectively. The present approach employs Witten's theory to deduce the ABCD IRF model invariants making use of the braid matrices of Cartan's type $A_{n-1}^{(1)}$, $B_n^{(1)}$, $C_n^{(1)}$, $D_n^{(1)}$ wherein the first case corresponds to Witten's theory. Furthermore the invariants related to non-standard representations of the different Lie groups may be determined. While Witten's interpretation is a 3D QFT treatment of Jones algebraic theory the present derivation establishes a 3D QFT version of Wenzl's[12,13,14], Turaev's[6] and Ge's[11] algebraic calculation of invariants of the classical Lie groups with skein relations of higher order.

2 DERIVATION OF THE ABCD IRF MODEL INVARIANTS

2.1 Chern-Simons theory and invariants

The basic object of interest are the expectation values of Wilson lines in Chern-Simons gauge theory with gauge Lie group G of Cartan's type $A_{n-1}^{(1)}$, $B_n^{(1)}$, $C_n^{(1)}$ or $D_n^{(1)}$ at level k. One considers a link L in S^3 consisting of a disjoint union of circles C oriented and labeled with a choice of representation R of G. The R_i are elements of the finite set $\mathcal{R}$ of representations of G that are heighest weight integrable representations of the loop group $\mathcal{L}G$ at level k. The (unnormalized) expectation value of L is given by Witten's partition function

$$\langle L \rangle = Z(M, L) = \int \mathcal{D}A \, e^{i\mathcal{L}_{CS}} \mathcal{O}(A) , \qquad (1)$$

where $\mathcal{L}_{CS}$ is the Chern-Simons Lagrangian and the observables $\mathcal{O}(A)$ are given by the Wilson lines

$$\mathcal{O}(A) := \prod_i W_{R_i}(C_i) = \prod_i \text{Tr}_{R_i} P \exp \oint_{C_i} A . \qquad (2)$$

Following Witten's approach $Z(M, L)$ can be evaluated employing an algorithm for unknotting knots. The link L is embedded in the manifold M in such a way that one crossing remains in a part M_R and the rest of the link is located in the part M_L each consisting of a three ball with boundary S^2 with four marked points $a_1, \ldots, a_4$ connected by two Wilson lines as illustrated in Fig. 1.

According to Witten there exist physical Hilbert spaces $\mathcal{H}_R$ and $\mathcal{H}_L$ associated with the boundaries of M_L and M_R. In the case of $G = SU(n)$ and the Wilson lines being in the n dimensional representation of $SU(n)$ called R the Hilbert spaces are two dimensional.

In the general case with G of Cartan's type $A_{n-1}^{(1)}, B_n^{(1)}, C_n^{(1)}$ or $D_n^{(1)}$ the Wilson lines are taken in the corresponding fundamental representation R of the Lie group G. The dimension of the associated Hilbert spaces $\mathcal{H}_{RRRR}$ can be deduced calculating the decomposition of $R \otimes R$ into irreducible representations of G

$$R \otimes R = \bigoplus_{i=1}^{s} E_i . \tag{3}$$

While in the case of A_{n-1} one obtains a symmetrical and antisymmetrical irreducible representation, the Cartan Lie algebras $B_n^{(1)}, C_n^{(1)}$ and $D_n^{(1)}$ decompose into a symmetrical, antisymmetrical and scalar representation. The number of distinct irreducible representations being the dimension of the Hilbert space (cf. ref. 15) it can be concluded that the Hilbert spaces $\mathcal{H}_{S^2;RRRR}^L$ and $\mathcal{H}_{S^2;RRRR}^R$ are of dimension two and three for the classes $A_{n-1}^{(1)}$ and $B_n^{(1)}, C_n^{(1)}, D_n^{(1)}$ respectively.

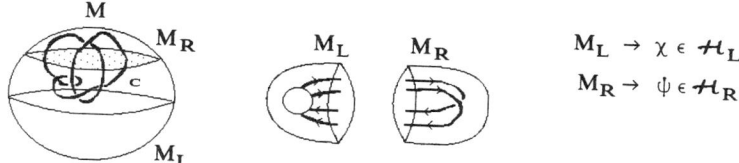

Figure 1. The 3-manifold M with the Wilson line C is cut into a simple piece M_R and complicated piece M_L.

An alternative point of view is to derive the dimension of the spaces from the theory of Moore and Seiberg[17,18]. Suppose that Ψ_1 and Ψ_2 are bases of the same Hilbert space $\mathcal{H}_{S^2;a_1,...a_4}$ and $\varepsilon_i, \tilde{\varepsilon}_i$ are the corresponding coupling tensor invariants of a graph in M (cf. rf. 19) given by

$$\Psi_1(m, \varepsilon_1, \varepsilon_2) = \sum_{\tilde{\varepsilon}_1, \tilde{\varepsilon}_2} K(m, \varepsilon_1, \varepsilon_2; n, \tilde{\varepsilon}_1, \tilde{\varepsilon}_2) \Psi_2(n, \tilde{\varepsilon}_1, \tilde{\varepsilon}_2) . \tag{4}$$

Then it follows employing the notation of Moore and Seiberg[17,18] that

$$K(m, \varepsilon_1, \varepsilon_2; n, \tilde{\varepsilon}_1, \tilde{\varepsilon}_2) = B_{m,n} \begin{bmatrix} a_1 & a_3 \\ a_2 & a_4 \end{bmatrix}_{\varepsilon_1, \varepsilon_2}^{\tilde{\varepsilon}_1, \tilde{\varepsilon}_2} \tag{5}$$

with B being the braiding matrix of the chosen Cartan's type. Since the braiding or monodromy matrices of Cartan's type $A_{n-1}^{(1)}$ and $B_n^{(1)}, C_n^{(1)}, D_n^{(1)}$ are two and three dimensional respectively the dimension of the Hilbert space may be determined.

Suppose that χ and Ψ are bases of the Hilbert spaces $\mathcal{H}_R$ and $\mathcal{H}_L$ respectively then the partition function $Z(M, L)$ is equal to the pairing

$$Z(M, L) = (\chi, \Psi) . \tag{6}$$

The dimension of the Hilbert spaces being two or three any three or four vectors obey a relation of linear dependence in the case of $A_{n-1}^{(1)}$ or $B_n^{(1)}, C_n^{(1)}, D_n^{(1)}$

$$\alpha\Psi + \beta\Psi_1 + \gamma\Psi_2 = 0 \quad \text{for} \quad A_{n-1}^{(1)}, \tag{7}$$

$$\alpha\Psi + \beta\Psi_1 + \gamma\Psi_2 + \delta\Psi_3 = 0 \quad \text{for} \quad B_n^{(1)}, C_n^{(1)}, D_n^{(1)}. \tag{8}$$

Here Ψ_1, Ψ_2, Ψ_3 are arbitrary vectors in the Hilbert space and $\alpha, \beta, \gamma, \delta$ are complex numbers. Hence in the following the $A_{n-1}^{(1)}$ will be referred to (8) with vanishing δ and Ψ_3.

A method to get additional vectors in $\mathcal{H}$ (cf. ref. 15) is to replace M_R in Fig. 1 by any other three manifolds X_1, X_2, X_3 with same boundary and suitable strings in X_i ($i = 1, \ldots 3$) and the Feynman integral (1) will generate new vectors in $\mathcal{H}$. The string configurations will be chosen of the following crossing types:

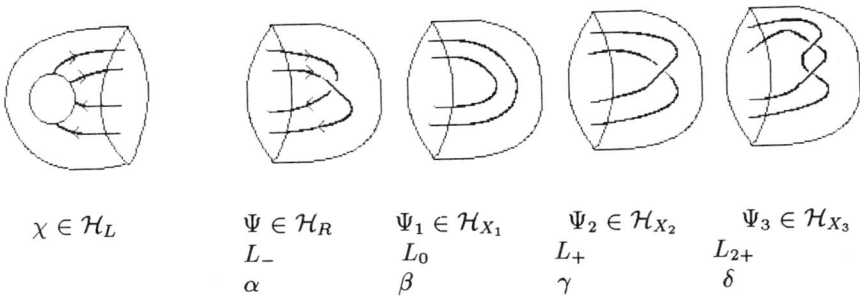

$\chi \in \mathcal{H}_L$	$\Psi \in \mathcal{H}_R$	$\Psi_1 \in \mathcal{H}_{X_1}$	$\Psi_2 \in \mathcal{H}_{X_2}$	$\Psi_3 \in \mathcal{H}_{X_3}$
L_-	L_0	L_+	L_{2+}
α	β	γ	δ

Figure 2. In order to obtain skein relations the 3-manifold M_R is replaced by different manifolds X_1, X_2 and X_3.

Taking the inner product of (8) with χ one obtains

$$\alpha(\chi, \Psi) + \beta(\chi, \Psi_1) + \gamma(\chi, \Psi_2) + \delta(\chi, \Psi_3) = 0. \tag{9}$$

This implies a recursive relation among the link expectation values in the form of a skein relation

$$\alpha Z(M, L_-) + \beta Z(M, L_0) + \gamma Z(M, L_+) + \delta Z(M, L_{2+}) = 0 \tag{10}$$

2.2 The monodromy operations

The monodromy operation B defined for each Lie group of the different Cartan types[17,20,21] can be used to generate the vectors Ψ_i ($i = 1, \ldots, 3$) from Ψ. B given in diagonal matrix representation possesses two and three eigenvalues λ_i in the case of $A_{n-1}^{(1)}$ and $B_n^{(1)}, C_n^{(1)}, D_n^{(1)}$, respectively, corresponding to the action in a two and three dimensional Hilbert space. Using the fact that every matrix obeys its own characteristic equation (by the Cayley Hamilton theorem) one deduces:

$$(\chi, [\prod_{i=1}^{3}(B - \lambda_i)]\Psi) = 0,$$

$$(\chi, [B^3 - (\lambda_1 + \lambda_2 + \lambda_3)B^2 + (\lambda_1\lambda_2 + \lambda_1\lambda_3 + \lambda_2\lambda_3)B - \lambda_1\lambda_2\lambda_3 I]\Psi) = 0.$$

Employing the relations

$$B\Psi = \Psi_1, \qquad B^2\Psi = \Psi_2, \qquad B^3\Psi = \Psi_3 \qquad (11)$$

one computes the coefficients in the skein relation of higher order from the eigenvalues of the braiding matrices in (11) for the different cases, similarly as in Witten's approach. The eigenvalues λ_i can be determined, e.g., from the Knizhnik-Zamolodchikov equation (cf. ref. 21) for the N-point correlation function Ψ

$$\kappa \frac{\partial}{\partial z_i}\Psi = \sum_{j \neq i} \frac{t_i - t_j}{z_i - z_j}\Psi, \qquad (i = 1, \ldots, n) \qquad (12)$$

and from the conformal weights

$$h = \frac{1}{\kappa}(\Lambda, \Lambda + 2\delta), \qquad (13)$$

where Λ is the heighest weight of an irreducible representation and δ is the sum of all positive roots of the corresponding algebra. This yields

$$\lambda = \delta_{sym} \exp\left[\pi i(\sum_{i \in K} h_i - h_{E_i})\right], \qquad (14)$$

where h_Y is the conformal weight of the state $Y = E_i$, K the number of transformed external lines and $\delta_{sym} = \pm 1$, where the + and - sign differentiates, respectively, between the cases that the irreducible representations E_i appear symmetrically (⊟) or antisymmetrically (⊟) in the elementary representation R (□). According to ref. 21 with $i = 1, 2, 3$ being the symmetric, antisymmetric and scalar representations respectively (13) and (14) yield the values listed in tables 1 and 2 for the convenience of the reader.

Agreement with the usual notation in the context of conformal theories is achieved using the well known relation between the parameter κ of the Knizhnik-Zamolodchikov equation and the level k of the representation.

$$\kappa = -(k + C_v), \qquad (15)$$

where C_v is the dual Coxeter number. In the case of $SU(n)$, $C_v = n$, and $\kappa = -(k+n)$ yields the notation used in Witten's approach[15].

Table 1. Conformal weights for the classical Lie groups with dimension d.

	d	δ^1_{sym}	δ^2_{sym}	δ^3_{sym}	h_R	h_1	h_2	h_3
A_{n-1}	$n-1$	1	-1	-	$-\frac{n^2-1}{2n\kappa}$	$-\frac{n^2+n-2}{n\kappa}$	$-\frac{n^2-n-2}{n\kappa}$	-
B_n	n	1	-1	1	$-\frac{n}{\kappa}$	$-\frac{2n+1}{\kappa}$	$-\frac{2n-1}{\kappa}$	0
C_n	n	1	-1	-1	$-\frac{2n+1}{2\kappa}$	$-\frac{2n+2}{\kappa}$	$-\frac{2n}{\kappa}$	0
D_n	n	1	-1	1	$-\frac{2n-1}{2\kappa}$	$-\frac{2n}{\kappa}$	$-\frac{2n-2}{\kappa}$	0

Table 2. The eigenvalues of the monodromy matrices for the classical Lie groups.

	λ_1	λ_2	λ_3
A_{n-1}	$\exp[\frac{\pi i}{\kappa}\frac{n-1}{n}]$	$-\exp[-\frac{\pi i}{\kappa}\frac{n+1}{n}]$	—
B_n	$\exp[\frac{\pi i}{\kappa}]$	$-\exp[-\frac{\pi i}{\kappa}]$	$\exp[-\frac{\pi i}{\kappa}(2n)]$
C_n	$\exp[\frac{\pi i}{\kappa}]$	$-\exp[-\frac{\pi i}{\kappa}]$	$\exp[-\frac{\pi i}{\kappa}(2n+1)]$
D_n	$\exp[\frac{\pi i}{\kappa}]$	$-\exp[-\frac{\pi i}{\kappa}]$	$\exp[-\frac{\pi i}{\kappa}(2n-1)]$

2.3 Evaluation of the invariants

The coefficients of the skein relation (8) can be found from the relations (10) and (11) written in the form

$$B^3\Psi - (\lambda_1 + \lambda_2 + \lambda_3)B^2\Psi + (\lambda_1\lambda_2 + \lambda_1\lambda_3 + \lambda_2\lambda_3)B\Psi - \lambda_1\lambda_2\lambda_3 I\Psi = 0 \quad (16)$$

and applied to the B_n, C_n and D_n algebras

$$\begin{aligned}
\alpha &= -\lambda_1\lambda_2\lambda_3 ,\\
\beta &= \lambda_1\lambda_2 + \lambda_1\lambda_3 + \lambda_2\lambda_3 ,\\
\gamma &= -(\lambda_1 + \lambda_2 + \lambda_3) ,\\
\delta &= 1 .
\end{aligned} \quad (17)$$

The coefficients in the case of A_{n-1} evaluated by Witten are obtained as follows:

$$\lambda_3 := 0 \quad \text{and} \quad \alpha := \beta ,$$
$$\beta := \gamma \quad \text{and} \quad \gamma := \delta .$$

In order to agree with the notation used in knot literature employing standard framing, the Wilson lines in the manifolds X_i ($i = 1, 2, 3$) must be adjusted by 1, 2 and 3-fold Dehn twists. This causes a multiplication of the coefficients β, γ, δ with $\exp[-2\pi i h_R], \exp[-4\pi i h_R], \exp[-6\pi i h_R]$, respectively. The multiplication with an irrelevant common factor and the q variable substitution

$$q = \exp\left[\frac{-2\pi i}{\kappa}\right] \quad (18)$$

yields the skein relations for the different Cartan type algebras corresponding to the ABCD IRF model invariants given in table 3.

These results agree exactly with the original defining relations of the ABCD IRF model invariants of Akutsu, Degutchi and Wadati[9] ((6.53) and (6.55)) given there in the form

$$\begin{aligned}
\alpha(L_+) &= (1-q)q^{\frac{n}{2}-\frac{1}{2}}\alpha(L_0) + q^n\alpha(L_-) &&\text{for } A_{n-1} ,\\
\alpha(L_{2+}) &= (1-q+\beta)e^{-i\omega[2\lambda+\omega(\sigma-1)]}\alpha(L_+) \\
&\quad +(q+\beta q - \beta)e^{-2i\omega[2\lambda+\omega(\sigma-1)]}\alpha(L_-) \\
&\quad -q\beta e^{-3i\omega[2\lambda+\omega(\sigma-1)]}\alpha(L_0) &&\text{for } B_n, C_n, D_n ,
\end{aligned}$$

where $q = e^{-2i\omega}$; $\beta = \sigma\exp[-i(2\lambda + \omega(\sigma-1))]$;

Table 3. The ABCD IRF model invariants

IRF invariant			Skein relation coefficients
A_{n-1}	α	L_-	q^n
	β	L_0	$q^{\frac{n}{2}-\frac{1}{2}} - q^{\frac{n}{2}+\frac{1}{2}}$
	γ	L_+	-1
B_n	α	L_-	$-q^{4n}$
	β	L_0	$q^{2n} - q^{3n-\frac{1}{2}} + q^{3n+\frac{1}{2}}$
	γ	L_+	$q^{n-\frac{1}{2}} - q^{n+\frac{1}{2}} + q^{2n}$
	δ	L_{2+}	-1
C_n	α	L_-	q^{4n+2}
	β	L_0	$q^{2n+1} - q^{3n+2} + q^{3n+1}$
	γ	L_+	$q^n - q^{n+1} + q^{2n+1}$
	δ	L_{2+}	-1
D_n	α	L_-	$-q^{4n-2}$
	β	L_0	$q^{2n-1} - q^{3n-1} + q^{3n-2}$
	γ	L_+	$q^{n-1} - q^n + q^{2n-1}$
	δ	L_{2+}	-1

$$\sigma = 1 \qquad \lambda = (2n-1)\frac{\omega}{2} \qquad \text{for } B_n,$$
$$\sigma = -1 \qquad \lambda = (n+1)\omega \qquad \text{for } C_n,$$
$$\sigma = 1 \qquad \lambda = (n-1)\omega \qquad \text{for } D_n.$$

In the case of A_{n-1} multiplication by $q^{-\frac{n}{2}}$ yields a HOMFLY type polynomial (generalized two variable polynomial). The case $n = 2$ yields the Jones polynomial and $n \to 0$ the Alexander polynomial. The difference of a factor of -1 in L_0 complies with an irrelevant renormalization of the invariants corresponding both to the same group representation in algebraic theory.

2.4 The normalization of the invariants

The normalization of the invariants corresponds to the normalized expectation value of a link L defined by

$$\langle L \rangle = \frac{Z(M,L)}{Z(M)}, \qquad (19)$$

where as usually $M = S^3$. For a collection of l unlinked, unknotted Wilson lines C_i one now may obtain[15]:

$$\langle L \rangle = \prod_{i=1}^{l} \langle C_i \rangle. \qquad (20)$$

The normalization may be evaluated by way of calculating the expectation value of an unknot U recursively. Searching for a suitable representation of the Hopf link H, the unknot U and the unknotted rings R one finds from the skein relation

$$\delta L_{2+} + \gamma L_+ + \alpha L_- + \beta L_0 = 0 , \qquad (21)$$

illustrated in Fig.3 for two specific cases, the system of equations

$$\delta H + (\gamma + \alpha)U + \beta R = 0 , \qquad (22)$$
$$\alpha H + (\delta + \beta)U + \gamma R = 0 . \qquad (23)$$

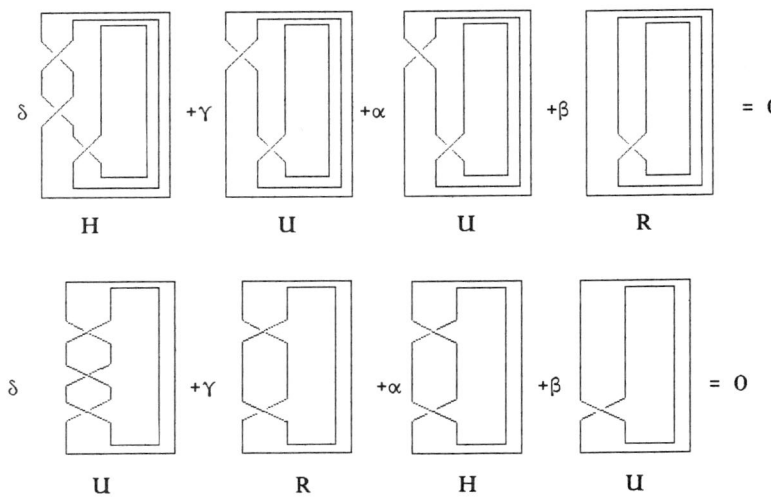

Figure 3. Two different applications of the skein relation to several links in order to calculate the normalization of the invariants.

The rings labeled by R are two unlinked and unknotted Wilson lines and (20) yields

$$\langle C^2 \rangle = \langle C \rangle^2 \text{ or } R = U^2 . \qquad (24)$$

The normalization for the ABCD IRF model invariant is then the solution of the system of equations, given by

$$\langle O \rangle = U = -\frac{\delta^2 + \delta\beta - \alpha(\gamma + \alpha)}{\delta\gamma - \alpha\beta} . \qquad (25)$$

The case of the two dimensional A model with $\delta \to 0$ reduces to the well known normalization introduced by Witten[15] (4.13):

$$\langle C \rangle = U = -\frac{\alpha + \gamma}{\beta} . \qquad (26)$$

2.5 The invariants of the non-standard representations

The previous approach showed how to deduce known IRF model invariants using standard representations of the classical Lie groups $A^{(1)}_{n-1}, B^{(1)}_n, C^{(1)}_n$ and $D^{(1)}_n$. An interesting generalization consists in regarding at the so called non-standard representations introduced by Ge and Xue[22].

These non-standard representations are sequences of new solutions of the spectral parameter-independent Yang-Baxter equation, wherein the coefficients of the Kauffman diagrams depend on the possible labeling (cf. ref. 11). According to Li and Ge[23] the eigenvalues of the corresponding monodromy matrices may be given in the following form:

For A_{n-1} : $\quad \lambda_1 = q$ $\qquad$ for B_n : $\quad \lambda_1 = q$
$\qquad\qquad\quad \lambda_2 = -q^{-1}$ $\qquad\qquad\qquad \lambda_2 = -q^{-1}$
$\qquad\qquad\qquad\qquad\qquad\qquad\qquad\qquad \lambda_3 = \delta_1 q^{-\mu+\delta_1}$

for C_n : $\quad \lambda_1 = q$ $\qquad$ for D_n : $\quad \lambda_1 = q$
$\qquad\quad\ \lambda_2 = -q^{-1}$ $\qquad\qquad\qquad \lambda_2 = -q^{-1}$
$\qquad\quad\ \lambda_3 = q^{-\mu+1}$ $\qquad\qquad\quad \lambda_3 = -\delta_1 q^{-\mu-\delta_1}$.

The parameters μ and δ_1 may be calculated from the possible sets of labelings of the diagrammatic Yang-Baxter equation.

The corresponding invariants may now be determined as in the case of the standard representation applying using Witten's Chern-Simons theory. The results are given in table 4.

The link invariants derived in this way are equivalent to those obtained by Ge and Li[23] using statistical mechanics theory with transformed Markov traces. The results agree with those of the standard representation theory modulo an irrelevant rescaling when determining μ and δ_1 for the different Lie group types. For example in the standard representation case of $SU(2)$, $\mu = 2$, $\delta_1 = 1$ and the rescaling $q' = q^{\frac{1}{2}}$ yields the Jones polynomial. Considering the case of B_3, $\mu = 3$, $\delta_1 = 1$ the invariants turn out to be equivalent to the 3-state vertex polynomials introduced by Akutsu, Degutchi and Wadati[9].

Table 4. The invariants of the non-standard representation.

Link invariant			Skein relation coefficients
A_{n-1}	α	L_-	$q^{-2\mu}$
	β	L_0	$q^{1-\mu} - q^{-1-\mu}$
	γ	L_+	-1
B_n	α	L_-	$-q^{-4\mu+4}$
	β	L_0	$q^{-2\mu+2} - q^{-3\mu+4} + q^{-3\mu+2}$
	γ	L_+	$q^{-\mu+2} - q^{-\mu} + q^{-2\mu+2}$
	δ	L_{2+}	-1
C_n	α	L_-	$\delta_1 q^{-4\mu-4\delta_1}$
	β	L_0	$\delta_1(q^{-3\mu-3\delta_1-1} - q^{-3\mu-3\delta_1+1} - \delta_1 q^{-2\mu-2\delta_1})$
	γ	L_+	$q^{-\mu-\delta_1+1} - q^{-\mu-\delta_1-1} - \delta_1 q^{-2\mu-2\delta_1}$
	δ	L_{2+}	-1
D_n	α	L_-	$-\delta_1 q^{-4\mu+4\delta_1}$
	β	L_0	$\delta_1(q^{-3\mu+3\delta_1-1} - q^{-3\mu+3\delta_1+1} + \delta_1 q^{-2\mu+2\delta_1})$
	γ	L_+	$q^{-\mu+\delta_1+1} - q^{-\mu+\delta_1-1} - \delta_1 q^{-2\mu+2\delta_1}$
	δ	L_{2+}	-1

3 CONCLUSIONS AND OUTLOOK

The present report supplies an intrinsically 3D definition of the ABCD IRF model invariants and their non-standard representation correspondents with skein relations of higher order in analogy with Witten's pioneering work relating QFT and the Jones polynomial. This approach is a further small step to a 3D understanding of the real nature of links derived usually by projection or algebraic methods. The presented polynomial invariants of higher order as well as the N state vertex polynomial invariants with $N \geq 3$ are more powerful than e.g. the Jones and the HOMFLY invariants permitting for example to distinguish between the famous Birman pair illustrated in Fig.4 and defined by:

$$L_1 = (b_1 b_2 b_1)^4 b_1^{-12} b_2^6, \text{ and } L_2 = b_1^{-6} b_2^{12} . \qquad (27)$$

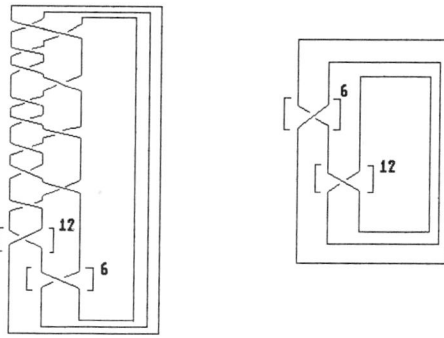

Figure 4. The Birman pair.

Note that the Jones polynomial does not distinguish between L_1 and L_2, whereas the higher dimensional skein relations will provide for them a systematic classification possibility (cf. ref. 9). A possible generalization is viewed as the possibility of topological surgery on M of $Z(M, L)$ generating new and more efficient polynomial invariants of higher order. If $\tilde{M}$ is a 3-manifold (with a Wilson line in the R_i representation) obtained from the 3-manifold M (with a Wilson line in the R_j representation) by surgery with a diffeomorphism K of the boundary, then well known topology axioms provide the generalized surgery formula of Witten

$$Z(\tilde{M}, R_i) = \sum_j K_i{}^j Z(M, R_j) . \qquad (28)$$

In the case of $G = SU(2)$, $\tilde{M} = S^3$ and $M = S^2 \times S^1$, K becomes the known modular group diffeomorphism S and surgery gives rise to the possibility to calculate $Z(S^3)$ from $Z(S^2 \times S^1, R_j)$. $Z(S^3, L(\bigcup_{i \in J} R_i))$ may be calculated successively as for example

$$Z(S^3, L(R_i, R_j)) = \sum_k S_i{}^k Z(S^2 \times S^1, R_k, R_j) = S_{ij} \qquad (29)$$

In the general case of G being of Cartan's type $A_{n-1}^{(1)}, B_n^{(1)}, C_n^{(1)}$ or $D_n^{(1)}$ the surgery diffeomorphism K_i^j may essentially be obtained using Kac's formula (cf. ref .24). Although no explicit results may be calculated a determination not only of $Z(S^3)$ but also successive generalizations from $Z(M,L)$ of the ABCD IRF model invariants may essentially be obtained.

An alternative 3D viewpoint of Cotta-Ramusino et al.[26] and Smolin[25] allows one to compute the skein coefficients in the case of $G = SU(n)$ in a large coupling approximation $k \to \infty$. The method basically consists in calculating an infinitesimal deformation of the corresponding Wilson lines in the form of a small loop generating the HOMFLY invariant. An examination of the ABCD IRF model invariants could be an interesting alternative. Furthermore the exceptional Lie groups G_2, F_4, E_6, E_7 and E_8 may be examined.

In conclusion the 3D interpretation of the ABCD IRF model invariants provides, besides the mathematical advantage of unifying the different link polynomials and generating new invariants, a further small step in the understanding of the connection of knot literature and QFT.

REFERENCES

[1] L.H. Kauffman, *Annals of Mathematics Study 115*, Princeton University Press, 1987.
[2] V.F. Jones, *Bull. Am. Math. Soc.* 12 (1985) 103.
[3] V.F. Jones, *Pac. J. of Math.* 137 (1989) 311.
[4] L.H. Kauffman, *Contemp. Math.* 78 (1988) 263.
[5] J. Hoste, A. Ocneanu, K. Millet, P. Freyd, W. Lickorish, D. Yetter, *Bull. Am. Math. Soc.* 12 (1985) 239.
[6] V. Turaev, *Invent. Math.* 92 (1988) 527.
[7] Y. Wu, *J. of Knot Theory and its Ramifications* 1 (1992) 47.
[8] Y. Akutsu, M. Wadati, *Commun. Math. Phys.* 117 (1988) 243.
[9] Y. Akutsu, T. Degutchi, M. Wadati, *Phys. Reports* 180 (1989) 247.
[10] L.H. Kauffman, *J. of Knot Theory and its Ramifications* 1 (1992) 59.
[11] M. Ge, L. Wang, K. Xue, Y. Wu, Advanced Series in Math. Physics, Volume 9.
[12] H. Wenzl, *Invent. Math.* 92 (1988) 349.
[13] H. Wenzl, *Pac. J. of Math.* 145 (1990) 153.
[14] H. Wenzl, *Commun. Math. Phys.* 133 (1990) 383.
[15] E. Witten, *Comm. Math. Phys.* 121 (1989) 351.
[16] E. Witten, *Nucl. Phys.* B311 (1988) 46.
[17] G. Moore, N. Seiberg, *Phys. Lett.* B212 (1988) 451.
[18] G. Moore, N. Seiberg, *Commun. Math. Phys.* 123 (1989) 177.
[19] E. Witten, *Nucl. Physics* B322 (1989) 629.
[20] T. Kohno, *Ann. Inst. Fourier Grenoble* 37(4) (1987) 139.
[21] Y. Akutsu, T. Degutchi, M. Wadati, *J. Phys. Soc. of Japan* 58 (1989) 1153.
[22] M. Ge, K. Xue, Preprint ITP-SB-90-20 Stony Brook, 1990.
[23] Y. Li, M. Ge, *J. of Physics* A 24 (1991) 4241.
[24] V. Kac, "Infinite Dimensional Lie Algebras", Cambridge University Press 1985.
[25] L. Smolin, *Mod. Phys. Lett.* A 4 (1990) 1091.
[26] P. Cotta-Ramusino, E. Guadagnini, M. Martellini, M. Mintchev, *Nucl. Phys.* B330 (1990) 557.

ON COMBINATORIAL THREE-MANIFOLD INVARIANTS

G. Felder and O. Grandjean

Mathematik, ETH-Zentrum, 8092 Zürich, Switzerland

1 INTRODUCTION

In a recent paper[1], Turaev and Viro gave a set of axioms for certain "initial data", and a procedure to construct combinatorial invariants for three-dimensional manifolds out of these data and a triangulation of the manifold. Then they showed that examples of initial data satisfying these axioms are provided by $6j$-symbols of the quantum group $U_q(sl_2)$, when q is a root of unity. This construction has been further analysed in refs. 2 and 3.

It is clear that the Turaev-Viro axioms have to be modified to include more general types of known invariants defined in terms of triangulations, such as invariants associated to quantum groups of higher rank[4], and the Dijkgraaf-Witten invariants[5-9].

Indeed, the axioms of Turaev and Viro use the facts that all representations are self-conjugate, and that multiplicities in the Clebsch-Gordan series are at most one, which are sl_2 peculiarities.

The aim of this paper is to generalize the axiomatic system of Turaev and Viro to include non-conjugate representations and multiplicities. The inspiring example is that of $6j$-symbols for finite groups. We also show that the Dijkgraaf-Witten model provides an other class of examples, and thereby give a simple direct proof that their partition function gives an invariant.

It should be emphasized that our axioms provide invariants for oriented three-manifolds, whereas the initial data obeying the Turaev-Viro axioms do not distinguish orientation but are also applicable to non-orientable manifolds.

This work is based on the ETH diploma thesis[10] of the second author.

2 CONSTRUCTION OF THE INVARIANTS

2.1 Initial data

We call initial data a set consisting of the following objects.

1. J is a finite set and $\text{adm}(J^3) \subset J^3$ is a set of ordered triples of elements of J, called admissible triples.

2. K is a commutative ring with unit and with involution $K \to K$, $w \mapsto \bar{w}$. The group of invertible elements of K is denoted by K^*. The element $\bar{w}$ is called conjugate of w.

3. $N : \mathrm{adm}(J^3) \to \mathbf{N}^*$, $(A, B, C) \mapsto N_{AB}{}^C$ is a map. The number $N_{AB}{}^C$ is called multiplicity of the admissible triple (A, B, C).

4. $W : J \to K^*$, $A \mapsto w_A$ is a map. The element $w_A^2 \in K^*$ is called dimension of $A \in J$.

5. $\mathrm{adm}(J^3 \times \mathbf{N}^*)$ is the subset of $J^3 \times \mathbf{N}^*$ defined as follows:
$$(A, B, C; i) \in \mathrm{adm}(J^3 \times \mathbf{N}^*) \quad \text{iff} \quad (A, B, C) \in \mathrm{adm}(J^3), \ i \leq N_{AB}{}^C.$$

6. $\mathrm{adm}(J^6 \times (\mathbf{N}^*)^4)$ is the subset of $J^6 \times (\mathbf{N}^*)^4$ defined as follows:
$$(A, B, C, D, E, F; i, j, k, l) \in \mathrm{adm}(J^6 \times (\mathbf{N}^*)^4)$$
iff $(A, B, C; i)$, $(C, D, E; j)$, $(B, D, F; k)$, $(A, F, E; l) \in \mathrm{adm}(J^3 \times \mathbf{N}^*)$.

7. $|\ ::::\ |\ ::\ | : \mathrm{adm}(J^6 \times (\mathbf{N}^*)^4) \to K$ is a map

$$(A, B, C, D, E, F; i, j, k, l) \mapsto \begin{vmatrix} A & B & C \\ D & E & F \end{vmatrix} \begin{matrix} i & k \\ j & l \end{matrix}.$$

The image of an element of $\mathrm{adm}(J^6 \times (\mathbf{N}^*)^4)$ is called symbol of this element.

2.2 Numbered and coloured triangulations

Let M be a compact oriented three-manifold and $\mathcal{T}$ a triangulation of M. A numbering of $\mathcal{T}$ is an injective map from the set of vertices of $\mathcal{T}$ to $\mathbf{N}$. A numbering of a triangulation induces an orientation of the edges, which can be indicated by an arrow going from the vertex with lower number to the vertex with higher number. Two numberings are called equivalent if they induce the same orientation on all 1-simplices. Although all subsequent construction only refer to equivalence classes of numberings, it is often convenient for the formulation to use numberings.

A 3-simplex of a numbered triangulation is said to have positive orientation if there is an orientation preserving homeomorphism from it to the standard affine 3-simplex with standard numbering of vertices, which maps vertices to vertices, and preserves the ordering of the vertices (See Fig. 1). Otherwise it is said to have negative orientation. Equivalent numberings of $\mathcal{T}$ give rise to the same set of 3-simplices with positive orientation. A colouring of $\mathcal{T}$ is a map which assigns to each edge an element of J, and to each triangle a natural number. An admissible colouring is a colouring such that for each triangle the condition of Fig. 2 is satisfied. The numbers at the vertices indicate the ordering of the vertices induced by the numbering. Thus 0, 1, 2 can be replaced by i, j, k, with $i < j < k$. The set of admissible colourings of the numbered triangulation $\mathcal{T}$ is denoted $\mathrm{adm}(\mathcal{T})$. It only depends on the equivalence class of the numbering.

positive orientation   negative orientation

Figure 1: Orientation of 3-simplices.

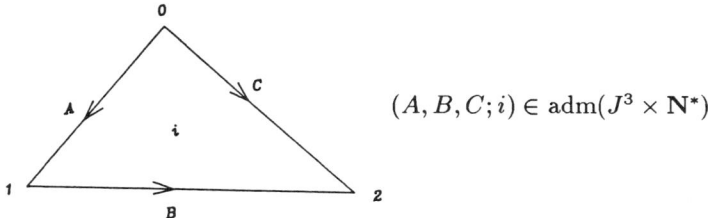

$(A, B, C; i) \in \mathrm{adm}(J^3 \times \mathbf{N}^*)$

Figure 2. Admissible colouring of a 2-simplex.

2.3 Graphical representation of the initial data

We introduce a graphical representation of initial data, which will enable us to express more clearly the conditions of the next subsection.

Let T be a 3-simplex of $\mathcal{T}$, $E_1, \ldots, E_6$ the edges of T, $\Gamma_1, \ldots, \Gamma_4$ the faces of T, and ϕ the restriction of an admissible colouring to T. To these data we assign a symbol or the conjugate of a symbol depending on whether T is positively or negatively oriented, according to the rule depicted in Fig. 3. As before, the numbering displayed indicates the ordering induced by the numbering of the vertices of T. We denote by T^ϕ the coloured 3-simplex, and $|T^\phi|$ the corresponding (conjugate) symbol.

2.4 Conditions on initial data

We are interested in initial data satisfying certain conditions. These conditions are best described graphically. The graphical representations involve coloured 3-simplices sharing faces. In order to recover the expressions corresponding to the pictures, we just

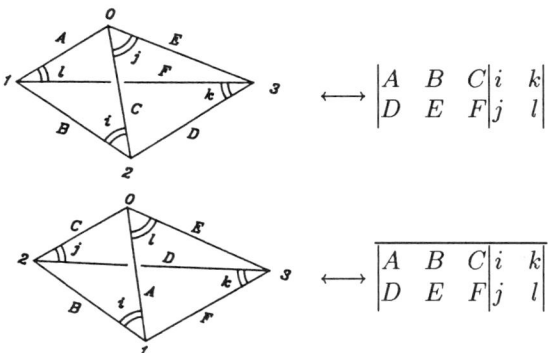

Figure 3. Correspondence between 3-simplices and symbols.

write down the product of the symbols defined by the 3-simplices. We then sum over colours associated to shared faces, and to edges not belonging to the boundary.

The initial data are said to satisfy condition (∗) if the following equation is true for all numberings and admissible colourings:

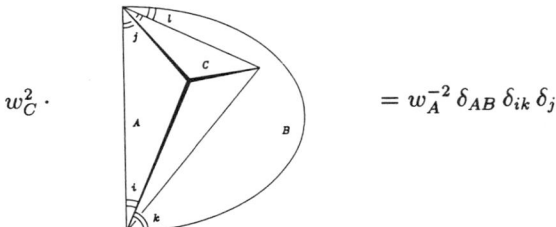

This picture represents two 3-simplices, one of which is deformed. These two simplices share two faces, five edges, and four vertices. The B edge belongs to the deformed simplex and the A edge to the other. We take the sum over C and over the multiplicity indices associated to the faces that share the C-edge. We notice that one simplex is positively oriented and the other is negatively oriented. We give an example of an equation described by this picture:

$$\sum_{C,m,n} w_C^2 \begin{vmatrix} A & D & E \\ C & F & H \end{vmatrix} \begin{matrix} i & n \\ m & j \end{matrix} \begin{vmatrix} B & D & E \\ C & F & H \end{vmatrix} \begin{matrix} k & n \\ m & l \end{matrix} = w_A^{-2} \delta_{AB} \delta_{ik} \delta_{jl}.$$

Following our conventions, the summation indices take only values for which the symbols are defined. All other equations described by this picture have the same form. The left-hand side contains the product of two symbols, one of which is conjugate, times the dimension of the summed element of $J(C)$. The condition (∗) consists of six equations.

The initial data are said to satisfy condition (∗∗) if the following equation is true for at least one numbering and all admissible colourings. The numbering and colouring have to be identical for simplices appearing on both sides of the equation.

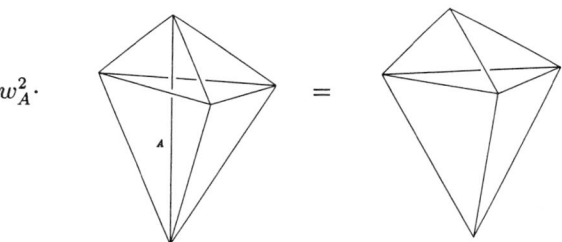

This picture gives one equation for each inequivalent choice of numbering, with a total of ten independent equations. With one choice of numbering, the equation is

$$\sum_{A,i,j,k} w_A^2 \begin{vmatrix} B & C & D \\ E & F & A \end{vmatrix} \begin{matrix} a & i \\ b & j \end{matrix} \begin{vmatrix} H & I & B \\ A & F & J \end{matrix} \begin{matrix} c & k \\ j & d \end{matrix} \begin{vmatrix} I & C & K \\ E & J & A \end{vmatrix} \begin{matrix} e & i \\ f & k \end{matrix} =$$

$$\sum_l \begin{vmatrix} H & I & B \\ C & D & K \end{vmatrix} \begin{matrix} c & e \\ a & l \end{matrix} \begin{vmatrix} H & K & D \\ E & F & J \end{vmatrix} \begin{matrix} l & f \\ b & d \end{matrix}.$$

Lemma 2.1 *If the initial data satisfy condition (∗), then the ten equations described by condition (∗∗) are all equivalent.*

The initial data are said to satisfy condition $(***)$ if there is an element $w \in K^*$ such that for all $A \in J$ the following equation holds

$$\sum_{B,C} N_{AB}{}^C w_B^2 w_C^2 = \sum_{B,C} N_{CA}{}^B w_B^2 w_C^2 = \sum_{B,C} N_{BC}{}^A w_B^2 w_C^2 = w^2 w_A^2.$$

2.5 Invariance theorem

Let M be a compact oriented 3-manifold with boundary and $\mathcal{T}$ a numbered triangulation of M. Let a be the number of vertices of $\mathcal{T}$, e the number of vertices of $\mathcal{T}$ which lie on the boundary ∂M, $E_1,\ldots E_b$ the number of edges of $\mathcal{T}$ ordered in such a way that the first f of them lie on ∂M and lastly $T_1, \ldots, T_d$ the 3-simplices of $\mathcal{T}$. The triangulation $\mathcal{T}$ of M induces a triangulation of ∂M which is denoted by $\partial \mathcal{T}$. A colouring of $\partial \mathcal{T}$ is a map which assigns an element of J to each edge of $\partial \mathcal{T}$, and a natural number to each 2-simplex of $\partial \mathcal{T}$. A colouring of $\partial \mathcal{T}$ is said to be admissible if the condition of Fig. 2 holds for each 2-simplex of $\partial \mathcal{T}$. The set of admissible colourings is denoted by $\mathrm{adm}(\partial \mathcal{T})$. Notice that an admissible colouring of $\mathcal{T}$ restricts to an admissible colouring of $\partial \mathcal{T}$. For each $\alpha \in \mathrm{adm}(\partial \mathcal{T})$, we denote by $\mathrm{adm}(\alpha, \mathcal{T})$ the set of admissible colourings of $\mathcal{T}$ whose restriction to the boundary coincides with α, or, in other words, which extend α. An admissible colouring $\phi \in \mathrm{adm}(\mathcal{T})$ defines an admissible colouring of each T_i, and we associate a (conjugate) symbol $|T_i^\phi|$.

We define

$$|M|_\phi := w^{-2a+e} \prod_{r=1}^f w_{\phi(E_r)} \prod_{s=f+1}^b w_{\phi(E_r)}^2 \prod_{t=1}^d |T_t^\phi| \in K \quad, \forall \phi \in \mathrm{adm}(\mathcal{T})$$

$$\Omega_M(\alpha) := \sum_{\phi \in \mathrm{adm}(\alpha, \mathcal{T})} |M|_\phi, \quad \forall \alpha \in \mathrm{adm}(\partial \mathcal{T}).$$

The quantity $\Omega_M(\alpha)$ is called partition function.

Theorem 2.2 *Let M be a compact oriented 3-manifold, S a numbered triangulation of ∂M, and α an admissible colouring of S. If the initial data satisfy conditions $(*)$, $(**)$, $(***)$, then $\Omega_M(\alpha)$ does not depend on the numbered triangulation $\mathcal{T}$ which extends S.*

This theorem is proved in the next subsections.

If the manifold M has empty boundary, we have $e = 0$ and $f = 0$. In that case, we define

$$|M| := \sum_{\phi \in \mathrm{adm}(\mathcal{T})} |M|_\phi.$$

Corollary 2.3 *Let M be a compact oriented manifold without boundary. If the initial data satisfy conditions $(*)$, $(**)$, $(***)$, then $|M|$ does not depend on the numbered triangulation $\mathcal{T}$.*

More generally, the general construction of[1] gives a "topological quantum field theory" in the sense of Witten and Atiyah.

2.5.1 Outline of the proof of the invariance theorem

The proof of the invariance theorem follows closely the proof of Turaev and Viro[1]. The only difference resides in the numbering of the triangulation, and the colouring of the 2-simplices. The main ingredient of the proof is the relative Alexander theorem proven in[1]. This theorem states that two triangulations of a compact 3-manifold M

which coincide on the boundary ∂M, can be transformed one into another by a finite sequence of Alexander moves which do not change the triangulation on the boundary, and their inverses. Thus it is sufficient to prove that $\Omega_M(\alpha)$ is invariant under Alexander moves. A simpler and more transparent criterion for invariance is obtained in the dual picture. Instead of considering triangulations one considers simple 2-polyhedra and simple graphs (see ref. 1). Simple 2-polyhedra and graphs are obtained from triangulations as follows. Let M be a compact oriented 3-manifold and $\mathcal{T}$ a triangulation of M. For each simplex σ of $\mathcal{T}$ we consider the barycentric star σ^*. The set $\mathcal{T}^* = \{\sigma^* | \sigma \in \mathcal{T}\}$ is a disjoint covering of the pair $(M, \partial M)$. The 2-skeleton $\mathcal{T}^*_{(2)} = \{\sigma^* | \sigma \in \mathcal{T}, \dim(\sigma) \geq 1\}$ of $\mathcal{T}^*$ is a simple 2-polyhedron, whose boundary $\Gamma = X \cap \partial M$ is a simple graph. In this representation Alexander moves can be decomposed into a finite sequence of more elementary moves: the three moves $\mathcal{L}$ (lune) $\mathcal{M}$ (Matveev) and $\mathcal{B}$ (bubble). However, in general, these moves transform 2-polyhedra obtained from triangulations to more general 2-polyhedra. This justifies the introduction of this concept.

The proof is organized as follows: we first define the notions of numbering and colouring for a restricted class of 2-polyhedra and simple graphs. If a 2-polyhedron is the 2-skeleton of the cell decomposition dual to a triangulation, the numberings and colourings of the two objects are in one-to-one correspondence. Then, we define a partition function for simple 2-polyhedra, which coincides with the partition function introduced above for triangulations. We then show that the partition function is invariant under Matveev moves, and that it does not depend on the choice of numbering of the internal vertices.

2.5.2 Numbering and colouring

Let $X \subset M$ be a 2-polyhedron such that $\partial X \subset \partial M$. Then the set of the natural strata of X and of the connected components of $M \setminus X$ is a disjoint covering of M. We set $\Gamma = \partial X = X \cap \partial M$.

A numbering of X (resp. Γ) is an injective map from the set of connected components of $M \setminus X$ (resp. $\partial M \setminus \Gamma$) to $\mathbf{N}$. If X is associated to a triangulation $\mathcal{T}$ of M, then we have a bijective correspondence between connected components of $M \setminus X$ (resp. $\partial M \setminus \Gamma$) and vertices of $\mathcal{T}$ (resp. $\partial \mathcal{T}$). Thus, we also have such a relation between numberings of X (resp. Γ) and numberings of $\mathcal{T}$ (resp. $\partial \mathcal{T}$). In what follows it will be assumed that X and Γ are numbered.

In order to define the notion of colouring, we use the natural stratification of X and Γ. A colouring of X is a map which assigns an element of J to each 2-stratum of X and a natural number to each internal (i.e., not belonging to Γ) 1-stratum of X. A colouring is admissible if it satisfies the condition of Fig. 4 for each 1-stratum. A colouring of Γ is a map which assigns an element of J to each 1-stratum of Γ and a natural number to each 0-stratum of Γ. A colouring is admissible if it satisfies the condition of Fig. 5 for each 0-stratum. An admissible colouring of X clearly induces an admissible colouring of Γ.

Just as we can associate a symbol to each 3-simplex of a coloured numbered triangulation, we can associate a symbol to each internal 0-stratum of a simple 2-polyhedron. The symbol is computed by constructing a coloured numbered tetrahedron for each internal 0-stratum, which coincides with the dual cell, in the case where the 2-polyhedron is the 2-skeleton of the dual of a triangulation. Fig. 6 shows the construction of a tetrahedron associated to an internal 0-stratum x of X. Firstly, we connect the 1-strata emerging from x. Then we consider the numbering and the colouring induced on the boundary. We get a tetrahedron T_x. Then we replace T_x by a tetrahedron $\hat{T}_x$, whose vertices, edges and faces correspond to faces, edges and vertices of T_x, respectively.

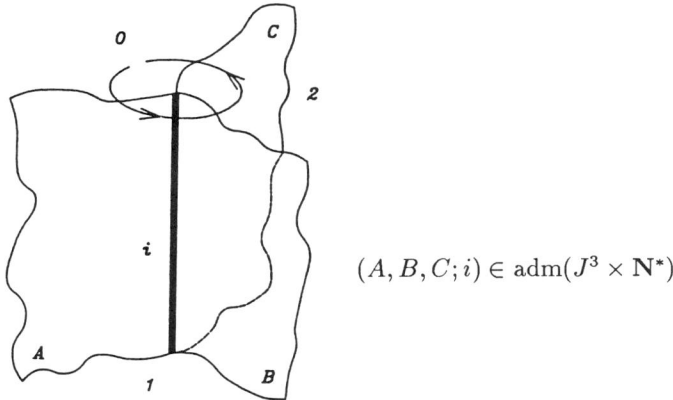

Figure 4. Admissible colouring of a simple 2-polyhedron.

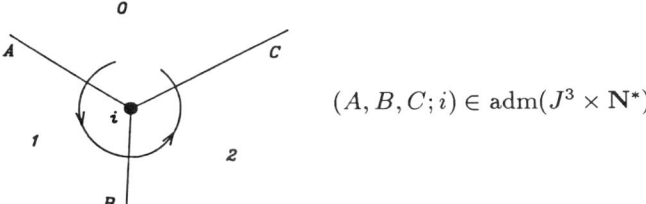

Figure 5. Admissible colouring of a simple graph.

Fig. 7 shows the correspondence between a 2-simplex of $\partial\mathcal{T}$ and a 0-stratum of Γ. The $\partial\mathcal{T}$ is drawn with solid lines, and Γ with dashed lines. In this figure, $i \in \mathbf{N}$ is the colouring of the 0-stratum and of the 2-simplex, A, B, $C \in J$ indicate the colouring of the 1-strata and 1-simplices they cross. With the help of Fig. 6 and 7, we see that there is a bijective relation between the (admissible) colourings of $\mathcal{T}$ (resp. $\partial\mathcal{T}$) and those of the corresponding simple 2-polyhedra X (resp. simple graph Γ).

2.5.3 Partition function

Let X be a numbered simple 2-polyhedron which defines a cell decomposition of M. Let $x_1, \ldots x_d$ be the vertices of $X \setminus \partial X$, $E_1, \ldots E_f$ the 1-strata of ∂X and $\Gamma_1, \ldots \Gamma_b$ the 2-strata of X. We denote by $\mathrm{adm}(X)$ (resp. $\mathrm{adm}(\partial X)$) the set of admissible colourings of X (resp. ∂X). For $\phi \in \mathrm{adm}(X)$ we denote by $\partial\phi \in \mathrm{adm}(\partial X)$ the colouring induced by ϕ on the boundary. We define the partition function for X.

$$|X|_\phi = w^{-2\chi(X)+\chi(\partial X)} \prod_{r=1}^{b} w_{\phi(\Gamma_r)}^{2\chi(\Gamma_r)} \prod_{s=1}^{f} w_{\partial\phi(E_s)}^{\chi(E_s)} \prod_{t=1}^{d} |\hat{T}_{x_t}^\phi|, \quad \forall \phi \in \mathrm{adm}(X),$$

$$\Omega_X(\alpha) = \sum_{\substack{\phi \in \mathrm{adm}(X) \\ \partial\phi = \alpha}} |X|_\phi, \quad \forall \alpha \in \mathrm{adm}(\partial X).$$

where χ is the Euler characteristic. If X is the 2-skeleton of the dual of a triangulation of M, we let $\alpha^* \in \mathrm{adm}(\partial X)$ be the colouring of ∂X corresponding to the admissible colouring $\alpha \in \partial\mathcal{T}$. Then we have the identity $\Omega_M(\alpha) = \Omega_X(\alpha^*)$ with proof as in ref. 1.

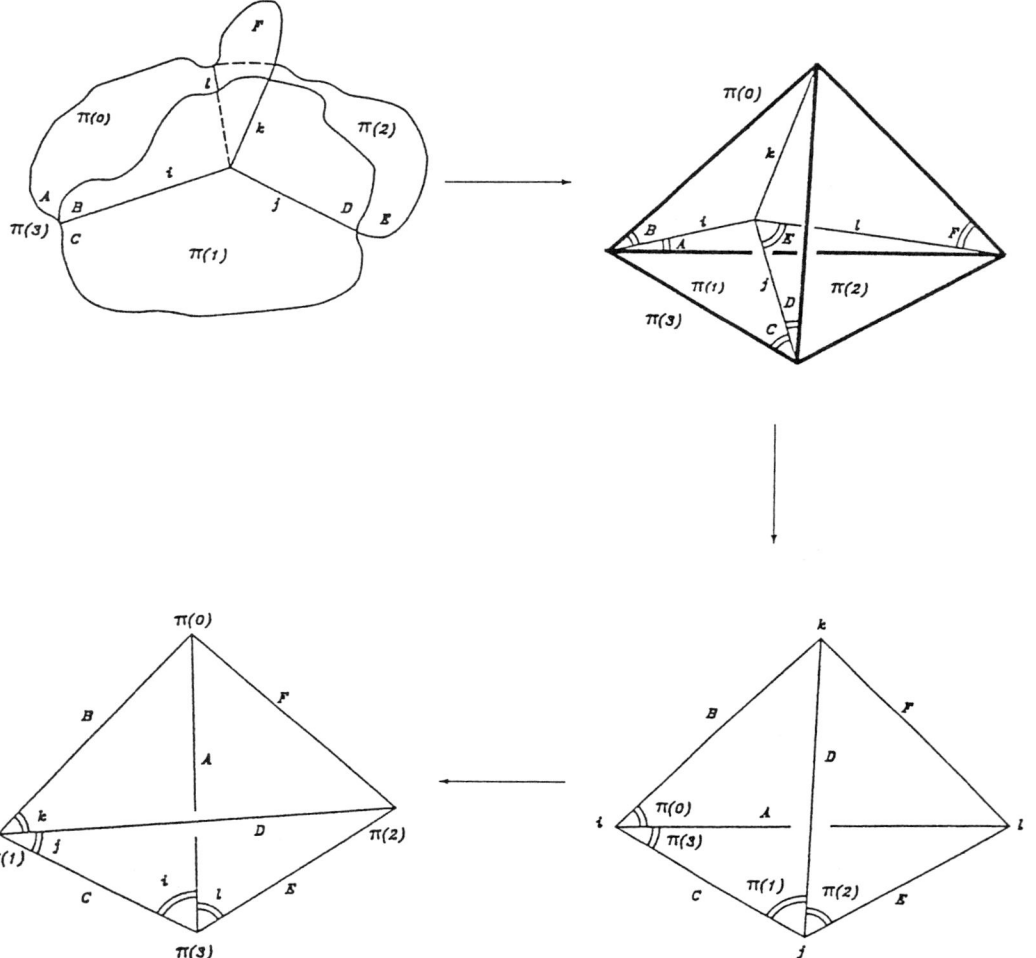

Figure 6. Correspondence between an internal 0-stratum of X and a tetrahedron. $\pi \in S_4$ defines the numbering.

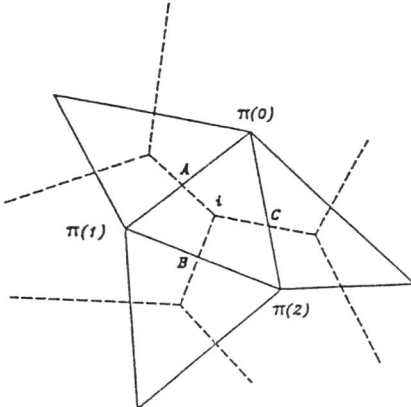

Figure 7. Correspondence between a 2-simplex of ∂T and a 0-stratum of Γ. $\pi \in S_3$ defines the numbering.

2.5.4 Invariance of the partition function under the moves $\mathcal{L}$, $\mathcal{M}$ and $\mathcal{B}$

Before proving the invariance of $\Omega_X(\alpha)$ we quote a lemma whose proof can be found in ref. 1.

Lemma 2.4 *Let X, Y and Z be simple 2-polyhedra such that*

i) *X defines a cell decomposition of M*

ii) *$X = Y \cup Z$*

iii) *Each stratum of $T = Y \cap Z$ is a stratum of ∂Y and ∂Z*

Then we have, for each $\beta \in \mathrm{adm}(\partial X)$,

$$\Omega_X(\beta) = \sum_{\alpha \in \mathrm{adm}(T)} \Omega_Y(\alpha \cup (\beta|_{Y \cap \partial X})) \, \Omega_Z(\alpha \cup (\beta|_{Z \cap \partial X})),$$

where $\alpha \cup (\beta|_{Y \cap \partial X})$ and $\alpha \cup (\beta|_{Z \cap \partial X})$ are the colourings of ∂Y and ∂Z induced by α and β, respectively.

Let X be a numbered simple 2-polyhedron which defines a cell decomposition of M. Notice that the moves $\mathcal{L}$, $\mathcal{M}$ and $\mathcal{B}$ do not modify the boundary ∂X (Fig. 8).

Proposition 2.5 *Let $\alpha \in \mathrm{adm}(\partial X)$. If the initial data satisfy condition (*), then $\Omega_X(\alpha)$ is invariant under $\mathcal{L}$.*

Proof: Let X' the simple 2-polyhedron obtained from X with the move $\mathcal{L}$. There are simple 2-polyhedra Y, Y' and Z such that $X = Y \cup Z$, $X' = Y' \cup Z$, $Y \cap Z = \partial Y \subset \partial Z$, $Y' \cap Z = \partial Y' = \partial Y \subset \partial Z$. The simple 2-polyhedra Y and Y' are depicted in Fig. 9 and 10. The permutation $\pi \in S_4$ indicates the numbering. In view of Lemma 2.4, we see that we just have to prove:

$$\Omega_Y(\beta) = \Omega_{Y'}(\beta), \; \forall \beta \in \mathrm{adm}(\partial Y) = \mathrm{adm}(\partial Y').$$

We consider the configurations of Fig. 9 and 10. We first compute $\Omega_{Y'}(\beta)$, where β is the colouring of the simple graph ∂Y (see Fig. 10).

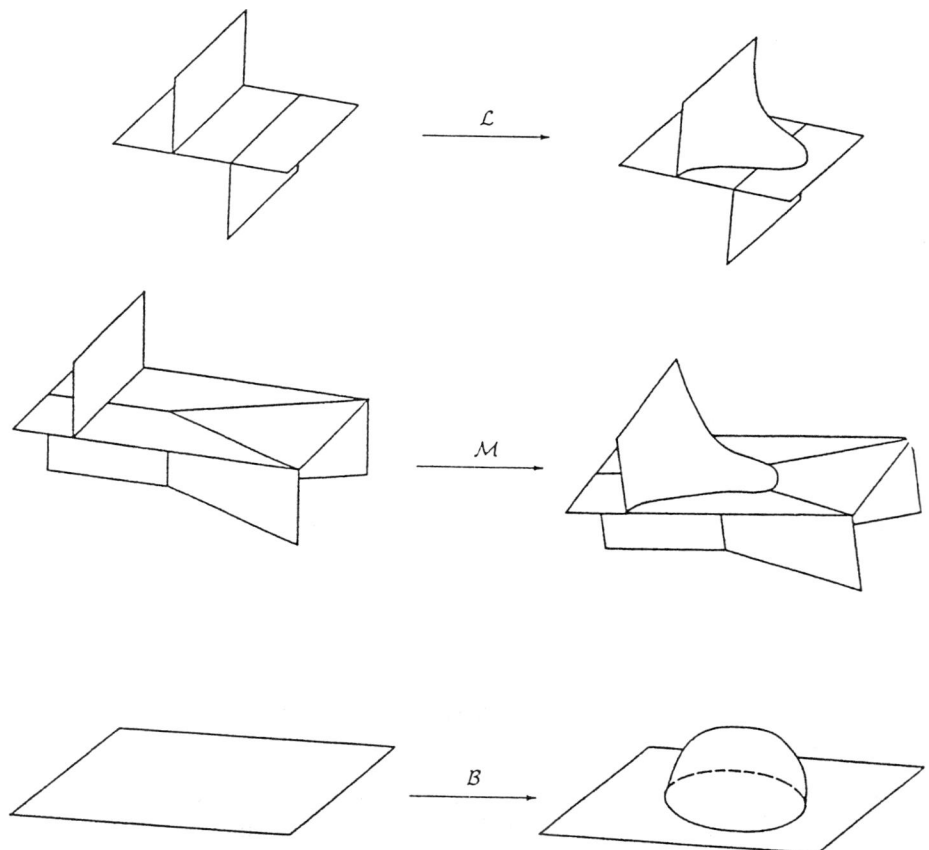

Figure 8. The moves $\mathcal{L}$, $\mathcal{M}$ and $\mathcal{B}$.

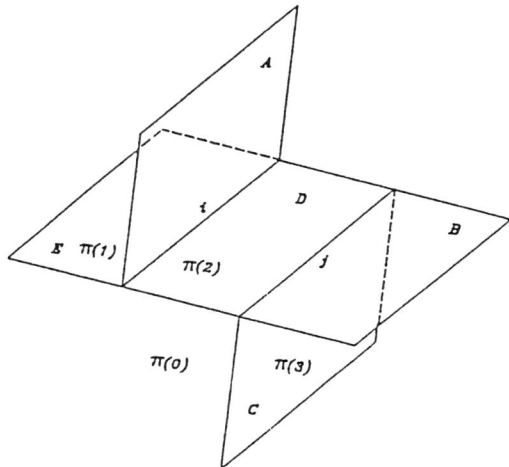

Figure 9. The numbered coloured 2-polyhedron Y.

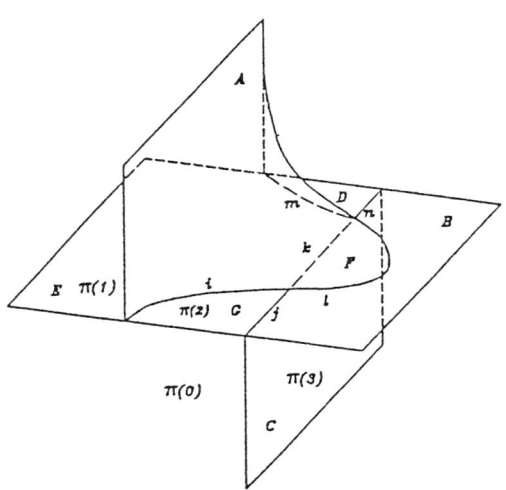

Figure 10. The numbered coloured 2-polyhedron Y'.

$$\Omega_{Y'}(\beta) = \sum_{\substack{\phi \in \mathrm{adm}(Y') \\ \partial\phi = \beta}} w^{-4} \, w_A \, w_B \, w_C \, w_D \, w_E \, w_G \, w_F^2 \, |\hat{T}_x^\phi| \, |\hat{T}_y^\phi|.$$

The sum over ϕ is in fact over F, k and l. If we construct $\hat{T}_x^\phi$ and $\hat{T}_y^\phi$ associated to the 0-strata x, y of Y', we obtain with condition $(*)$ the following equation.

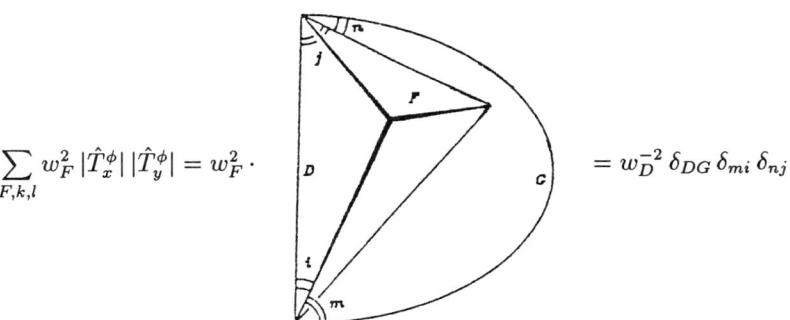

$$\sum_{F,k,l} w_F^2 \, |\hat{T}_x^\phi| \, |\hat{T}_y^\phi| = w_F^2 \cdot \quad\quad\quad = w_D^{-2} \, \delta_{DG} \, \delta_{mi} \, \delta_{nj}$$

and we get the result

$$\Omega_{Y'}(\beta) = \delta_{DG} \, \delta_{mi} \, \delta_{nj} \, w^{-4} \, w_A \, w_B \, w_C \, w_E.$$

Now we compute $\Omega_Y(\beta)$. There are no admissible colourings of Y inducing β on ∂Y unless $D = G$, $j = n$ and $i = m$. If this condition is fulfilled, we have

$$\Omega_Y(\beta) = w^{-4} \, w_A \, w_B \, w_C \, w_E.$$

□

Proposition 2.6 *Let $\alpha \in \mathrm{adm}(\partial X)$. If the initial data satisfy condition $(**)$, then $\Omega_X(\alpha)$ is invariant under $\mathcal{M}$.*

Proof: Let X' the simple 2-polyhedron obtained from X with the move $\mathcal{M}$. There are simple 2-polyhedra Y, Y' and Z such that $X = Y \cup Z$, $X' = Y' \cup Z$, $Y \cap Z = \partial Y \subset \partial Z$, $Y' \cap Z = \partial Y' = \partial Y \subset \partial Z$. The simple 2-polyhedra Y and Y' are depicted in Fig. 11 and 12. The permutation $\pi \in S_5$ indicates the numbering. In view of Lemma 2.4, we have to prove:

$$\Omega_Y(\beta) = \Omega_{Y'}(\beta) \,, \quad \forall \beta \in \mathrm{adm}(\partial Y) = \mathrm{adm}(\partial Y').$$

We consider the configurations of Fig. 11 and 12. We first compute $\Omega_Y(\beta)$, where β is the colouring of the simple graph ∂Y (see Fig. 11).

$$\Omega_Y(\beta) = \sum_{\substack{\phi \in \mathrm{adm}(Y) \\ \partial\phi = \beta}} w^{-5} \, w_A \, w_B \, w_C \, w_D \, w_E \, w_F \, w_G \, w_H \, w_I \, |\hat{T}_x^\phi| \, |\hat{T}_y^\phi|.$$

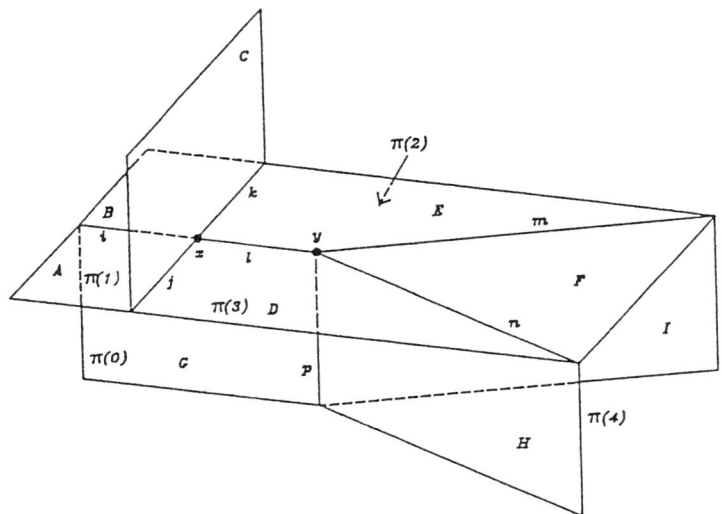

Figure 11. The numbered coloured 2-polyhedron Y.

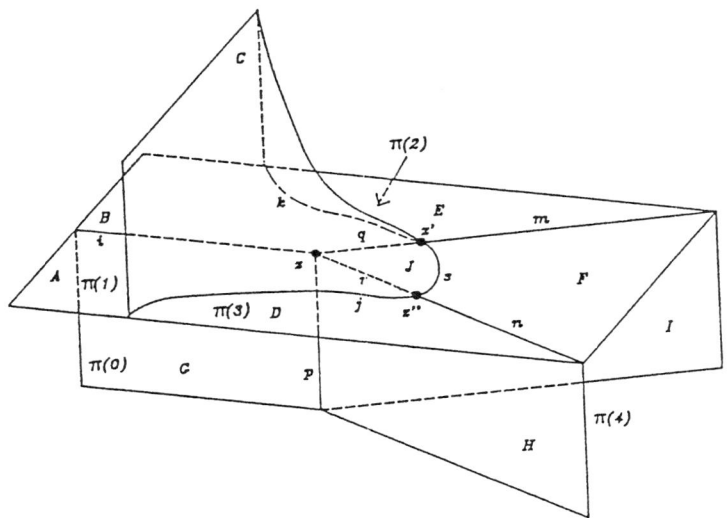

Figure 12. The numbered coloured 2-polyhedron Y'.

The sum over ϕ is a sum over l. If we construct $\hat{T}_x^\phi$ and $\hat{T}_y^\phi$ associated to the 0-strata x, y of Y, we obtain the following equation.

$$\Omega_Y(\beta) = w^{-5}\, w_A\, w_B\, w_C\, w_D\, w_E\, w_F\, w_G\, w_H\, w_I\, \cdot$$

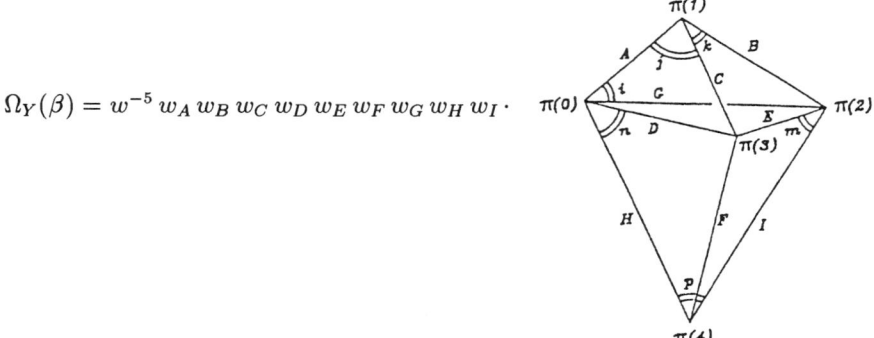

The partition function for Y' is given by

$$\Omega_{Y'}(\beta) = \sum_{\substack{\phi \in \mathrm{adm}(Y') \\ \partial\phi = \beta}} w^{-5}\, w_A\, w_B\, w_C\, w_D\, w_E\, w_F\, w_G\, w_H\, w_I\, w_J^2 \prod_{i=1}^{3} |\hat{T}_{z_i}^\phi|.$$

The sum over ϕ is a sum over J, q, r and s. Constructing the tetrahedra $|\hat{T}_{z_1}^\phi|$, $|\hat{T}_{z_2}^\phi|$ and $|\hat{T}_{z_3}^\phi|$ leads to the following relation:

$$\sum_{J,q,r,s} w_J^2 \prod_{i=1}^{3} |\hat{T}_{z_i}^\phi| = w_J^2 \cdot$$

We see that condition (**) implies that $\Omega_Y(\beta)$ equals $\Omega'_Y(\beta)$. □

Proposition 2.7 *Let* $\alpha \in \mathrm{adm}(\partial X)$. *If the initial data satisfy condition* $(***)$, *then* $\Omega_X(\alpha)$ *is invariant under* $\mathcal{B}$.

Proof: Let X' the simple 2-polyhedron obtained from X with the move $\mathcal{M}$. There are simple 2-polyhedra Y, Y' and Z such that $X = Y \cup Z$, $X' = Y' \cup Z$, $Y \cap Z = \partial Y \subset \partial Z$, $Y' \cap Z = \partial Y' = \partial Y \subset \partial Z$. The simple 2-polyhedra Y and Y' are depicted in Fig. 13. The permutation $\pi \in S_3$ indicates the numbering. In view of Lemma 2.4, we have to prove:

$$\Omega_Y(\beta) = \Omega_{Y'}(\beta)\,,\ \forall \beta \in \mathrm{adm}(\partial Y) = \mathrm{adm}(\partial Y').$$

We consider the configurations of Fig. 13. The partition functions are

$$\Omega_Y(\beta) = w^{-2}\, w_A^2,$$
$$\Omega_{Y'}(\beta) = \sum_{\substack{\phi \in \mathrm{adm}(Y') \\ \partial\phi = \beta}} w^{-4}\, w_B^2\, w_C^2.$$

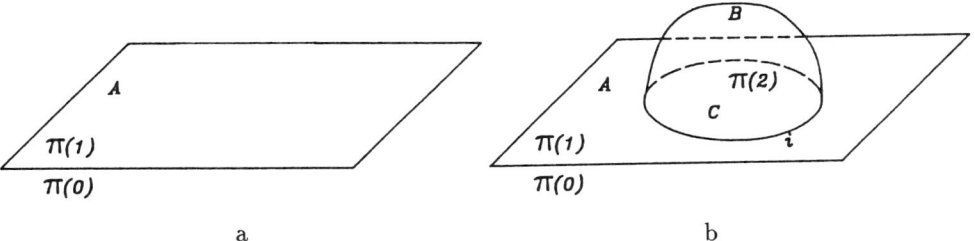

Figure 13. The numbered coloured 2-polyhedra Y and Y'.

The sum over ϕ is a sum over B, C and i. The claimed identity follows from condition $(***)$. $\square$

2.5.5 Independence of numbering

Let T be a triangulation of the compact oriented 3-manifold M, $x_1, \ldots, x_d$ the vertices of T ordered in such a way that the first e lie on the boundary ∂M. Let ν a numbering of T: $\nu(x_i)$ is the number of the vertex x_i. By going to an equivalent numbering it can be assumed that the image of ν is the set $\{1, \ldots, d\}$.

Case with empty boundary. We first consider the simpler case where $\partial M = \emptyset$. If ν' is another numbering of T with image $\{1, \ldots, d\}$, then $\nu' \circ \nu^{-1}$ is a permutation, and to each permutation in S_d there corresponds a numbering ν'. Any permutation can be written as a product of transpositions of consecutive numbers. Thus, in order to prove the invariance of $|M|$ with respect to the numbering, it is sufficient to show that $|M|$ does not change if we permute two consecutive values of ν, i.e., if we replace ν by $\nu' = \pi \circ \nu$, where $\pi(j) = j+1$, $\pi(j+1) = j$ for some j, and $\pi(i) = i$ if $i \notin \{j, j+1\}$.

If j and $j+1$ label vertices not bounding the same edge, there is nothing to prove, since the orientation of the edges is not changed. If the vertices numbered j and $j+1$ bound the same edge, we can do an Alexander move centered at this edge. This has the effect that the two vertices are not connected by an edge any more. Thus we can exchange j and $j+1$ without changing $|M|$. Then we can do an inverse Alexander move to get back the original triangulation with the new numbering.* Fig. 14 shows this in the case of three 3-simplices sharing an edge.

Case with boundary. The proof for the case with non empty boundary involves only a slight modification. We denote by $\partial\nu : \{x_1, \ldots, x_e\} \to \mathbf{N}$ the numbering of ∂T induced by ν and by $\nu_{\text{int}} : \{x_{e+1}, \ldots, x_a\} \to \mathbf{N}$ the numbering of the internal vertices.

Lemma 2.8 *The value of $\Omega_M(\alpha)$, where $\alpha \in \text{adm}(\partial T)$, is the same for all numberings of T which extend the numbering of ∂T.*

Proof: We first prove the result for two numberings μ and μ' such that $\partial\mu = \partial\mu'$, and $\text{Im}(\mu_{\text{int}}) = \text{Im}(\mu'_{\text{int}})$. Then the proof goes exactly as in the case without boundary, except when one of the numbers to be permuted is on one or more 1-simplices whose other extremity is on the boundary. In this case we make an Alexander move on each of these edges before the permutation and the inverse moves after the permutation.

We now prove the general case. Let μ and μ' be numberings of T such that $\partial\mu = \partial\mu'$. We set
$$n = \max\{\max_i \mu(x_i), \max_i \mu'(x_i)\}.$$

---

*This argument is due to Mauro Crivelli.

Let $m_{e+1}, \ldots, m_a$ be the values of μ_{int}, in ascending order. We can replace m_a by $n+a-e$, possibly with the help of an Alexander move, as in the first part of the proof, if the vertex is connected by an edge to the boundary. Likewise, we can replace m_{a-1} by $n+a-e-1$, and so on. These substitutions keep $\Omega_M(\alpha)$ unchanged. If we apply the same procedure to μ' we reduce the problem to the special case considered in the first part of the proof. □

Lastly, we want to specify in what sense $\Omega_M(\alpha)$ depends on the numbering of $\partial\mathcal{T}$. Two numberings are said to be equivalent if they induce the same orientation on all 1-simplices of $\partial\mathcal{T}$.

Proposition 2.9 *The value of $\Omega_M(\alpha)$ depends only on the equivalence class of the numbering of $\mathcal{T}$.*

Proof: Let $\partial\nu, \partial\nu'$ be two equivalent numberings of $\partial\mathcal{T}$. Choose a numbering μ of $\mathcal{T}$ extending $\partial\nu$. We saw in the proof of the preceding lemma that we can replace μ by a numbering whose internal numbers are all greater than those of $\partial\nu$ and $\partial\nu'$. After this, we can replace $\partial\nu$ by $\partial\nu'$, without changing $\Omega_M(\alpha)$. □

This concludes the proof of the invariance theorem.

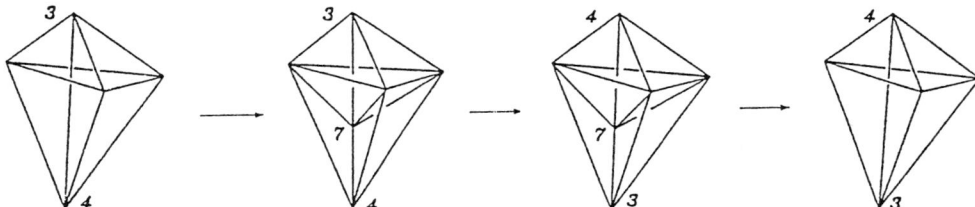

Figure 14. The permutation of two consecutive numbers

2.6 Equivalent initial data

The correct way of viewing symbols is to think of them as components of tensors of type (2,2). The next proposition shows that invariants do not depend on the basis chosen to represent them.

Definition: Two sets of initial data are called equivalent if all data are equal except possibly the symbols, $|:::|::|_1$, $|:::|::|_2$, and there are unitary matrices $s(A,B,C) \in U(N_{AB}{}^C, K)$ such that

$$\left| \begin{matrix} A & B & C \\ D & E & F \end{matrix} \right| \begin{matrix} i & k \\ j & l \end{matrix} \right|_2$$
$$= \sum_{mnpq} \left| \begin{matrix} A & B & C \\ D & E & F \end{matrix} \right| \begin{matrix} m & p \\ n & q \end{matrix} \right|_1 s(A,B,C)_{mi} s(C,D,E)_{nj} \overline{s(B,D,F)}_{pk} \overline{s(A,F,E)}_{ql}.$$

It is clear that this notion of equivalence is indeed an equivalence relation.

Proposition 2.10 *If a set of initial data obeys conditions* $(*)$, $(**)$, *and* $(***)$, *then also equivalent initial data do. Equivalent initial data give rise to the same invariants for closed oriented manifold.*

Proof: The proof of $(*)$, $(**)$ is best formulated using the graphical representation of 2.3. The main observation is that if we replace in these formulas the symbols by equivalent symbols related by s, we get, for each face shared by two tetrahedra, a factor

$$\sum_k \overline{s(A,B,C)}_{ik} s(A,B,C)_{jk} = \delta_{ij}$$

where A, B, and C are the colours of the edges. This immediately shows $(**)$. To prove $(*)$ one has to apply the unitarity relation once more.

To prove that the invariants are the same for equivalent data, one applies the same argument, using the fact that each face of the triangulation is shared precisely by two tetrahedra. □

One can then similarly show that equivalent initial data give rise to equivalent topological quantum field theories.

3 EXAMPLES

Here we give two examples of initial data giving rise to invariants of three manifolds, and topological field theories. They are both constructed out of finite group and are given in terms of the tensor category of representation of a (quasi-) Hopf algebra. In the first example, the Hopf algebra is the group algebra of a finite group, and in the second example (the Dijkgraaf-Witten model) we consider the commutative algebra of functions on a finite group, with standard coproduct but with twisted associativity isomorphism. Other examples, not discussed here, are given by quantum groups at root of unity: see ref. 1 for the sl_2 case and the general construction of ref. 4.

3.1 6j-symbols for finite groups

In this example, the set J indexes a complete system of irreducible $*$-representations of the group algebra $\mathbf{C}[G]$ of a finite group G, and the symbols of the initial data are the $6j$-symbols of $\mathbf{C}[G]$. Then the hypothesis of the invariance theorem are satisfied.

Firstly, we introduce some notations: let G be a finite group and $\mathbf{C}[G]$ its group algebra. The group algebra is an involutive Hopf algebra with coproduct and involution given by:

$$\Delta : \mathbf{C}[G] \longrightarrow \mathbf{C}[G] \otimes \mathbf{C}[G]$$
$$\sum_{g \in G} a(g)\, g \longmapsto \sum_{g \in G} a(g)\, g \otimes g$$

$$* : \mathbf{C}[G] \longrightarrow \mathbf{C}[G]$$
$$\sum_{g \in G} a(g)\, g \longmapsto \sum_{g \in G} \overline{a(g)}\, g^{-1}$$

where $\overline{a(g)}$ denotes the complex conjugate of $a(g)$. The algebra $\mathbf{C}[G]$ is semi-simple and finite dimensional. Each representation (we consider only finite dimensional representations) of $\mathbf{C}[G]$ is equivalent to a $*$-representation. Thus, we can consider a complete system of irreducible $*$-representations of $\mathbf{C}[G]$ indexed by a finite set J:

$$\mathcal{R} := \{(\rho_A, V_A, <\cdot,\cdot>_{V_A}) | A \in J\}$$

where for each $A \in J$, V_A is a vector space of finite dimension n_A, ρ_A is an algebra homomorphism from $\mathbf{C}[G]$ to $L(V_A)$ the algebra of linear operators on V_A, and $<\cdot,\cdot>_{V_A}$ is an inner product with respect to which ρ_A is a $*$-homomorphism. Moreover, for each $A \in J$, we choose an orthonormal basis $\{e_\alpha^A\}_{\alpha=1,\ldots,n_A}$ of V_A.

The tensor product of representations is defined through the coproduct of $\mathbf{C}[G]$:

$$\begin{aligned}\rho_A \otimes \rho_B : \mathbf{C}[G] &\longrightarrow L(V_A \otimes V_B) \\ a &\longmapsto (\rho_A \otimes \rho_B)(a) := (\rho_A \otimes \rho_B)\Delta(a)\end{aligned}$$

This tensor product is associative since Δ is coassociative. As the representations of $\mathbf{C}[G]$ are completely reducible, we can define multiplicity coefficients $N_{AB}{}^C$ with the following equation:

$$\rho_A \otimes \rho_B \simeq \bigoplus_{C \in J} N_{AB}{}^C \rho_C$$

We have two decompositions for the tensor product of three representations:

$$\rho_A \otimes \rho_B \otimes \rho_C \simeq \bigoplus_{D,E} N_{AB}{}^D N_{DC}{}^E \rho_E \simeq \bigoplus_{F,E} N_{BC}{}^F N_{AF}{}^E \rho_E$$

Since the decomposition in isotypic components is unique, we get:

$$\sum_D N_{AB}{}^D N_{DC}{}^E = \sum_F N_{BC}{}^F N_{AF}{}^E$$

It follows that the three following spaces are isomorphic:

$$\mathrm{Hom}_{\mathbf{C}[G]}(V_E, V_A \otimes V_B \otimes V_C) \qquad (1)$$

$$\bigoplus_D \mathrm{Hom}_{\mathbf{C}[G]}(V_D, V_A \otimes V_B) \otimes \mathrm{Hom}_{\mathbf{C}[G]}(V_E, V_D \otimes V_C) \qquad (2)$$

$$\bigoplus_F \mathrm{Hom}_{\mathbf{C}[G]}(V_F, V_B \otimes V_C) \otimes \mathrm{Hom}_{\mathbf{C}[G]}(V_E, V_A \otimes V_F). \qquad (3)$$

In the following, each inner product on a space $L(V, W)$, where V and W are inner product spaces, is defined in this way:

$$(\varphi, \psi) \longmapsto <\varphi, \psi> = \frac{1}{\dim V} \sum_{i=1}^{\dim V} <\phi(e_i), \psi(e_i)>_W$$

where $\{e_i\}_{i=1,\ldots,\dim V}$ is an orthonormal basis of V.

Now, we pass to the construction of the $6j$-symbols. For each triple $(A, B, C) \in J^3$ we choose an orthonormal basis $\{\varphi_{ABC}^i\}_{i=1,\ldots,N_{AB}{}^C}$ of $\mathrm{Hom}_{\mathbf{C}[G]}(V_C, V_A \otimes V_B)$. Since the spaces (1), (2) and (3) are isomorphic, we can construct two orthonormal bases of $\mathrm{Hom}_{\mathbf{C}[G]}(V_E, V_A \otimes V_B \otimes V_C)$, namely:

$$\{(\varphi_{ABD}^i \otimes \mathrm{id})\varphi_{DCE}^j | D \in J, 1 \leq i \leq N_{AB}{}^D, 1 \leq j \leq N_{DC}{}^E\}$$

$$\{(\mathrm{id} \otimes \varphi_{BCF}^k)\varphi_{AFE}^l | F \in J, 1 \leq k \leq N_{BC}{}^F, 1 \leq l \leq N_{AF}{}^E\}$$

The $6j$-symbols are proportional to matrix elements of the corresponding basis transformation:

$$\begin{Bmatrix} A & B & D \\ C & E & F \end{Bmatrix} \begin{vmatrix} i & k \\ j & l \end{vmatrix} := \frac{1}{\sqrt{n_D n_F}} < (\varphi^i_{ABD} \otimes \mathrm{id})\varphi^j_{DCE}, (\mathrm{id} \otimes \varphi^k_{BCF})\varphi^l_{AFE} >$$

The $6j$-symbols do not depend on the bases e^A_α of the spaces V_A but on the bases φ^i_{ABC}. We can now define the initial data:

1. The set J indexes a complete system of irreducible $*$-representations of $\mathbf{C}[G]$. A triple $(A, B, C) \in J^3$ is admissible if $N_{AB}{}^C \neq 0$.

2. We set $K = \mathbf{C}$ and the involution is the complex conjugation.

3. The number $N_{AB}{}^C$ is defined as the multiplicity of ρ_C in the tensor product $\rho_A \otimes \rho_B$.

4. The map $W : J \to \mathbf{C}^*$ is given by $A \mapsto \sqrt{n_A}$.

5. The set $\mathrm{adm}(J^3 \times \mathbf{N}^*)$ indexes the bases of the spaces $\mathrm{Hom}_{\mathbf{C}[G]}(V_C, V_A \otimes V_B)$.

6. The set $\mathrm{adm}(J^6 \times (\mathbf{N}^*)^4)$ is defined as earlier.

7. The symbols are the $6j$-symbols.

It turns out that these initial data satisfy the conditions $(*)$, $(**)$ and $(***)$. For these initial data, the distinguished element w of condition $(***)$ is the square root of the order of the group: $\sqrt{|G|}$. Different choices of orthonormal bases φ^i_{ABC} lead to equivalent initial data.

3.2 The Dijkgraaf-Witten model

Take J to be a finite group of order $|J|$, $\mathrm{adm}(J^3)$ the set of triples (A, B, C) such that $C = AB$, with $N_{AB}{}^C = 1$ and $w_A = 1$ for all $A \in J$. Then $\mathrm{adm}(J^6 \times (\mathbf{N}^*)^4)$ consists of elements of the form

$$(A, B, AB, C, ABC, BC; 1, 1, 1, 1).$$

We set

$$\alpha(A, B, C) = \begin{Bmatrix} A & B & AB \\ C & ABC & BC \end{Bmatrix} \begin{vmatrix} 1 & 1 \\ 1 & 1 \end{vmatrix} \in K.$$

The conditions $(*)$, $(**)$, $(***)$ are then equivalent to

$(*)$ $\overline{\alpha(A, B, C)} \alpha(A, B, C) = 1$,

$(**)$ $\alpha(AB, C, D)\alpha(A, B, CD) = \alpha(B, C, D)\alpha(A, BC, D)\alpha(A, B, C)$,

$(***)$ $w^2 = |J|$.

In other words, α is a group three-cocycle with values in $U(1, K) = \{x \in K | x\bar{x} = 1\}$, the group of unitary elements of K. Equivalent initial data are given by cocycles differing by coboundaries. Thus equivalence classes of initial data are labeled by $H^3(G, U(1, K))$.

We now show, if $K = \mathbf{C}$, how this model can be understood in terms of $6j$-symbols for a simple quasi-Hopf algebra. Let $\mathcal{A} = \mathbf{C}^G$ be the algebra of complex-valued functions on the finite group G. It is of course a commutative Hopf algebra, with $*$-structure, but we can make it into a quasi-Hopf algebra[11] by introducing the associativity isomorphism $\alpha \in \mathcal{A} \otimes \mathcal{A} \otimes \mathcal{A} = \mathbf{C}^{G \times G \times G}$. The pentagon equation[11] for α is indeed the cocycle condition. Note that this twist does not have any effect on the algebra or coalgebra structure of $\mathcal{A}$, but twists the tensor category of its representations. It turns out that $\mathcal{A}$ is a quasi-Hopf subalgebra of the DPR algebra[12]. Since α takes values in $U(1)$, we get a semisimple C^* tensor category (see ref. 13), and $6j$-symbols can be defined: let $\{V_A\}_{A \in J}$ a complete set of inequivalent irreducible representations. Then we have the canonical isomorphisms:

$$\bigoplus_D \mathrm{Hom}_{\mathcal{A}}(V_D, V_A \otimes V_B) \otimes \mathrm{Hom}_{\mathcal{A}}(V_E, V_D \otimes V_C) \xrightarrow{S} \mathrm{Hom}_{\mathcal{A}}(V_E, V_A \otimes (V_B \otimes V_C))$$

$$\downarrow \alpha$$

$$\bigoplus_F \mathrm{Hom}_{\mathcal{A}}(V_F, V_B \otimes V_C) \otimes \mathrm{Hom}_{\mathcal{A}}(V_E, V_A \otimes V_F) \xrightarrow{T} \mathrm{Hom}_{\mathcal{A}}(V_E, (V_A \otimes V_B) \otimes V_C).$$

The vertical arrow is composition with the associativity isomorphism α and $6j$-symbols can be defined as matrix elements of the isomorphism $T^* \alpha S$ for some choice of orthonormal bases of the spaces $\mathrm{Hom}_{\mathcal{A}}(V_A, V_B \otimes V_C)$. In our example, irreducible representations are one-dimensional, labeled by the elements of the group: $V_A = \mathbf{C}$, $\rho_A(f) = f(A)$, and we have the canonical isomorphism $V_A \otimes V_B \simeq V_{AB}$. Thus $\mathrm{Hom}(V_{AB}, V_A \otimes V_B) = \mathbf{C}$, with canonical orthonormal basis 1, and the $6j$-symbol is just α evaluated on the three independent group elements. Choosing different orthonormal bases amounts to replacing α by a cocycle in the same cohomology class.

ACKNOWLEDGEMENTS

We wish to thank M. Crivelli, P. Feng, T. Kerler and B. Mirzai for helpful comments and discussions.

REFERENCES

[1] V. G. Turaev and O. Y. Viro, *State sum of 3-manifold and quantum 6j-symbols*, Topology 31, 865-902 (1992).
[2] M. Karowski, W. Müller and R. Schrader: *State sum invariants of compact 3-manifolds with boundary and 6j-symbols*, J. Phys. A25, 4847-4860 (1992).
[3] K. Walker, *On Witten's 3-manifold invariant*, preprint 1992.
[4] B. Durhuus, H. P. Jacobsen and R. Nest, *Topological quantum field theories from 6j-symbols*, Copenhagen preprint 1991.
[5] R. Dijkgraaf and E. Witten, *Topological gauge theories and group cohomology*, Commun. Math. Phys. 129, 393-429 (1990).
[6] G. Kuperberg, *Involutory Hopf algebras and 3-manifold invariants*, Int. J. Math. 2, 41-46 (1991).
[7] D. S. Freed and F. Queen, *Chern-Simons theory with finite gauge group*, preprint 1991.
[8] D. Altschuler and A. Coste, *Quasi-quantum groups, knots, three-manifolds and topological field theory*, CERN preprint TH 6360/92.
[9] D. N. Yetter, *Topological field theory associated to finite groups and crossed G-sets*, J. Knot Theory and its Ramifications 1, 1-20 (1992).
[10] O. Grandjean, *Quelques invariants topologiques des variétés de dimension 3*, Diploma thesis, ETH Zurich (1992).
[11] V. G. Drinfeld, *On almost cocommutative Hopf algebras*, Leningrad Math. J. 1, 321–342 (1990).
[12] R. Dijkgraaf, V. Pasquier and Ph. Roche, in *Modern quantum field theory*, Tata Institute, Bombay 1990.
[13] T. Kerler, to appear in the proceedings of the Cargèse summer school, 1991.

SCHWINGER-DYSON EQUATION IN THREE-DIMENSIONAL SIMPLICIAL QUANTUM GRAVITY

Hirosi Ooguri

Research Institute for Mathematical Sciences
Kyoto University, Kyoto 606-01, Japan
and
Lyman Laboratory of Physics
Harvard University, Cambridge, MA 02138, USA

ABSTRACT

We study simplicial quantum gravity in three dimensions. Motivated by Boulatov's model which generates a sum over simplicial complexes weighted with the Turaev-Viro invariant, we introduce boundary operators in the simplicial gravity associated to compact orientable surfaces. An amplitude of the boundary operator is given by a sum over triangulations in the interior of the boundary surface. It turns out that the amplitude solves the Schwinger-Dyson equation even if we restrict the topology in the interior of the surface, so long as the surface is non-degenerate. We propose a set of factorization conditions on the amplitudes which singles out a solution associated to triangulations of S^3.

1. INTRODUCTION

In order to understand a quantum theory of gravity, we need to clarify what we mean by integrations over Riemannian metrics on a d-dimensional differentiable manifold M. One of the possibilities proposed by Regge [1] is to discretize the integrals and replace them by sums over smooth triangulations on M. In this approach, each triangulation T specifies a Riemannian metric on M in the following way. We impose the condition that an interior of each d-simplex in T is flat Euclidean and that the curvature has a support only on $(d-2)$-simplexes. Moreover all the edges in T are regarded as being straight and having the same geodesics length a. (There is another version of the theory where the edges are allowed to have different lengths [2], but we do not examine this possibility here.) These conditions are enough to define a Riemannian metric associated to the triangulation T of M. It is reasonable to expect that any metric on M can be approximated in this way by using sufficiently large number of simplexes and by taking the size a of each simplex to be infinitesimally small. It is then important to know what types of triangulations give relevant contributions to the summations. If smooth triangulations with infinitely many simplexes dominate the summations, it would make sense to *define* the integrals over metrics by the sums over triangulations. In two dimensions, this program has worked rather well and we have learned quite a lot about two-dimensional quantum gravities from this approach for the last few years. The purpose of this paper is to study the three-dimensional case.

Let us be more specific about what we mean by the sums over triangulations. In order to quantize the Einstein gravity, we would like to perform integrals over metrics $g_{\mu\nu}$ weighted with the exponential of the action given by $S[g] = \Lambda \int d^d x \sqrt{g} + \frac{1}{G} \int d^d x \sqrt{g} R$ where R is the scalar curvature. Suppose that the metric g is associated to some triangulation T of M. Since each d-simplex has the same volume proportional to a^d, the total volume $\int d^d x \sqrt{g}$ of M should be proportional to $a^d n_d(T)$ where $n_i(T)$ ($i = 0, 1, \ldots, d$) is the number of i-simplexes in T. The integral of the scalar curvature R on the other hand is expressed in the cases of $d = 2$ and 3 as follows. When $d = 2$, the curvature has a support on the vertices of T. Since the deficit angle around each vertex v is given by $(2\pi - \frac{\pi}{3} \tau_\triangle(v))$ where $\tau_\triangle(v)$ is the number of triangles containing v, we obtain

$$\int d^2 x \sqrt{g} R \sim \sum_{v:\text{vertices}} \left(1 - \frac{1}{6}\tau_\triangle(v)\right) = n_0(T) - \tfrac{1}{2} n_2(T).$$

Especially when the manifold M is closed, $3n_2(T) = 2n_1(T)$ since every edge on T is shared by exactly two triangles. In this case, the right-hand side in the above becomes $\chi = n_0(T) - n_1(T) + n_2(T)$, the Euler number of M. When $d = 3$, the curvature is concentrated on the edges of T and is characterized by the deficit angle around each edge e given by $(2\pi - \cos^{-1}(\frac{1}{3})\tau_t(e))$ where $\tau_t(e)$ denotes the number of tetrahedra containing e. Since each edge has the geodesic length a, the integral of the scalar curvature is given by

$$\int d^3 x \sqrt{g} R \sim a \sum_{e:\text{edges}} \left[2\pi - \cos^{-1}\left(\frac{1}{3}\right) \tau_t(e)\right]$$

$$= a \left[2\pi n_1(T) - 6 \cos^{-1}\left(\frac{1}{3}\right) n_3(T)\right].$$

Therefore, for both $d = 2$ and 3, the action $S[g]$ is expressed as a linear combination of $n_d(T)$ and $n_{d-2}(T)$.

$$S[g] = -\kappa n_{d-2}(T) + \beta n_d(T).$$

The partition function of the simplicial quantum gravity is then defined by

$$Z_d = \sum_T \frac{1}{C(T)} e^{\kappa n_{d-2}(T) - \beta n_d(T)}, \qquad (1)$$

where the sum is over all possible triangulations of M, and the factor $C(T)$ is the order of a symmetry of T if any. This factor is included in order to take into account the volume factor of the diffeomorphism group. First of all, we need to know if this summation is convergent at all. Since Z_d can be written as

$$Z_d = \sum_{n_d, n_{d-2}=0}^{\infty} Z_{n_d, n_{d-2}}(M) e^{\kappa n_{d-2} - \beta n_d}$$

where $Z_{n_d, n_{d-2}}(M)$ is a number of triangulations of M with given values of n_d and n_{d-2} weighted with the factor $C(T)$, we would like to know the asymptotic behaviour of $Z_{n_d, n_{d-2}}(M)$ for large values of n_d and n_{d-2}. In two dimensions, n_2 and n_0 are not independent and they are related as as $n_0 = \frac{1}{2} n_2 + \chi$. Especially when $\chi = 2$ (i.e. $M \simeq S^2$), the exact asymptotic behaviour of Z_{n_2, n_0} is known from the combinatorial study of ref. [3] as

$$Z_{n_2, n_0}(S^2) \sim n_2^{-\frac{7}{2}} e^{\beta_c n_2}, \quad (n_2 \to \infty, \ n_0 = \tfrac{1}{2} n_2 + 2), \quad e^{\beta_c} = \frac{2^4}{3^{3/2}}. \quad (2)$$

The sum over triangulations can also be evaluated by the matrix model [4], and the above formula for S^2 is reproduced by the sum over one-particle irreducible vacuum diagrams [4] in the large-N limit of the matrix model [5] [6] [7]. (The sum over connected vacuum diagrams also gives a similar formula with a different value of β_c; $e^{\beta_c} = 2 \cdot 3^{\frac{3}{4}}$.) When M is a surface of higher genus ($\chi \leq 0$), subleading corrections [8] to the large-N matrix model give [9]

$$Z_{n_2 n_0}(M) \sim n_2^{-\frac{5}{4}\chi - 1} e^{\beta_c n_2} \quad (n_2 \to \infty, \ n_0 = \tfrac{1}{2} n_2 + \chi).$$

Therefore the contribution to $Z_{d=2}$ from T with many triangles is expressed as

$$Z_{d=2} \sim e^{\kappa \chi} \sum_{n_2} n_2^{-\frac{5}{4}\chi - 1} e^{-(\tilde{\beta}-\beta_c) n_2} \sim \begin{cases} (\tilde{\beta} - \beta_c)^{\frac{5}{4}\chi} & \text{if } \chi \neq 0, \\ \log\left(\frac{1}{\tilde{\beta}-\beta_c}\right) & \text{if } \chi = 0, \end{cases}$$

where $\tilde{\beta} = \beta - \frac{1}{2}\kappa$, and it is convergent for $\tilde{\beta} > \beta_c$. Moreover as $\tilde{\beta}$ approaches to the critical value β_c, the contributions from triangulations with $n_2 \gg 1$ become more and more significant for $Z_{d=2}$ with $\chi \leq 0$ and for $\partial_{\tilde{\beta}}^3 Z_{d=2}$ with $\chi = 2$. Thus we have a good reason to believe that the program of the simplicial quantum gravity works well in two dimensions. Indeed this approach has turned out to be quite successful and its validity has been confirmed by comparisons of the matrix model computations with the results of the continuum field theories [10] [11] [12] [13] [14] [15] [16].

Recently several numerical results have been reported on the three-dimensional simplicial quantum gravity [17] [18] [19] [20] [21]. Their results seem to be compatible with the following bound on the number of triangulations of S^3

$$Z_{n_3}(S^3) = \sum_{n_1} Z_{n_3 n_1}(S^3) < C e^{\beta_c n_3} \quad (n_3 \to \infty), \quad (3)$$

for some constants C and β_c. In three dimensions, the Euler number $\chi = n_0 - n_1 + n_2 - n_3$ is zero for any triangulation of a closed manifold. We also have $n_2 = 2n_3$ which follows from the fact that any triangle in T is shared by exactly two tetrahedra. Therefore $n_3 = n_1 - n_0$, and n_3 should always be less than n_1. On the other hand, since each edge in T belongs to at least one tetrahedra, n_1 must be less than or equal

53

to $6n_3$. Namely $n_3 < n_1 \leq 6n_3$. Thus, if the exponential bound (3) is indeed correct, the sum over triangulations in eq. (1) would be bounded as

$$Z_{d=3} \leq \sum_{n_3,n_1} Z_{n_3 n_1} e^{-(\beta-6\kappa)n_3} < C \sum_{n_3} e^{-(\beta-6\kappa-\beta_c)n_3}$$

for $\kappa \geq 0$ and

$$Z_{d=3} < \sum_{n_3,n_1} Z_{n_3 n_1} e^{-(\beta-\kappa)n_3} < C \sum_{n_3} e^{-(\beta-\kappa-\beta_c)}$$

for $\kappa < 0$. The sum (1) would then be convergent at least for $\beta - 6\kappa > \beta_c$ ($\kappa \geq 0$) and for $\beta - \kappa > \beta_c$ ($0 > \kappa$). It is thus important to develop an analytical tool to examine the behaviour of $Z_{n_3}(S^3)$.

In two dimensions, the matrix model [4] has been found useful to study the simplicial quantum gravity, and there have been some attempts to extend it to three dimensions [22] [23] [24]. In this so-called tensor model, one considers and integral over rank-three tensors rather than matrices, and the perturbative expansion of the integral generates simplicial complexes, i.e. collections of tetrahedra whose boundary triangles are pairwisely identified. However the collections of tetrahedra generated in this model are not necessarily triangulations of smooth manifolds. In fact their Euler numbers are not always zero, but in general non-negative integers [22].

The basic reason for the failure of the tensor model is that the model does not contain as many parameters as one needs to distinguish various topologies in three dimensions. In two dimensions, a topology of a closed orientable surface is characterized by the number of handles, and two parameters in the model are enough to control both the topology and the volume n_2 of the surface. The matrix model indeed has two parameters, the size N of the matrix and the coupling constant g. In three dimensions, on the other hand, we need at least three parameters to control three independent variables n_3, n_1 and $\chi = n_0 - n_1 + n_2 - n_3$. The simplest version of the tensor model has only two parameters, the size N of the tensor and the coupling constant, and it is not possible to force χ to be zero.

Recently, Boulatov proposed an improved version of the tensor model [25] by incorporating a gauge symmetry of a finite group or a quantum group. If one takes the gauge group to be a finite group G, the order of G couples to the Euler number of the simplicial complex. For example, when G is the cyclic group Z_p, the partition function of the model is expressed as a sum over simplicial complexes T weighted with a factor $p^{\chi(T)}|H_1(T, Z_p)|$, where $|H_1(T; Z_p)|$ is the order of the first cohomology group of T with coefficients in Z_p. In three dimensions, the necessary and the sufficient condition for a simplicial complex T to be a manifold is $\chi(T) = 0$, and otherwise $\chi(T) > 0$. Thus, if one could make sense of the limit of $p \to 0$, the perturbative expansion of the model in this limit would be dominated by the sum over simplicial manifolds with $\chi(T) = 0$.

The $p \to 0$ limit of the Z_p model not only imposes $\chi = 0$ but gives an additional condition on topologies of simplicial manifolds, i.e. the first and the second homologies of the manifolds are restricted to be trivial [25]. Without any condition on the topology, the number of three-dimensional simplicial manifolds with a given number n_3 of tetrahedra grows at least factorially in n_3 [22], and the sum over simplicial manifolds becomes divergent. Thus one needs to impose restrictions on the topologies. It is not clear if the conditions on the homologies are enough to tame the factorial growth since there are many topologies in three dimensions with the same homologies as those of S^3. If not, the sum over triangulations generated by the Z_p

model would still be divergent, and we would need to develop a mechanism to impose further restrictions on the topologies.

In this paper, we examine the Boulatov's model associated to the quantum group $SU_q(2)$. In this model, the sum over simplicial complexes is weighted with $\Lambda_q^{\chi(T)} I_q(T)$ where

$$\Lambda_q = -\frac{2(k+2)}{(q^{1/2} - q^{-1/2})^2}, \quad q = e^{2\pi i/(k+2)}$$

and $I_q(T)$ is the Turaev-Viro invariant [26]. We define boundary operators of the model associated to compact orientable triangulated two-dimensional surfaces, and derive the Schwinger-Dyson equation for the amplitudes of the operators. It turns out that the Schwinger-Dyson equation takes the same form for any value of q when the surfaces involved are non-degenerate. Since the amplitude of the boundary operator in the Boulatov model is expressed a sum over triangulations in the interior of the boundary surface weighted with $\Lambda_q^\chi I_q$ depending on q, the independence of the Schwinger-Dyson equation on q suggests that we can restrict the topology of T in the interior of the surface without spoiling the equation. Indeed we find that, associated to an arbitrary closed orientable three-dimensional manifold, one can construct a solution to the equation. In order to single out a solution associated to triangulations of S^3, we introduce a set of factorization conditions for degenerate surfaces with which the equation can be solved uniquely and inductively.

In the case of the two-dimensional simplicial gravity, the Schwinger-Dyson equation can be derived from purely combinatorial considerations without using the matrix model. This is also true in the three-dimensional model studied here. However, in order to take into account the symmetry factor to each graph, it is more convenient and transparent to employ the perturbative expansion of the Boulatov model.

This paper is organized as follows. In Section 2, we briefly describe of the Schwinger-Dyson equation in the large-N matrix model and its combinatorial meaning. We then define the Boulatov's model in Section 3 and examine its properties. The boundary operator are defined in Section 4, and the Schwinger-Dyson equation is derived. In Section 5, we show that there is a solution to the equation associated to each closed orientable three-dimensional manifold. The factorization conditions on the amplitude which characterize the solution associated to S^3 are defined in Section 6. The Schwinger-Dyson equation combined with the factorization conditions is shown to have a unique solution.

2. SCHWINGER-DYSON EQUATION IN THE MATRIX MODEL

Let (M_{ij}) be a $N \times N$ hermitian matrix and consider the following integral.

$$Z = \int [dM] \exp(-S(M)) \qquad (4)$$

where

$$S(M) = \frac{1}{2} \text{tr} M^2 - \frac{g}{3\sqrt{N}} \text{tr} M^3.$$

by expanding the integral (4) in powers of g, Z is expressed as a sum over orientable closed trivalent graphs, each of which is dual to a triangulation of an orientable closed (but not necessarily connected) surface. A sum over closed and connected simplicial manifolds is obtained by taking the logarithm of Z as

$$\log Z = \sum_T \frac{1}{C(T)} \left(\frac{g}{\sqrt{N}}\right)^{n_2(T)} N^{n_0(T)} = \sum_T \frac{1}{C(T)} g^{n_2(T)} N^{\chi(T)}.$$

In particular the sum over triangulations of S^2 is given by the leading term in the large-N limit of $\log Z$.

The basic quantity we examine in this section is the expectation value of $\operatorname{tr} M^n$ at the large-N.

$$u_n = \lim_{N\to\infty} \frac{1}{N^{\frac{1}{2}n+1}} \langle \operatorname{tr} M^n \rangle = \lim_{N\to\infty} \frac{1}{N^{\frac{1}{2}n+1}} \cdot \frac{1}{Z} \int [dM] \operatorname{tr} M^n \exp(-S(M)).$$

The power of N is multiplied to $\langle \operatorname{tr} M^n \rangle$ so that u_n gives a finite value in the limit of $N \to \infty$. The Schwinger-Dyson equation for u_n [27], [28] is derived from

$$\begin{aligned} 0 &= \frac{1}{Z} \int [dM] \frac{\partial}{\partial M_{ij}} \left[(M^{n-1})_{ij} \exp(-S(M)) \right] \\ &= -\langle \operatorname{tr} M^n \rangle + \frac{g}{\sqrt{N}} \langle \operatorname{tr} M^{n+1} \rangle + \sum_{k=0}^{n-2} \langle \operatorname{tr} M^k \operatorname{tr} M^{n-k-2} \rangle. \end{aligned} \quad (5)$$

Here the integration by part is justified in each term in the perturbative expansion in g. In the limit of large-N, the last term in the right-hand side of eq. (5) factorizes as

$$\frac{1}{N^{\frac{1}{2}k+1}} \frac{1}{N^{\frac{1}{2}(n-k-2)+1}} \langle \operatorname{tr} M^k \operatorname{tr} M^{n-k-2} \rangle = u_k u_{n-k-2} + O(\frac{1}{N^2})$$

since only the planar graphs contribute. Thus we obtain a recurrence relation for u_n as

$$u_n = g u_{n+1} + \sum_{k=0}^{n-2} u_k u_{n-k-2}. \quad (6)$$

This is the Schwinger-Dyson equation in the large-N limit of the matrix model.

This equation has the following simple graphical interpretation. The expectation value u_n can be expressed as a sum over planar trivalent graphs with n external legs. The graphs are not necessarily connected, but they do not contain a closed disconnected part because of the factor $1/Z$ in u_n. Let us take one of the graphs T in the summation and pay attention to one of the external legs on it. The leg should be either connected to a trivalent vertex in T or to another external leg. In the former case, the rest will be a graph with $(n+1)$ external legs. If on the other hand the leg is connected to another external legs, the line connecting these two legs will separate the original planar graph into two disconnected graphs with k and $(n-k-2)$ external legs ($k = 1, 2, \ldots, n-2$). The two terms in the right-hand side of the Schwinger-Dyson equation correspond to these two possibilities.

It is also straightforward to show that the Schwinger-Dyson equation completely determines the power series expansion of u_n in g. By substituting

$$u_n(g) = \sum_{n_2=0}^{\infty} u_{n,n_2} g^{n_2}$$

into eq. (6), we obtain a recurrence relation with respect to n and n_2 as

$$u_{n,n_2} = u_{n+1,n_2-1} + \sum_{k=0}^{n-2} \sum_{s=0}^{n_2} u_{k,s} u_{n-k-2,n_2-s}. \quad (7)$$

Since $u_0(g)$ is normalized to be 1, the initial condition of the recurrence relation is $u_{0,n_2} = \delta_{0,n_2}$, with which u_{n,n_2} is determined inductively by solving (7) following the bibliographic ordering: $(n,n_2) \succ (n',n'_2)$ if $n_2 > n'_2$ or $n_2 = n'_2$, $n > n'$.

Since the nonlinear term in the Schwinger-Dyson equation (6) is the convolution of u_n's, it can be transformed into the integral equation

$$-1 + g\lambda^2 = P \int \frac{d\lambda'}{2\pi} \frac{\rho(\lambda')}{\lambda - \lambda'} \tag{8}$$

where

$$u_n = \int d\lambda \, \rho(\lambda) \lambda^n \tag{9}$$

and $P \int$ denotes the principal part of the integral. This equation can be solved using the technique of the Riemann-Hilbert correspondence [4] as

$$\rho(\lambda) = \frac{1}{\pi}[1 + \tfrac{1}{2}g(a+b) + g\lambda]\sqrt{(\lambda-a)(b-\lambda)} \quad (a \leq \lambda \leq b)$$
$$= 0 \qquad (\lambda < a \text{ or } b < \lambda)$$

where a and b are determined by

$$g(b-a)^2 + 2(a+b)[2+g(a+b)] = 0, \quad (b-a)^2[1+g(a+b)] = 1.$$

We can also compute the partition function (4) itself from the solution to the Schwinger-Dyson equation. By using the relation between u_3 and Z,

$$u_3 = \frac{1}{N^2\sqrt{N}}\langle \mathrm{tr}M^3 \rangle = \frac{3}{N^2}\frac{1}{Z}\frac{\partial Z}{\partial g} \tag{10}$$

we obtain

$$Z(g) = Z(0)\exp\left(\frac{N^2}{3}\int_0^g dg' u_3(g')\right).$$

The approach described here using the Schwinger-Dyson equation is in fact equivalent at the large-N to the standard computation [4] [8] using the eigenvalues λ_i ($i = 1, \ldots, N$) of M. By inverting the integral transformation (9), we can express $\rho(\lambda)$ in terms of u_n as

$$\rho(\lambda) = \int \frac{dk}{2\pi} e^{-ik\lambda} \sum_{n=0}^{\infty} \frac{1}{n!} u_n \lambda^n = \int \frac{dk}{2\pi} e^{-ik\lambda} \left\langle \frac{1}{N}\mathrm{tr}\exp\left(ik\frac{M}{\sqrt{N}}\right)\right\rangle$$
$$= \sum_{i=1}^{N} \frac{1}{N} \left\langle \delta\left(\lambda - \frac{\lambda_i}{\sqrt{N}}\right)\right\rangle.$$

Namely $\rho(\lambda)$ is the density of the eigenvalues of M.

3. BOULATOV'S MODEL

In this section, we describe the model due to Boulatov [25] and examine its properties. Let us consider the following integral

$$Z_q = \int [dM]\exp(-S(M)) \tag{11}$$

with $S(M)$ given by

$$S(M) = \frac{1}{6} \sum_{a_1,a_2,a_3=1}^{N} \sum_{\{j_1,j_2,j_3\}} \sum_{\{-j_k \leq m_k \leq j_k\}} |M^{j_1 j_2 j_3; a_1 a_2 a_3}_{m_1 m_2 m_3}|^2$$

$$- \frac{g\Lambda_q}{12} \sum_{a_1,\ldots,a_6=1}^{N} \sum_{\{j_1,\ldots,j_6\}} \sum_{\{-j_k \leq m_k \leq j_k\}} (-1)^{\sum_{k=1}^{6} j_k} \begin{Bmatrix} j_1 & j_2 & j_3 \\ j_4 & j_5 & j_6 \end{Bmatrix}_q \quad (12)$$

$$\times (-q)^{\sum_{k=1}^{6} m_k} M^{j_1 j_2 j_3; a_1 a_2 a_3}_{-m_1 -m_2 -m_3} M^{j_3 j_4 j_5; a_3 a_4 a_5}_{m_3 -m_4 m_5} M^{j_1 j_5 j_6; a_1 a_5 a_6}_{m_1 -m_5 m_6} M^{j_2 j_6 j_4; a_2 a_6 a_4}_{m_2 -m_6 m_4}$$

where

$$\begin{Bmatrix} j_1 & j_2 & j_3 \\ j_4 & j_5 & j_6 \end{Bmatrix}_q$$

is the Racah-Wigner $6j$-symbol of the quantum group $SU_q(2)$ [29] with $q = e^{2\pi i/(k+2)}$ and

$$\Lambda_q = -\frac{2(k+2)}{(q^{1/2} - q^{-1/2})^2}.$$

The integration variable $M^{j_1 j_2 j_3; a_1 a_2 a_3}_{m_1 m_2 m_3}$ obeys the conditions

(1) $M^{m_1 m_2 m_3}_{j_1 j_2 j_3; a_1 a_2 a_3} = 0$ unless $|j_1 - j_2| \leq j_3 \leq j_1 + j_2$ and $j_1 + j_2 + j_3 \leq k$,

(2) $\overline{M}^{j_1 j_2 j_3; a_1 a_2 a_3}_{m_1 m_2 m_3} = (-1)^{j_1+j_2+j_3}(-q)^{m_1+m_2+m_3} M^{j_1 j_2 j_3; a_1 a_2 a_3}_{-m_1 -m_2 m_3}$,

(3) $M^{j_1 j_2 j_3; a_1 a_2 a_3}_{m_1 m_2 m_3} = M^{j_2 j_3 j_1; a_2 a_3 a_1}_{m_2 m_3 m_1}$.

The perturbative expansion of this integral gives a sum over Feynman graphs each of which is dual to an oriented simplicial complex in three dimensions. A contribution of each simplicial complex T is evaluated by assigning a triplet of indices (a_e, j_e, m_e) to each edge e in T and by summing over the indices on all the edges with the conditions

$$|j_{e_1(\Delta)} - j_{e_2(\Delta)}| \leq j_{e_3(\Delta)} \leq j_{e_1(\Delta)} + j_{e_2(\Delta)}, \quad j_{e_1(\Delta)} + j_{e_2(\Delta)} + j_{e_3(\Delta)} \leq k$$

for every triangle Δ in T where $e_r(\Delta)$ $(r = 1, 2, 3)$ are the edges of Δ. We then obtain

$$\log Z_q = \sum_T \frac{1}{C(T)} (g\Lambda_q)^{n_3(T)} N^{n_1(T)} \Lambda_q^{n_0(T)-1} I_q(T),$$

where the sum $\sum_T$ is over connected simplicial complexes, and $I_q(T)$ is given by

$$I_q(T) = \frac{1}{\Lambda_p^{n_0(T)-1}} \sum_{\{j\}} \prod_{e:\text{edges}} (-1)^{2j_e} [2j_e + 1]_q$$

$$\times \prod_{t:\text{tetrahedra}} (-1)^{\sum_{k=1}^{6} j_{e_k}(t)} \begin{Bmatrix} j_{e_1}(t) & j_{e_2}(t) & j_{e_3}(t) \\ j_{e_4}(t) & j_{e_5}(t) & j_{e_6}(t) \end{Bmatrix}_q \quad (13)$$

where

$$[2j+1]_q = \frac{q^{j+1/2} - q^{-j-1/2}}{q^{1/2} - q^{-1/2}},$$

and $e_r(t)$ $(r = 1, \ldots, 6)$ are the six edges of the tetrahedron t.

It has been shown by Turaev and Viro [26] that $I_q(T)$ is a topological invariant of T, i.e. $I_q(T_1) = I_q(T_2)$ if T_1 and T_2 are combinatorially equivalent. Especially when T is a triangulation of an orientable manifold M, it has been shown in [30] [31] that $I_q(T)$ is equal to $|\tau_q(M)|^2$ where $\tau_q(M)$ is the Witten-Reshetikhin-Turaev invariant [32]. (The relation between $I_q(T)$ and $\tau_q(M)$ at $q \to 1$ has also been studied in [33] [34]. See also [35] and [36].) For example

$$I_q(S^3) = 1, \quad I_q(S^1 \times S^2) = \Lambda_q.$$

In general, it has been shown by Kohno [37] that $I_q(M)$ is bounded from above as

$$I_q(M) \leq \Lambda_q^{h(M)} \qquad (14)$$

where $h(M)$ is the Heegaard genus. Any three-dimensional manifold can be decomposed into a pair of handlebodies $\mathcal{H}_h$ of genus h glued together on their boundaries (the Heegaard splitting), and the Heegaard genus $h(M)$ is the minimum number of such h. Especially $h(M) = 0$ if and only if M is homeomorphic to S^3 and otherwise $h(M) \geq 1$.

By introducing a new variable μ by $N = \mu/\Lambda_q$, the partition function Z_q is expressed as

$$\log Z_q = \frac{1}{\Lambda_q} \sum_T \frac{1}{C(T)} g^{n_3(T)} \mu^{n_1(T)} \Lambda_q^{\chi(T)} I_q(T). \qquad (15)$$

Suppose that we could take the limit $q \to 0$ so that $\Lambda_q \to 0$ while maintaining the inequality (14). As we mentioned in the introduction, $\chi(T) \geq 0$ for any simplicial complex T in three dimensions and the equality holds if and only if T is a simplicial manifold. Because of the inequality (14), the $q \to 0$ limit of $\log Z_q$ would be dominated by a sum over simplicial manifolds with $\chi(T) = 0$ and $h(T) = 0$, i.e. $T \simeq S^3$. Unfortunately, it is not clear how to make sense of this limit since the model is defined only when q is a root of unity. In the following, we will develop an alternative mechanism to impose the condition $T \simeq S^3$ without using the $q \to 0$ limit.

4. SCHWINGER-DYSON EQUATION IN THREE DIMENSIONS

In the case of the matrix model, the integral over the hermitian matrix M in (1) can be reduced to an integral over the eigenvalues λ_i ($i = 1, \ldots, N$) of M

$$\int [dM] \exp(-S(M)) = \int \prod_{i=1}^N d\lambda_i \prod_{i<j} (\lambda_i - \lambda_j)^2 \exp\left(-\sum_i (\frac{1}{2}\lambda_i^2 - \frac{g}{3\sqrt{N}}\lambda_i^3)\right),$$

and the powerful methods of the WKB approximation [4] and the orthogonal polynomials [8] have been used to successfully analyze the model. This approach, however, does not seem to have an obvious generalization to three dimensions. It is not clear if the integral over $M_{m_1 m_2 m_3}^{j_1 j_2 j_3; a_1 a_2 a_3}$ could be significantly reduced by introducing some variables similar to the eigenvalues.

In Section 2, we saw that the large-N matrix model can also be studied by using the Schwinger-Dyson equation for u_n, the expectation value of $\text{tr} M^n$. In fact u_n is the integral transform (9) of $\rho(\lambda)$, the density of the eigenvalue. This is not too surprising since both the WKB approximation and the Schwinger-Dyson equation at large-N make use of the classical equation of motion of the model. The point here is that the method of the Schwinger-Dyson equation is applicable to the three-dimensional case.

In case of the matrix model, the expectation value of $\mathrm{tr} M^n$ is expressed as a sum over planar graphs with n external legs. Thus $\mathrm{tr} M^n$ may be regarded as an operator which creates a circle with n edges in each of the planar graphs. The natural thing to consider in three dimensions should then be an operator which creates a closed connected triangulated surface Σ. Indeed we can construct such an operator as follows. We attach a triplet of indices (j_e, m_e, a_e) to each edge e on the surface Σ and associate $M^{j_1 j_2 j_3; a_1 a_2 a_3}_{m_1 m_2 m_3}$ ($j_r = j_{e_r(\Delta)}$, $m_r = \pm m_{e_r(\Delta)}$, $a_r = a_{e_r(\Delta)}$) to each triangle Δ. The signs ($\pm$) of m_r for a pair of triangle sharing the edge e are chosen to be opposite. We then define a polynomial $F_\Sigma(j; M)$ of M by summing over $\{m\}$ and $\{a\}$ as

$$F_\Sigma(j; M) = \sum_{\{m\},\{a\}} \prod_{e:\mathrm{edges}} \frac{(-1)^{j_e}}{2} \left[(-q)^{m_e} + (-q)^{-m_e}\right] \qquad (16)$$
$$\times \prod_{\Delta:\mathrm{triangles}} M^{j_{e_1}(\Delta) j_{e_2}(\Delta) j_{e_3}(\Delta) a_{e_1}(\Delta) a_{e_2}(\Delta) a_{e_3}(\Delta)}_{\pm m_{e_1}(\Delta) \pm m_{e_2}(\Delta) \pm m_{e_3}(\Delta)}.$$

By taking an expectation value of F_Σ, we obtain a function $\Psi_\Sigma(j)$ of the indices $\{j\}$ on Σ as

$$\Psi_\Sigma(j) = \langle F_\Sigma(j; M) \rangle = \frac{1}{Z} \int [dM] F_\Sigma(j; M) \exp(-S(M)). \qquad (17)$$

We have not yet summed over the indices $\{j\}$.

We saw in eq.(13) that the weight $I_q(T)$ for the simplicial complex T in in the summation (15) is given by a sum over $\{j\}$ on the edges of T. We may regard $I_q(T)$ as a partition function of a lattice statistical model on T. In such a topological lattice model, we can introduce a notion of physical states on the closed triangulated surface Σ as follows [26] [33] [34]. We start with a space V_Σ spanned by functions $\Phi(j)$ of $\{j\}$ on Σ and call it a space of states. In order to define physical states in V_Σ, we consider a three-dimensional simplicial complex T_{Σ_0, Σ_1} which is homeomorphic to $\Sigma \times [0,1]$ and whose boundaries $\Sigma \times \{0\}$ and $\Sigma \times \{1\}$ are triangulated as Σ_0 and Σ_1. We then define a map P_{Σ_0, Σ_1} from V_{Σ_0} to V_{Σ_1} by

$$P_{\Sigma_0, \Sigma_1} : \Phi(j) \in V_{\Sigma_0} \to \sum_{\{j'\}} \Phi(j') P_{\Sigma_0, \Sigma_1}(j', j) \in V_{\Sigma_1}$$

where $P_{\Sigma_0, \Sigma_1}(j', j)$ is defined by attaching an index J_e on each edge e on T_{Σ_0, Σ_1} and by summing over $\{J\}$ as

$$P_{\Sigma_0, \Sigma_1}(j', j) = \frac{\Lambda_q^{n_0(\Sigma_1)}}{\Lambda_q^{n_0(T_{\Sigma_0, \Sigma_1})}} \sum_{\{J\}} \prod_{e \in \Sigma \times \{0\}} \delta(J_e, j_e)$$
$$\times \prod_{e \in \Sigma \times \{1\}} \left(\frac{\delta(J_e, j'_e)}{(-1)^{2j'_e}[2j'_e + 1]_q}\right) \prod_{e \in T_{\Sigma_0, \Sigma_1}} (-1)^{2J_e}[2J_e + 1]_q \qquad (18)$$
$$\times \prod_{t \in T_{\Sigma_0, \Sigma_1}} (-1)^{\sum_{k=1}^{6} J_{e_k}(t)} \left\{ \begin{array}{ccc} J_{e_1}(t) & J_{e_2}(t) & J_{e_3}(t) \\ J_{e_4}(t) & J_{e_5}(t) & J_{e_6}(t) \end{array} \right\}_q.$$

Because of the topological invariance, P_{Σ_0, Σ_1} is independent of the choice of the triangulation in the interior of T_{Σ_0, Σ_1} and satisfies [26]

$$\sum_{\{j'\}} P_{\Sigma_0, \Sigma_1}(j, j') P_{\Sigma_1, \Sigma_2}(j', j'') = P_{\Sigma_0, \Sigma_2}(j, j'') \qquad (19)$$

where Σ_0, Σ_1 and Σ_2 are all homeomorphic to Σ. Especially when $\Sigma_0 = \Sigma_1 = \Sigma$, $P_{\Sigma,\Sigma}$ acts as a projection operator on V_Σ. The physical subspace $V_\Sigma^{(\text{phys})}$ is defined as a subspace of V_Σ projected out by $P_{\Sigma,\Sigma}$ as

$$V_\Sigma^{(\text{phys})} = \{\Phi \in V_\Sigma : \sum_{\{j'\}} \Phi(j') P_{\Sigma,\Sigma}(j',j) = \Phi(j)\}.$$

Because of the property (19), the operator P_{Σ_0,Σ_1} defines an invertible map from a physical state in V_{Σ_1} onto a physical state in V_{Σ_0}. Therefore $V_{\Sigma_0}^{(\text{phys})}$ and $V_{\Sigma_1}^{(\text{phys})}$ are isomorphic.

Since $\Psi_\Sigma(j)$ given by (17) is also a function of $\{j\}$, we may consider its pairing with a physical state Φ on Σ as

$$(\Phi, \Psi_\Sigma) = \Lambda_q^{\frac{1}{2}n_2(\Sigma) + h(\Sigma)} \sum_{\{j\}} \Phi(j) \Psi_\Sigma(j) \tag{20}$$

where $h(\Sigma)$ is the genus of Σ. The factor $\Lambda_q^{\frac{1}{2}n_2(\Sigma) + h(\Sigma)}$ will simplify the equations in the following. As we shall see below, $\Phi_\Sigma(j)$ itself is not a physical state, and the pairing (Φ, Ψ_Σ) depends on the triangulation Σ of the surface. The Schwinger-Dyson equation describes the dependence of (Φ, Ψ_Σ) on Σ.

Among physical states on Σ, especially important are the Hartle-Hawking type states [34] defined as follows. A handlebody $\mathcal{H}_h$ of genus h is obtained by cutting out $2h$ pairs of discs from a boundary of a three-dimensional ball B^3 and by identifying them pairwisely in opposite orientations. The boundary of $\mathcal{H}_h$ is then a closed orientable surface of genus h. The boundaries of the discs become simple homology cycles on the boundary of $\mathcal{H}_h$ which do not intersect with each other and which are contractible in $\mathcal{H}_h$. They are called the meridians of the handlebody. Let us consider a simplicial manifold $\mathcal{H}_\Sigma^{(\alpha)}$ homeomorphic to a handlebody of genus h whose boundary is triangulated as on Σ ($h(\Sigma) = h$) and whose meridians are given by $\{\alpha_1, \ldots, \alpha_h\}$ where each α_i is chosen to be a sequence of edges on Σ. The Hartle-Hawking type state $\Phi^{(\alpha)}(j)$ associated to the handlebody is defined by

$$\Phi^{(\alpha)}(j) = \frac{\Lambda_q^{n_0(\Sigma)}}{\Lambda_q^{n_0(\mathcal{H}_\Sigma^{(\alpha)})}} \sum_{\{J\}} \prod_{e \in \Sigma} \left(\frac{\delta(J_e, j_e)}{(-1)^{2j_e}[2j_e+1]_q} \right) \prod_{e \in \mathcal{H}_\Sigma^{(\alpha)}} (-1)^{2J_e}[2J_e+1]_q$$

$$\times \prod_{t \in \mathcal{H}(\{\alpha\})} (-1)^{\sum_{k=1}^6 J_{e_k(t)}} \left\{ \begin{array}{ccc} J_{e_1(t)} & J_{e_2(t)} & J_{e_3(t)} \\ J_{e_4(t)} & J_{e_5(t)} & J_{e_6(t)} \end{array} \right\}_q. \tag{21}$$

Because of the topological invariance, $\Phi^{(\alpha)}(j)$ defined in this way is independent of the choice of the triangulation of the interior of the handlebody, and it gives a physical state on Σ.

By expanding integral (17) in powers of g, the pairing of Ψ_Σ with the Hartle-Hawking type state $\Phi^{(\alpha)}$ is expressed as

$$(\Phi^{(\alpha)}, \Psi_\Sigma) = \Lambda_q^{\frac{1}{2}n_2(\Sigma) + h(\Sigma)} \sum_{\{j\}} \Phi^{(\alpha)}(j) \langle F(j; M) \rangle$$

$$= \Lambda_q^{\frac{1}{2}n_2(\Sigma) + h(\Sigma)} \sum_{T: \partial T = \Sigma} \left(\frac{C(\Sigma)}{C(T)} \right) (g\Lambda_q)^{n_3(T)} \left(\frac{\mu}{\Lambda_q} \right)^{n_1(T)} \tag{22}$$

$$\times \Lambda_q^{n_0(T)-1} I_q(\mathcal{H}_\Sigma^{(\alpha)} \cup_\Sigma (-T))$$

$$= \sum_{T: \partial T = \Sigma} \left(\frac{C(\Sigma)}{C(T)} \right) g^{n_3(T)} \mu^{n_1(T)} \Lambda_q^{\chi(T) - \frac{1}{2}\chi(\Sigma)} I_q(\mathcal{H}_\Sigma^{(\alpha)} \cup_\Sigma (-T))$$

where the sum $\sum_T$ is over simplicial complexes whose boundary is Σ and $I_q(\mathcal{H}_\Sigma^{(\alpha)} \cup_\Sigma (-T))$ is the Turaev-Viro invariant (13) for the simplicial complex obtained by gluing $\mathcal{H}_\Sigma^{(\alpha)}$ and T together on their boundaries after inverting the orientation of T. As in the case of the matrix model, the dual graph of T is not necessarily connected though it does not contain a closed disconnected part. Especially a triangle on the boundary Σ of T may be attached to another triangle on Σ. For any simplicial complex T bounded by Σ, its Euler number $\chi(T)$ is greater than or equal to $\frac{1}{2}\chi(\Sigma)$, and the equality holds if and only if $\mathcal{H}_\Sigma^{(\alpha)} \cup (-T)$ is a simplicial manifold. If we could take the limit $q \to 0$, the right-hand side in the above would become a sum over T such that $\chi(T) = \frac{1}{2}\chi(\Sigma)$ and $\mathcal{H}_\Sigma^{(\alpha)} \cup_\Sigma (-T) \simeq S^3$. Although this limit itself is not well-defined, we will find a set of factorization conditions which together with the Schwinger-Dyson equation imposes the similar conditions on T.

In order to derive the Schwinger-Dyson equation, we choose a triangle Δ_0 on Σ and consider an open surface Σ_{Δ_0} obtained by removing the triangle Δ_0 from Σ. Associated to Σ_{Δ_0}, we define a boundary operator $F_{\Sigma_{\Delta_0}}(j; M)_{m_1 m_2 m_3}^{a_1 a_2 a_3}$ by

$$F_\Sigma(j; M)_{m_1 m_2 m_3}^{a_1 a_2 a_3} = \sum_{\{m\},\{a\}} \prod_{r=1}^{3} \delta(m_r, m_{e_r(\Delta_0)}) \delta(a_r, a_{e_r(\Delta_0)})$$

$$\times \prod_{e:\text{edges}} \frac{(-1)^{j_e}}{2} [(-q)^{m_e} + (-q)^{-m_e}]$$

$$\times \prod_{\Delta:\text{triangles}} M_{\pm m_{e_1(\Delta)} \pm m_{e_2(\Delta)} \pm m_{e_3(\Delta)}}^{j_{e_1(\Delta)} j_{e_2(\Delta)} j_{e_3(\Delta)}; a_{e_1(\Delta)} a_{e_2(\Delta)} a_{e_3(\Delta)}}.$$

As in the case of the matrix model, we make use of the identity

$$\frac{1}{Z} \int [dM] \sum_{\{m\},\{a\}} \frac{\partial}{\partial M_{m_1 m_2 m_3}^{j_1 j_2 j_3; a_1 a_2 a_3}} \left[F_{\Sigma_{\Delta_0}}(j; M)_{m_1 m_2 m_3}^{a_1 a_2 a_3} \exp(-S(M)) \right] = 0$$

where we have set $j_{e_r(\Delta_0)} = j_r$ ($r = 1, 2, 3$). By performing the derivative with respect to M, we obtain

$$\Psi_\Sigma(j) = g\Lambda_q \sum_{j_4, j_5, j_6} (-1)^{\sum_{k=1}^{6} j_k} \begin{Bmatrix} j_1 & j_2 & j_3 \\ j_4 & j_5 & j_6 \end{Bmatrix} \Psi_{\Sigma_{\widehat{\Delta}_0}}(j)$$

$$+ \sum_{\{m\},\{a\}} \left\langle \frac{\partial F_{\Sigma_{\Delta_0}}(j; M)_{m_1 m_2 m_3}^{a_1 a_2 a_3}}{\partial M_{m_1 m_2 m_3}^{j_1 j_2 j_3; a_1 a_2 a_3}} \right\rangle \tag{23}$$

where $\Sigma_{\widehat{\Delta}_0}$ is a triangulated surface obtained from Σ by decomposing Δ_0 into three small triangles, and j_4, j_5, j_6 are spins on the three additional edges on $\Sigma_{\widehat{\Delta}_0}$. If it were not for the last term in the right-hand side of eq.(23), Ψ_Σ would transform like a physical state under the change of triangulation $\Sigma \to \Sigma_{\widehat{\Delta}_0}$.

Since each M in $F_{\Sigma_{\Delta_0}}(j; M)$ corresponds to a triangle Δ on Σ_{Δ_0}, the last term $\langle \frac{\partial}{\partial M} F_{\Sigma_{\Delta_0}}(j; M) \rangle$ in the right-hand side of (23) can be expressed a sum over closed surfaces Σ_{Δ,Δ_0} obtained by removing Δ and Δ_0 from Σ and by identifying their edges. Let us denote the number of common edges of the two triangles by $E(\Delta, \Delta_0)$ and the number of common vertices by $V(\Delta, \Delta_0)$. We can classify the position of Δ on Σ_{Δ_0} according to the values of E and V as follows.

(1) When $E(\Delta, \Delta_0) \neq 0$ and $V(\Delta, \Delta_0) = E(\Delta, \Delta_0) + 1$, Δ is in the neighborhood of Δ_0 on Σ. In this case, Σ_{Δ,Δ_0} has the same topology as Σ.

(2) When $E(\Delta, \Delta_0) = 0$ and $V(\Delta, \Delta_0) = 0, 1$, the removal of the triangles will create a new handle on Σ and the edges of Δ_0 becomes a homology cycle around the handle.

(3) When $V(\Delta, \Delta_0) \geq E(\Delta, \Delta_0) + 2$, the triangles Δ and Δ_0 wrap around a cycle on Σ, and Σ_{Δ,Δ_0} is degenerate around the cycle.

Let us consider a generic situation when $E(\Delta, \Delta_0) + 1 \geq V(\Delta, \Delta_0)$ for any triangle Δ on Σ_{Δ_0}, which includes the cases (1) and (2). We also assume here that the surface Σ is not degenerate in the neighborhood of Δ_0. The cases involving degenerate surfaces will be examined in the next section.

In the generic situation, by multiplying an arbitrary physical state Φ on both sides of eq.(23) and by summing over $\{j\}$, we obtain

$$(\Phi, \Psi_\Sigma) = g(\Phi, \Psi_{\Sigma_{\Delta_0}}) + \sum_{\Delta; E(\Delta,\Delta_0)\neq 0} \mu^{E(\Delta,\Delta_0)}(\Phi, \Psi_{\Sigma_{\Delta,\Delta_0}})r$$
$$+ \sum_{\Delta; E(\Delta,\Delta_0)=0} (i[\Phi]_{\Delta,\Delta_0}, \Psi_{\Sigma_{\Delta,\Delta_0}}). \tag{24}$$

The second term in the right-hand side is a sum over Δ in the neighborhood of Δ_0, and Σ_{Δ,Δ_0} is of the same topology as Σ. On the other hand, $h(\Sigma_{\Delta,\Delta_0}) = h(\Sigma) + 1$ in the last term, and i_{Δ,Δ_0} is defined by

$$i[\Phi]_{\Delta,\Delta_0}(j) = \sum_{\{j_{e_r(\Delta)}\}_{r=1}^3} \left(\prod_{r=1}^3 \delta(j_{e_r(\Delta)}, j_{e_r(\Delta_0)}) \right) \Phi(j).$$

It is easy to show that this gives a map from $V_\Sigma^{(phys)}$ into $V_{\Sigma_{\Delta,\Delta_0}}^{(phys)}$.

Especially when Φ is the Hartle-Hawking type state $\Phi^{(\alpha)}$, $i[\Phi^{(\alpha)}]_{\Delta,\Delta_0}$ is also the Hartle-Hawking type state associated to a handlebody whose boundary is Σ_{Δ,Δ_0} and whose meridians are $\{\alpha_1, \ldots, \alpha_{i(\Sigma)}, \partial\Delta_0\}$

$$i[\Phi^{(\alpha)}]_{\Delta,\Delta_0}(j) = \Phi^{(\alpha \cup \{\partial\Delta_0\})}(j) \tag{25}$$

where $\partial\Delta_0$ is the boundary of Δ_0. The equation (24) for $\Phi = \Phi^{(\alpha)}$ can then be written as

$$\langle\Sigma^{(\alpha)}\rangle = g\langle\Sigma_{\Delta_0}^{(\alpha)}\rangle + \sum_{\Delta; E(\Delta,\Delta_0)\neq 0} \mu^{E(\Delta,\Delta_0)}\langle\Sigma_{\Delta,\Delta_0}^{(\alpha)}\rangle + \sum_{\Delta; E(\Delta,\Delta_0)=0} \langle\Sigma_{\Delta,\Delta_0}^{(\alpha\cup\{\partial\Delta_0\})}\rangle \tag{26}$$

where

$$\langle\Sigma^{(\alpha)}\rangle = (\Phi^{(\alpha)}, \Psi_\Sigma) = \Lambda_q^{\frac{1}{2}n_2(\Sigma)+h(\Sigma)} \sum_{\{j\}} \Phi^{(\alpha)}(j)\langle F_\Sigma(j;M)\rangle. \tag{27}$$

We call this the Schwinger-Dyson equation for the amplitude $\langle\Sigma^{(\alpha)}\rangle$ of the triangulated surface Σ with the choice $\{\alpha_i\}$ of simple homology cycles. We see that, when Σ_{Δ,Δ_0} is non-degenerate, the Schwinger-Dyson equation (26) does not depend on the value of q explicitly.

5. RESTRICTION ON TOPOLOGIES

5-1. General Case

We have found that the Schwinger-Dyson equation (26) takes the same form for any value of q as far as the surface Σ in the equation is non-degenerate. On the other hand, the amplitude $\langle \Sigma^{(\alpha)} \rangle$ which solves the equation is expanded as a sum over T weighted with $\Lambda_q^{\chi(T)-\frac{1}{2}\chi(\Sigma)} I_q(\mathcal{H}_\Sigma^{(\alpha)} \cup_\Sigma (-T))$ as in eq. (22). Depending on the topology of T, the weight takes various functional forms in q. Since the Schwinger-Dyson equation for non-degenerate surfaces is linear in $\langle \Sigma^{(\alpha)} \rangle$, a sum over a restricted class of triangulations T such that the weight takes the same functional form should also give a solution to the equation.

Here we claim a stronger statement that, for each choice of a compact orientable three-dimensional manifold M, a sum over triangulations T such that

$$\mathcal{H}_\Sigma^{(\alpha)} \cup_\Sigma (-T) \simeq M \tag{28}$$

satisfies the Schwinger-Dyson equation. Being a triangulation of the manifold, T should also satisfy $\chi(T) = \frac{1}{2}\chi(\Sigma)$. Let us define

$$\langle \Sigma^{(\alpha)} \rangle_{|M} = \sum_{T: \mathcal{H}_\Sigma^{(\alpha)} \cup_\Sigma (-T) \simeq M} \left(\frac{C(\Sigma)}{C(T)} \right) g^{n_3(T)} \mu^{n_1(T)} = \sum_{n_3, n_1} Z_{n_3, n_1}(\Sigma; \alpha)_{|M} g^{n_3} \mu^{n_1} \tag{29}$$

where the sum Σ_T is over triangulations T satisfying the condition (28), and $Z_{n_3,n_1}(\Sigma; \alpha)_{|M}$ is the number of such triangulations with $n_3(T) = n_3$, $n_1(T) = n_1$. We are going to show that this gives a solution to the Schwinger-Dyson equation.

By definition, for each triangulation T counted in the above, $\mathcal{H}_\Sigma^{(\alpha)} \cup_\Sigma (-T)$ give a triangulation of M. Let us pay attention to one of the triangle Δ_0 on the boundary Σ of T. In the simplicial complex T, the triangle Δ_0 may be either attached to a tetrahedron or identified with another triangle Δ on Σ_{Δ_0}.

If Δ_0 is attached to a tetrahedron in T, we can remove the tetrahedron from T and give it to the handlebody $\mathcal{H}_\Sigma^{(\alpha)}$, without changing the triangulation of M. After removing the tetrahedron, T becomes a triangulation T' consisting of $(n_3 - 1)$ tetrahedron whose boundary is $\Sigma_{\widehat{\Delta}_0}$. Obviously T' obeys $\chi(T') = \frac{1}{2}\chi(\Sigma_{\widehat{\Delta}_0})$ if $\chi(T) = \frac{1}{2}\chi(\Sigma)$. The handlebody $\mathcal{H}_\Sigma^{(\alpha)}$ becomes $\mathcal{H}_{\Sigma_{\widehat{\Delta}_0}}^{(\alpha)}$ after adding the tetrahedron on it. Since the triangulation of M remains unchanged, T' should satisfy

$$\mathcal{H}_{\Sigma_{\widehat{\Delta}_0}}^{(\alpha)} \cup_{\Sigma_{\widehat{\Delta}_0}} (-T') \simeq M. \tag{30}$$

Such T' is counted in $Z_{n_3-1,n_1}(\Sigma_{\widehat{\Delta}_0}; \alpha)_{|M}$. Conversely, from any triangulation T' satisfying the above, we can construct a triangulation T such that the condition (28) holds. However the correspondence between T and T' is not necessarily one to one when T and T' have non-trivial symmetries. We shall see below that this is taken care of by the factors $C(\Sigma)/C(T)$ and $C(\Sigma_{\widehat{\Delta}_0})/C(T')$.

Let $G(\Sigma)$ be the symmetry of Σ under interchange of the triangles on it, and $G(\Sigma_{\Delta_0})$ be the invariant subgroup of $G(\Sigma)$ which keeps the position of the triangle Δ_0 fixed. Then $[G(\Sigma) : G(\Sigma_{\Delta_0})] = C(\Sigma)/C(\Sigma_{\Delta_0})$ gives the number of the triangles on Σ which are conjugate to Δ_0 under the action of $G(\Sigma)$. Suppose that $m_1(T)$ of them are attached to tetrahedra in the interior of T, and $m_2(T)$ of them are attached to triangles on Σ_{Δ_0} so that $m_1(T) + m_2(T) = C(\Sigma)/C(\Sigma_{\Delta_0})$. The contribution $C(\Sigma)/C(T)$ of the triangulation T to $Z_{n_3,n_1}(\Sigma;\alpha)|_M$ can then be divided into two parts as

$$\frac{C(\Sigma)}{C(T)} = C(\Sigma_{\Delta_0}) \cdot \left(\frac{m_1(T)}{C(T)} + \frac{m_2(T)}{C(T)}\right).$$

Let us show that the sum

$$C(\Sigma_{\Delta_0}) \cdot \sum_T \left(\frac{m_1(T)}{C(T)}\right) \tag{31}$$

over triangulations T with satisfy eq. (28) and $n_3(T) = n_3$, $n_1(T) = n_1$ is equal to $Z_{n_3-1,n_1}(\Sigma_{\widehat{\Delta}_0};\alpha)|_M$. If we remove a tetrahedron attached to one of the $m_1(T)$ triangles conjugate to Δ_0, we obtain a triangulation T' consisting of $(n_3 - 1)$ tetrahedra. It may happen that there is more than one ways to remove a tetrahedron from T to obtain the same triangulation T'. Let $N(T \to T')$ be the number of ways to obtain T' from T. Obviously $m_1(T) = \sum_{T'} N(T \to T')$, and the summation given by eq. (31) is equal to

$$C(\Sigma_{\Delta_0}) \cdot \sum_T \sum_{T'} \left(\frac{N(T \to T')}{C(T)}\right).$$

The number $N(T \to T')$ can be expressed as

$$N(T \to T') = \frac{C(T)}{C(T,T')}$$

where $C(T,T')$ is the order of the symmetry of T which keeps the position of the tetrahedron at $T \setminus T'$ fixed. Therefore eq. (31) can be expressed as

$$C(\Sigma_{\Delta_0}) \cdot \sum_T \sum_{T'} \frac{1}{C(T,T')}$$

On the other hand,

$$Z_{n_3-1,n_1}(\Sigma_{\widehat{\Delta}_0};\alpha)|_M = C(\Sigma_{\Delta_0}) \cdot \sum_{T'} \left(\frac{C(\Sigma_{\widehat{\Delta}_0})/C(\Sigma_{\Delta_0})}{C(T')}\right)$$

$$= C(\Sigma_{\Delta_0}) \cdot \sum_{T'} \sum_T \left(\frac{N(T \leftarrow T')}{C(T')}\right)$$

where $N(T \leftarrow T')$ is the number of ways to construct T from T' by attaching a tetrahedron on the boundary of T'. The number $N(T \leftarrow T')$ can also be expressed as

$$N(T' \leftarrow T) = \frac{C(T')}{C(T,T')}.$$

Combining these, we obtain

$$Z_{n_3-1,n_1}(\Sigma_{\widehat{\Delta}_0};\alpha)|_M = C(\Sigma_{\Delta_0}) \cdot \sum_{T'} \sum_T \frac{1}{C(T,T')} = C(\Sigma_{\Delta_0}) \cdot \sum_T \left(\frac{m_1(T)}{c(T)}\right). \tag{32}$$

Next we consider the case when Δ_0 on the boundary of T is identified with another triangle Δ on Σ_{Δ_0}. If Δ is in the neighborhood of Δ_0 on Σ, i.e. $E(\Delta, \Delta_0) \neq 0$ and $V(\Delta, \Delta_0) = E(\Delta, \Delta_0) + 1$, we can remove the triangle $\Delta = \Delta_0$ from T to obtain T' whose boundary is $\Sigma_{\Delta, \Delta_0}$. In this case, the topologies of the boundaries are the same, i.e. $\chi(\Sigma) = \chi(\Sigma_{\Delta, \Delta_0})$. Since T' is obtained from T by removing one triangle, $E(\Delta, \Delta_0)$ edges and $(V(\Delta, \Delta_0) - 2)$ vertices, the Euler number of T' remains the same as that of T; $\chi(T') = \chi(T) - 1 + E(\Delta, \Delta_0) - (V(\Delta, \Delta_0) - 2) = \chi(T)$. Therefore $\chi(T) = \frac{1}{2}\chi(\Sigma)$ implies $\chi(T') = \frac{1}{2}\chi(\Sigma_{\Delta, \Delta_0})$. In this case, the identification of the two triangles transforms the handlebody $\mathcal{H}_\Sigma^{(\alpha)}$ into $\mathcal{H}_{\Sigma_{\Delta, \Delta_0}}^{(\alpha)}$, and the triangles $\Delta = \Delta_0$ is now in the interior of $\mathcal{H}_{\Sigma_{\Delta, \Delta_0}}^{(\alpha)}$. By construction, the triangulation T' satisfies $\mathcal{H}_{\Sigma_{\Delta, \Delta_0}}^{(\alpha)} \cup_{\Sigma_{\Delta, \Delta_0}} (-T') \simeq M$.

On the other hand, when Δ is not in the neighborhood of Δ_0, i.e. $E(\Delta, \Delta_0) = 0$ and $V(\Delta, \Delta_0) = 0, 1$, the removal of the triangle $\Delta = \Delta_0$ from T creates a hole in T. The Euler number of the boundary surface then decreases by two, $\chi(\Sigma_{\Delta, \Delta_0}) = \chi(\Sigma) - 2$. On the other hand, the numbers of the edges and the vertices of T' remain the same as those of T. Therefore $\chi(T') = \chi(T) - 1 = \frac{1}{2}\chi(\Sigma_{\Delta, \Delta_0})$. The handlebody $\mathcal{H}_\Sigma^{(\alpha)}$ acquires a new handle after the removal of $\Delta = \Delta_0$, and Δ_0 becomes a meridian disc of the handle. Therefore $\mathcal{H}_{\Sigma_{\Delta, \Delta_0}}^{(\alpha \cup \{\partial \Delta_0\})} \cup_{\Sigma_{\Delta, \Delta_0}} (-T') \simeq M$.

As in the case of eq. (32), we can estimate the contribution of these triangulations to $Z_{n_3, n_1}(\Sigma; \alpha)|_M$, taking the symmetry factor $C(\Sigma)/C(T)$ into account, and obtain

$$C(\Sigma_{\Delta_0}) \cdot \sum_T \left(\frac{m_2(T)}{c(T)}\right) = \sum_{\Delta: E(\Delta, \Delta_0) \neq 0} Z_{n_3, n_1 - E(\Delta, \Delta_0)}(\Sigma_{\Delta, \Delta_0}; \alpha)|_M$$
$$+ \sum_{\Delta: E(\Delta, \Delta_0) = 0} Z_{n_3, n_1}(\Sigma_{\Delta, \Delta_0}; \alpha \cup \{\partial \Delta_0\})|_M. \tag{33}$$

Combining this with eq. (32), $Z_{n_3, n_1}(\Sigma; \alpha)|_M$ is expressed as

$$Z_{n_3, n_1}(\Sigma; \alpha) = Z_{n_3 - 1, n_1}(\Sigma_{\widehat{\Delta}_0}; \alpha) + \sum_{\Delta: E(\Delta, \Delta_0) \neq 0} Z_{n_3, n_1 - E(\Delta, \Delta_0)}(\Sigma_{\Delta, \Delta_0}; \alpha)$$
$$+ \sum_{\Delta: E(\Delta, \Delta_0) = 0} Z_{n_3, n_1}(\Sigma_{\Delta, \Delta_0}; \alpha \cup \{\partial \Delta_0\}). \tag{34}$$

Therefore $\langle \Sigma^{(\alpha)} \rangle|_M$ solves the Schwinger-Dyson equation, for any choice of M.

5-2. Triangulations of Handlebodies

We have seen that the Schwinger-Dyson equation (34) holds for any closed orientable manifold M. Especially when $M = S^3$, there is a nice characterization of triangulations T counted in $Z_{n_3, n_1}(\Sigma; \alpha)|_{S^3}$.

If we remove a handlebody from S^3, the rest is also a handlebody which is described as follows. Let us first consider the case when Σ is isomorphic to a torus. In this case, it is convenient to choose a canonical homology basis $\{\alpha, \beta\}$ such that α is the meridian of the handlebody (solid torus) $\mathcal{H}_\Sigma^{(\alpha)}$ and that α and β intersects once with each other transversely. The cycle β is called the longitude of the handlebody. The choice of the longitude is unique upto α, i.e. $\beta + n\alpha$ is also the longitude of $\mathcal{H}_\Sigma^{(\alpha)}$.

Three-dimensional manifolds with genus-one Heegaard splittings are classified as

$$\mathcal{H}_\Sigma^{(\alpha)} \cup_\Sigma (-\mathcal{H}_\Sigma^{(n\alpha+\beta)}) \simeq S^3,$$
$$\mathcal{H}_\Sigma^{(\alpha)} \cup_\Sigma (-\mathcal{H}_\Sigma^{(n\alpha)}) \simeq S^1 \times S^2,$$
$$\mathcal{H}_\Sigma^{(\alpha)} \cup_\Sigma (-\mathcal{H}_\Sigma^{(n\alpha+m\beta)}) \simeq L_{m,n}, \quad (m \neq 0,1),$$

where $L_{m,n}$ is the lens space. Therefore, if $\mathcal{H}_\Sigma^{(\alpha)} \cup_\Sigma (-T) \simeq S^3$, T must be a triangulation of a handlebody whose meridian (longitude) is the longitude (meridian) of $\mathcal{H}_\Sigma^{(\alpha)}$.

This is also the case for $h(\Sigma) \geq 2$. Corresponding to the meridians $\{\alpha_i\}_{i=1}^{h(\Sigma)}$ of the handlebody $\mathcal{H}_\Sigma^{(\alpha)}$, we choose simple homology cycles $\{\beta_i\}$ on Σ such that α_i and β_i ($i = 1, \ldots, h(\Sigma)$) make a canonical homology basis, i.e. they have the following intersection properties.

$$\alpha_i \cdot \beta_j = \delta_{ij}, \quad \alpha_i \cdot \alpha_j = 0, \quad \beta_i \cdot \beta_j = 0.$$

As in the case of torus, the cycles $\{\beta_i\}$ are called the longitudes of $\mathcal{H}_\Sigma^{(\alpha)}$. It is known in general that the Heegaard splitting of S^3 is unique upto the stable equivalence [38], and T satisfying $\mathcal{H}_\Sigma^{(\alpha)} \cup_\Sigma (-T)$ should be a triangulation of a handlebody whose meridians (longitudes) are the longitudes (meridians) of $\mathcal{H}_\Sigma^{(\alpha)}$.

The sum over triangulations of the handlebody given by $\langle \Sigma^{(\alpha)} \rangle_{|S^3}$ is closely related to the sum over triangulations of S^3. Before explaining the relation, it is instructive to examine the amplitude $\langle \Sigma^{(\alpha)} \rangle$ given by eq.(27) and its relation to Z_q. Let us consider the case when Σ is homeomorphic to S^2 and is given by a single tetrahedron. For the single tetrahedron, the Hartle-Hawking type state Φ is given by the $6j$-symbol and $n_2 = 4$. The boundary operator $\Lambda_q^2 \sum_j \Phi(j) F(j; M)$ for the tetrahedron is then equal to the quartic term in the action $S(M)$ upto the factor $\frac{1}{12} g \Lambda_q$. Therefore, as in the case of the matrix model (10), the partition function Z_q and the amplitude of the tetrahedron are related as

$$\langle\text{tetrahedron}\rangle = 12\Lambda_q \frac{1}{Z}\frac{\partial Z}{\partial g} = \sum_T \left(\frac{12}{C(T)}\right) n_3(T) g^{n_3(T)-1} \mu^{n_1(T)} \Lambda_q^{\chi(T)} I_q(T).$$

On the other hand, $\langle\text{tetrahedron}\rangle$ is also expanded as a sum over simplicial complexes T bounded by the tetrahedron as in eq. (22). Thus the number of closed simplicial complexes T weighted with $\Lambda_q^{\chi(T)} I_q(T)$ and the symmetry factor $1/C(T)$ is equal to $12 n_3$ times the number of open simplicial complexes T such that $\partial T = (\text{tetrahedron})$ weighted with $\Lambda_q^{\chi(T)-2} I_q((\text{tetrahedron}) \cup (-T))$ and the symmetry factor $(12/C(T))$.

A similar relation holds between $\langle\Sigma^{(\alpha)}\rangle_{|S^3}$ and the sum over triangulations of S^3. When Σ is a single tetrahedron, the amplitude is a sum over triangulations of a three-dimensional ball B^3 whose boundary is the tetrahedron.

$$\langle\text{tetrahedron}\rangle_{|S^3} = \sum_{n_3,n_1} Z_{n_3,n_1}(\text{tetrahedron})_{|S^3} \cdot g^{n_3} \mu^{n_1}.$$

Here $Z_{n_3,n_1}(\text{tetrahedron})_{|S^3}$ is the number of triangulations of the interior of the tetrahedron with n_3 tetrahedra and n_1 edges weighted with the symmetry factor

($12/C(T)$). The factor 12 is the order of the symmetry of the tetrahedron. We can show that this is related to $Z_{n_3,n_1}(S^3)$ as

$$Z_{n_3-1,n_1}(\text{tetrahedron})_{|S^3} = 12 n_3 Z_{n_3,n_1}(S^3). \tag{35}$$

Any triangulation of B^3 is obtained by removing a tetrahedron from a triangulation of S^3, and conversely any triangulation of S^3 is given by gluing a tetrahedron to a triangulation of B^3. For each triangulation T of S^3, there are $n_3(T)$ distinct triangulations of B^3 as far as T does not have a symmetry. On the other hand, if T has a non-trivial symmetry, different choices of tetrahedra in $T \simeq S^3$ related by the symmetry give the same triangulation of B^3. The symmetry factors in $Z_{n_3,n_1}(\text{tetrahedron})$ and $Z_{n_3,n_1}(S^3)$ compensate for the overcounting due to the symmetries, and eq. (35) holds.

6. FACTORIZATION

In the last section, we have shown that $\langle \Sigma^{(\alpha)} \rangle_{|S^3}$ gives a solution to the Schwinger-Dyson equation when $E(\Delta, \Delta_0) + 1 \geq V(\Delta, \Delta_0)$ for any Δ on Σ and when the surface Σ in the equation is not degenerate. In this section, we will define a set of factorization conditions on $\langle \Sigma^{(\alpha)} \rangle_{|S^3}$ so that the Schwinger-Dyson equation becomes applicable even when it involves degenerate surfaces. We will then show that the Schwinger-Dyson equation combined with the factorization conditions has a unique solution. Thus they characterize the sum over triangulations T such that $\mathcal{H}_\Sigma^{(\alpha)} \cup_\Sigma (-T) \simeq S^3$.

Let us first consider a case when the surface Σ is degenerate. When the pinching cycle is homologically trivial on Σ, the surface is actually separated into two components Σ_1 and Σ_2 touching to each other at the vertex. In this case, T must also consists of two disconnected components T_1 and T_2 such that $\partial T_i = \Sigma_i$ ($i = 1, 2$). The Schwinger-Dyson equation remains applicable in this case if we supplement a condition that Δ in the summation in the right-hand side of eq. (34) is restricted to be those on the same boundary component as Δ_0 does. Especially when we can choose the meridians of the handlebody $\mathcal{H}_\Sigma^{(\alpha)}$ such that

$$\{\alpha_1, \ldots, \alpha_{h(\Sigma)}\} = \{\alpha_1^{(1)}, \ldots, \alpha_{h(\Sigma_1)}^{(1)}\} \cup \{\alpha_1^{(2)}, \ldots, \alpha_{h(\Sigma_2)}^{(2)}\} \tag{36}$$

where $\alpha_i^{(1)} \in \Sigma_1$ and $\alpha_i^{(2)} \in \Sigma_2$, the amplitude $\langle \Sigma^{(\alpha)} \rangle_{|S^3}$ factorizes as

$$\langle \Sigma^{(\alpha)} \rangle_{|S^3} = \langle \Sigma_1^{(\alpha^{(1)})} \rangle_{|S^3} \langle \Sigma_2^{(\alpha^{(2)})} \rangle_{|S^3}.$$

In this case, it is obvious that $\Delta \in \Sigma_i$ should be imposed in eq. (34) when $\Delta_0 \in \Sigma_i$ ($i = 1, 2$).

It is instructive to compare this behaviour of the amplitude with that of $\langle \Sigma^{(\alpha)} \rangle = (\Phi^{(\alpha)}, \Psi_\Sigma)$. When Δ and Δ_0 wrap around the homologically trivial cycle γ and when the longitudes $\{\alpha_i\}$ are separated as in eq. (36), i_{Δ,Δ_0} defined by (25) gives a map from $V_\Sigma^{(\text{phys})}$ into $V_{\Sigma_1}^{(\text{phys})} \otimes V_{\Sigma_2}^{(\text{phys})}$ as

$$i[\Phi^{(\alpha)}]_{\Delta,\Delta_0} = \Phi^{(\alpha^{(1)})} \Phi^{(\alpha^{(2)})}.$$

The contribution of the degenerate surface Σ_{Δ,Δ_0} to the Schwinger-Dyson equation is then

$$\langle \Sigma^{(\alpha)}_{\Delta,\Delta_0} \rangle = \langle \Sigma^{(\alpha_1)}_1 \Sigma^{(\alpha_2)}_2 \rangle$$
$$= \Lambda_q^{\frac{1}{2}n_2(\Sigma_1)+h(\Sigma_1)} \Lambda_q^{\frac{1}{2}n_2(\Sigma_2)+h(\Sigma_2)}$$
$$\times \sum_{\{j^{(1)}\},\{j^{(2)}\}} \Phi^{(\alpha^{(1)})}(j^{(1)}) \Phi^{(\alpha^{(2)})}(j^{(2)}) \Psi_{\Sigma_1+\Sigma_2}(j^{(1)},j^{(2)})$$
$$= \langle \Sigma^{(\alpha^{(1)})}_1 \rangle \langle \Sigma^{(\alpha^{(2)})}_2 \rangle + \Lambda_q \cdot \sum_{T: \partial T = \Sigma_1 + \Sigma_2} \left(\frac{C(\Sigma_1)C(\Sigma_2)}{C(T)} \right) g^{n_3(T)} \mu^{n_1(T)}$$
$$\times \Lambda_q^{\chi(T)-\frac{1}{2}\chi(\Sigma_1)-\frac{1}{2}\chi(\Sigma_2)} I_q(\mathcal{H}^{(\alpha^{(1)})}_{\Sigma_1} \cup_{\Sigma_1} (-T) \cup_{\Sigma_2} \mathcal{H}^{(\alpha^{(2)})}_{\Sigma_2}).$$

In the first term in the right-hand side, the sum is over simplicial complexes consisting of two disconnected components T_1 and T_2 each of which is attached to Σ_1 and Σ_2 respectively, and the second term is a sum over connected simplicial complexes. We see that the second term is multiplied by the extra-factor of Λ_q. This is interpreted as follows. If T remains connected after pinching around the cycle γ, T should contain a homotopy cycle linking with γ. The extra-factor Λ_q is associated with this homotopy cycle. The factorization condition may be regarded as the $q \to 0$ limit of this equation for $\langle \Sigma^{(\alpha)} \rangle$ where the second term is to be suppressed by the factor of Λ_q.

When the surface Σ is degenerate around a homologically non-trivial cycle, the sum over Δ in the Schwinger-Dyson equation runs over all triangles on Σ_{Δ_0} as far as Δ_0 is not on the degeneration point. When one of the vertices of Δ_0 is on the degeneration point, we must supplement the following rule: The sum over triangles in the Schwinger-Dyson equation should not include those in the neighborhood of Δ_0 and on the other side of the degeneration point across the degenerating cycle. The reason for this is the following. If we take a triangle Δ' which shares the degeneration point with Δ_0 and which is on the other side of the point, the identification of Δ' and Δ_0 does not change the Euler number of Σ. The number of triangles in T, on the other hand, decreases by one if Δ' and Δ_0 were identified in T and removed from T. Thus the equality $\chi(T) = \frac{1}{2}\chi(\Sigma)$ cannot be sustained. On the other hand, if we take Δ to be other than those triangles, there is a triangulation T such that $\mathcal{H}^{(\alpha)}_\Sigma \cup (-T) \simeq S^3$ and Δ and Δ_0 are identified. Thus the Schwinger-Dyson equation holds with this supplementary condition.

So far, we have examined the situation when Σ is already degenerate. When $V(\Delta,\Delta_0) \geq E(\Delta,\Delta_0) + 2$, the triangles Δ and Δ_0 wrap around a cycle γ on Σ and the surface will develop additional degeneration if we identify Δ and Δ_0 and remove them. If it is possible to construct a triangulation T in which the two triangles Δ and Δ_0 are attached to each other, T should be counted in $\langle \Sigma^{(\alpha)} \rangle_{|S^3}$. The Schwinger-Dyson equation (26) derived for non-degenerate surfaces becomes applicable in this case if we take into account the contribution of the degenerate surface Σ_{Δ,Δ_0}. On the other hand, if there is no such triangulation in which Δ is identified with Δ_0, the surface Σ_{Δ,Δ_0} should not be included the Schwinger-Dyson equation.

When the cycle γ is homologically trivial, it is always possible to construct a triangulation T in which Δ and Δ_0 are identified. Thus the term $\langle \Sigma^{(\alpha)}_{\Delta,\Delta_0} \rangle$ should be included in the Schwinger-Dyson equation. In this case, the surface Σ_{Δ,Δ_0} is pinched around at the common vertex of Δ and Δ_0, and the Schwinger-Dyson equation can be iterated with the supplementary conditions introduced in the above.

On the other hand, when the cycle γ is homologically non-trivial on Σ, the triangles Δ and Δ_0 wrap around a handle on Σ. In this case, the handle is pinched on the surface Σ_{Δ,Δ_0}. If there is a triangulation T in which the triangles Δ and Δ_0 are identified with each other, there must be a disc in T whose boundary is the cycle γ. After removing the triangle $\Delta = \Delta_0$ from T, we obtain T' which gives a triangulation of a handlebody whose boundary is Σ_{Δ,Δ_0}. The surface Σ_{Δ,Δ_0} is pinched at the common vertex of Δ and Δ_0. The complement of T' in S^3 should also be a handlebody $\mathcal{H}^{(\alpha)}_{\Sigma_{\Delta,\Delta_0}}$ which is pinched around the cycle γ. The process $\mathcal{H}^{(\alpha)}_\Sigma \cup (-T) \to \mathcal{H}^{(\alpha)}_{\Sigma_{\Delta,\Delta_0}} \cup (-T')$ then gives the reduction of the Heegaard splitting of S^3 which is the inverse of the stabilization [38]. For this to be possible, there must be a disc in the handlebody $\mathcal{H}^{(\alpha)}_\Sigma$ whose boundary is on Σ and intersects once with γ transversely. If this is the case, we can choose the meridians of $\mathcal{H}^{(\alpha)}_\Sigma$ so that $\alpha_1 \cdot \gamma = 1$ and $\alpha_i \cdot \gamma = 0$ ($i = 2, \ldots, h(\Sigma)$). The interior of $\mathcal{H}^{(\alpha_1,\ldots,\alpha_{h(\Sigma)})}_{\Sigma_{\Delta,\Delta_0}}$ is then homeomorphic to that of $\mathcal{H}^{(\alpha_2,\ldots,\alpha_{h(\Sigma)})}_{\Sigma'_{\Delta,\Delta_0}}$, where $\Sigma'_{\Delta,\Delta_0}$ is obtained from Σ_{Δ,Δ_0} by splitting the common vertex of Δ and Δ_0 into two so that the pinched handle is removed; $h(\Sigma'_{\Delta,\Delta_0}) = h(\Sigma_{\Delta,\Delta'}) - 1$. In this case, a term $\langle \Sigma'^{(\alpha)}_{\Delta,\Delta_0} \rangle_M$ must be included in the Schwinger-Dyson equation. Conversely, if it is not possible to find such a disc in $\mathcal{H}^{(\alpha)}_\Sigma$, there is no triangulation in which Δ and Δ_0 are identified with each other. In this case, there should be no term in the Schwinger-Dyson equation corresponding to the surface Σ_{Δ,Δ_0}.

With these supplementary conditions, we can apply the Schwinger-Dyson equation even when Σ is degenerate or when there is a triangle Δ such that $V(\Delta, \Delta_0) \geq E(\Delta, \Delta_0) + 2$. As in the case of the matrix model, we can show that the Schwinger-Dyson equation combined with these conditions can be solved inductively. The Schwinger-Dyson equation expressed as in eq.(34) is a recurrence relation for $Z_{n_3,n_1}(\alpha; \Sigma)$ along the bibliographical ordering: $(n_3, \Sigma) \succ (n'_3, \Sigma')$ if $n_3 > n'_3$ or $n_3 = n'_3$ and $n_2(\Sigma) > n_2(\Sigma')$. Because of the factorization conditions, we can trace back the Schwinger-Dyson equation to the expectation value of the null surface $\langle 1 \rangle$ which is normalized to be one. Thus the Schwinger-Dyson equation combined with the factorization conditions has a unique solution. The solution gives the sum over triangulations of handlebodies, and it is related to $Z_{n_3,n_1}(S^3)$ as in eq. (35).

ACKNOWLEDGEMENTS

I am grateful to Jan Ambjørn, Bergfinnur Durhuus, Tohru Eguchi, Antal Jevicki and Takao Matsumoto for discussions and comments. I would also like to thank Michael Atiyah and Peter Goddard for their hospitality at the Isaac Newton Institute for Mathematical Sciences, University of Cambridge, where part of this work was done. This research is supported by Grant-in-Aid for Scientific Research on Priority Areas 231 "Infinite Analysis" from the Ministry of Education, Science and Culture of Japan.

REFERENCES

[1] T. Regge, *Nuovo Cimento* **19** (1961) 558.
[2] For a review on recent numerical results in this approach, see H. Hamber, Nucl. Phys. (Proc. Suppl.) **25A** (1992) 150.
[3] W.T. Tutte, *Can. J. Math.* **14** (1962) 21.
[4] E. Brézin, C. Itzykson, G. Parisi and J.-B. Zuber, *Commun. Math. Phys.* **59** (1978) 35.
[5] T. Eguchi and H. Kawai, *Phys. Lett.* **114B** (1982) 247.

[6] F. David, *Nucl. Phys.* **B257** (1985) 45
[7] V.A. Kazakov, *Phys. Lett.* **150B** (1985) 282.
[8] D. Bessis, C. Itzykson and J.-B. Zuber, *Adv. Appl. Math.* **1** (1980) 109 .
[9] I. Kostov and M.L. Metha, *Phys. Lett.* **189B** (1987) 118.
[10] V.G. Knizhnik, A.M. Polyakov and A.B. Zamolodchikov, *Mod. Phys. Lett.* **A3** (1988) 819.
[11] F. David, *Mod. Phys. Lett.* **A3** (1988) 1651; J. Distler and H. Kawai, *Nucl. Phys* **B321** (1989) 509.
[12] M. Douglas and S. Shenker, *Nucl. Phys.* **B335** (1990) 635; D.J. Gross and A.A. Migdal, *Phys. Rev. Lett.* **64** (1990) 127; E. Brezin and V.A. Kazakov, *Phys. Lett.* **B236** (1990) 144.
[13] E. Witten, *Nucl. Phys.* **B340** (1990) 281; R. Dijkgraaf and E. Witten, *Nucl. Phys.* **B342** (1990) 342.
[14] R. Dijkgraaf, E. Verlinde and H. Verlinde, *Nucl. Phys.* **B348** (1991) 435; E. Verlinde and H. Verlinde, *Nucl. Phys.* **B348** (1991) 457.
[15] M. Fukuma, H. Kawai and R. Nakayama, *Int. J. Mod. Phys.* **A6** (1991) 1385.
[16] G. Moore, N. Seiberg and M. Staudacher, *Nucl. Phys.* **B362** (1991) 665; G. Moore and N. Seiberg, *Int. J. Mod. Phys.* **A7** (1991) 190.
[17] J. Ambjørn and S. Varsted, *Phys. Lett.* **266B** (1991) 285.
[18] M.E. Agishtein and A.A. Migdal, *Mod. Phys. Lett.* **A6** (1991) 1863.
[19] D.V. Boulatov and A. Krywicki *Mod. Phys. Lett.* **A6** (1991) 3005.
[20] J. Ambjørn and S. Varsted, *Nucl. Phys.* **B373** (1992) 557.
[21] J. Ambjørn, D.V. Boulatov, A. Krywicki and S. Varsted, *Phys. Lett.* **276B** (1992) 432.
[22] J. Ambjørn, B. Durhuus and T. Jonsson, *Mod. Phys. Lett.* **A6** (1991) 1133.
[23] N. Sasakura, *Mod. Phys. Lett.* **A6** (1991) 2613.
[24] M. Godfrey and M. Gross, *Phys. Rev.* **D43** (1992) R1749.
[25] D.V. Boulatov, *Mod. Phys. Lett.* **A7** (1992) 1629.
[26] V.G. Turaev and O.Y. Viro, *Toplogy* **31** (1992) 865.
[27] S.R. Wadia, *Phys. Rev.* **D24** (1981) 970.
[28] A.A. Migdal, *Phys. Rep.* **102** (1983) 199.
[29] A.N. Kirillov and N.Yu. Reshetikhin, in *Infinite Dimensional Lie Algebras and Groups,* edited by V.G. Kac (World Scientific, 1989).
[30] V.G. Turaev, *C. R. Acad. Sci. Paris,* t. 313, Série I (1991) 395; *"Topology of Shadow"*, preprint (1991).
[31] K. Walker, *"On Witten's 3-Manifold Invariants"*, preprint (1990).
[32] E. Witten, *Commun. Math. Phys.* **121** (1989) 351; N.Y. Reshetikhin and V.G. Turaev, *Invent. Math.* **103** (1991) 547.
[33] H. Ooguri and N. Sasakura, *Mod. Phys. Lett.* **A6** (1991) 3591.
[34] H. Ooguri, *Nucl. Phys.* **B382** (1992) 276.
[35] F. Archer and R. Williams, *Phys. Lett.* **B273** (1991) 438.
[36] S. Mizoguchi and T. Tada, *Phys. Rev. Lett.* **68** (1992) 1795.
[37] T. Kohno, *Topology* **31** (1992) 203.
[38] F. Waldhausen, *Topology* **7** (1968) 195.

OBSERVABLES IN THE KONTSEVICH MODEL

P. Di Francesco

Service de Physique Théorique de Saclay*
F-91191 Gif sur Yvette Cedex, France

ABSTRACT

Kontsevich introduced a hermitian random matrix model to compute the generating function of intersection numbers of the moduli space of (punctured) Riemann surfaces. He showed that this generating function is also a τ-function for the Korteveg–de Vries (KdV) hierarchy of differential equations. This model is fundamentally different from the usual double scaling limit of random matrix models known to yield analogous τ-functions. Our aim is to clarify the notion of "observables" in both pictures, as related to KdV time evolutions. As a result we prove two conjectures by Kontsevich and Witten about the form of these observables, which involve polynomial matrix averages.

INTRODUCTION

The matrix model formulation of two dimensional quantum gravity coupled to matter (with central charge c less than 1) enables to express the physical properties of the solutions in a very simple and elegant way [1]. Let us recall the original definition. One starts with a $N \times N$ hermitian matrix model, with partition function:

$$\mathcal{Z}(\mathcal{V}, N) = \int dM \exp -N \mathrm{tr} \mathcal{V}(M) , \qquad (1)$$

where $dM = (2\pi)^{-N^2/2} \prod_{i<j} d\mathrm{Re}(M_{ij}) \, d\mathrm{Im}(M_{ij}) \prod_i dM_{ii}$ is the Haar measure over hermitian matrices, and $\mathcal{V}$ some polynomial potential. Roughly speaking the Feynman diagrammatic expansion of (1) simulates all possible polygon decompositions of Riemann surfaces (i.e. gravitational fluctuations) while the precise form of the potential indicates which type of matter interactions are involved. The partition function

---

* Laboratoire de la Direction des Sciences et de la Matière du Commissariat à l'Energie Atomique.

(1) can be calculated explicitly using orthogonal polynomial techniques, but the main progress was to consistently define the so called double scaling limits of the orthogonal polynomial solution. On the one hand the large N limit was already known to concentrate on surfaces with spherical topology, whereas a large N expansion would give access order by order to surfaces with higher genera [2]. The idea is to combine a large N limit with a critical limit in which the parameters of the potential are tuned to make it (multi)critical as N grows. This has the advantage of capturing the whole genus expansion into a single asymptotic series of a rescaled variable x (the renormalized cosmological constant), kept fixed while N is sent to infinity and the potential is taken to a (multi)critical value. The orthogonal polynomial solution becomes in this limit an ordinary differential equation for the double scaled string susceptibility $u(x) = \partial_x^2 \log \tau(x)$, where $\log \tau(x)$ is the double scaled free energy:

$$\log \tau(x) = \lim_{\substack{N \to \infty \\ \mathcal{V} \to \mathcal{V}^*; x \text{ fixed}}} -\frac{1}{2N^2} \log \mathcal{Z}(\mathcal{V}, N) . \qquad (2)$$

If instead of taking a special m–critical potential $\mathcal{V}_m^*$, we consider a linear combination of them for various m's with coefficients t_m, we end up with a function $u(x, t_i) = \partial_x^2 \log \tau(x, t_i)$. The double-scaled solution is then simply characterized by the fact that (i) $\tau(x, t_i)$ is a τ-function for the Korteveg–de Vries (KdV) hierarchy and (ii) $\tau(x, t_i)$ satisfies an ordinary differential equation (the so called "string equation"). Upon introducing the differential operator $Q = d^2 + u(x, t_i)$, where d stands for ∂_x, and the fractional powers of Q truncated to their differential piece $Q_+^{m+1/2}$, the two above properties become:

$$\begin{aligned}(i) & \quad \partial_{t_m} Q = [Q_+^{m+1/2}, Q] , \\ (ii) & \quad P = \sum_{j>0}(j+1/2)t_j Q_+^{j+1/2} \; ; \; [P, Q] = 1 .\end{aligned} \qquad (3)$$

The string equation (ii) can be viewed as an interpolating equation of motion between the m–critical points $t_i = \delta_{i,m}$, which correspond to $c = 1 - 3(2m-1)^2/(2m+1)$ conformal matter coupled to 2D quantum gravity. More generally (multi)critical (multi)matrix models are known to interpolate between conformal points with $c < 1$, the KdV flows (i) being replaced by generalized KdV flows and (ii) modified accordingly [3] (Q is now a differential operator of order p, for a model of $p-1$ matrices, and one considers all its fractional powers of the form $Q_+^{k/p}$ where k is not a multiple of p).

The KdV flows (i) and their generalizations enable therefore to move in the space of $c < 1$ matter coupled to 2D gravity, along RG trajectories. The latter are obtained in ordinary conformal theory by adding to the action perturbations by relevant operators, it is therefore tempting to identify the "dressed" operators of the conformal theory coupled to gravity as dual to the "KdV times" t_i. More precisely, the insertion of a dressed operator ϕ_m in a correlator will be generated by differentiation w.r.t. t_m: $\partial_{t_m} \langle ... \rangle = \langle \phi_m ... \rangle$. This definition of observables was successfully applied to one and more matrix models [4] and shown to confirm predictions from the continuum theory such as KPZ scaling dimensions of dressed operators [5], Liouville correlators [6], etc... Going back to the original matrix model for a while, we see that these observables correspond to the insertion of very specific polynomials $\mathcal{V}_m^*(M)$ into the defining integral. On the other hand, such polynomial insertions correspond in the Feynman expansion to the insertion of sources with vertices of a well defined order, or in the dual picture to the creation of microscopic holes in the (discretized) Riemann surface. In fact, the properties (i) and (ii) can be rephrased into equations of motion (or loop equations) for these so called loop operators, and take the form of Virasoro constraints [7] $L_m \tau = 0$, $m \geq -1$, L_m certain Virasoro generators constructed in terms of bilinears of t_i and ∂_{t_i}. The simplest of those observables is the "puncture" operator dual to $t_0 = x$, the renormalized cosmological constant.

Meanwhile after introducing topological gravity [8] and uncovering its relations to KdV hierarchies [9], Witten conjectured that the one matrix model partition function $\tau(x, t_i)$ could also be interpreted as the generating function for intersection numbers of the moduli space of (punctured) Riemann surfaces [10]. These intersection numbers $\langle \sigma_0^{n_0} .. \sigma_p^{n_p} \rangle$ have a precise definition as integrals over a compactification of the moduli space of Riemann surfaces of genus g with n marked points of exterior powers of the first Chern class of the line bundle defined by the cotangent spaces at these points. We will not emphasize the topological aspect of this model here and refer the reader to ref. [10] for more details.

M. Kontsevich made this statement even deeper by introducing yet another matrix model of a very different nature [11], enabling to compute these intersection numbers directly. By interpreting his cell decomposition of the moduli space of (punctured) Riemann surfaces in terms of "fat graphs" he was able to write directly an ad–hoc hermitian matrix model whose connected partition function is exactly the generating function for the intersection numbers. Let $\Lambda = diag(\lambda_1, \lambda_2, \ldots, \lambda_N)$ be a real diagonal matrix, the partition function reads:

$$\Xi_N(\Lambda) = \frac{\int dY \exp \operatorname{tr}(iY^3/6 - \Lambda Y^2/2)}{\int dY \exp -\operatorname{tr}(\Lambda Y^2/2)} . \qquad (4)$$

Kontsevich established that when expressed in the variables

$$t_j = -(2j-1)!! \operatorname{tr}(\Lambda^{-2j-1}) \qquad (5)$$

this function is an asymptotic series whose truncation to terms of degree less than N is a universal polynomial of the $t.$'s. It admits a universal $N \to \infty$ limit $\Xi(t.)$, whose connected piece $\log \Xi(t.)$ is equal to the generating function $F(t.)$ for intersection numbers:

$$F(t.) = \sum_{n_j} \prod_j \frac{t_j^{n_j}}{n_j!} \langle \sigma_0^{n_0} \ldots \sigma_p^{n_p} \rangle . \qquad (6)$$

Using this correspondence, Witten was able to show using topological arguments [12] that $\Xi(t.)$ satisfies the properties (i) and (ii) of (3), and is therefore equivalent to $\tau(x, t_i)$ $(x = t_0)$[1]. Some generalizations of the Kontsevich model were also introduced and shown to satisfy generalized KdV time evolutions [11] [15]. Their topological interpretation was given in [16].

The obvious question in view of this equivalence is what about the observables? We just clarified their interpretation in the (double scaled) one matrix model as dual to the KdV times. Their topological counterpart is given by the first Chern class of the line bundle of the cotangent spaces at marked points. But Witten noticed on a few examples that the disconnected partition function $\Xi(t.)$ could also give rise to interesting objects by differentiation w.r.t. KdV times. In a number of cases, he

---

[1] See also [13] and [14] for alternative proofs.

found that the action of some differential polynomials of the KdV times $R(\partial_{t.})$ on $\Xi(t.)$ could be rewritten as a kind of polynomial expectation value in the form:

$$R(\partial_{t.})\,\Xi(t.) = \frac{\int d\mu_\Lambda(Y)\,P(Y)\,\exp\,\text{tr}(iY^3/6)}{\int d\mu_\Lambda(Y)} = \ll P(Y) \gg , \qquad (7)$$

where $P(Y)$ is a certain polynomial of traces of odd powers of Y (odd traces for short), and $d\mu_\Lambda(Y) = dY \exp -\text{tr}(\Lambda Y^2/2)$ is the natural Gaussian measure of the problem. This led him to the conjecture that there exists a general mapping $R \to P$ defined on $\mathbb{C}[\partial_{t_0}, \partial_{t_1}, \ldots]$. Let us first discuss a few implications of this fact. First of all if the mapping can be made explicit, this gives in principle a straightforward way of computing any intersection number using also (6). Another important consequence is the definition of yet another kind of observables in the topological model. Those are very much like the ones of the one matrix model before the double scaling limit, and correspond to the insertion of vertices with a well defined number of legs in the Feynman diagrammatic expansion of $\Xi(t.)$. The explicit mapping $R \to P$ yields rules for computing correlations of such observables.

The rest of the lecture will be dedicated to the proof of this conjecture and the explicit construction of the mapping $R \to P$ (see also [17] for a more detailed version). We will need a few definitions and preliminaries, and will first prove a weaker statement on Gaussian polynomial averages (sect.2), due to Kontsevich, who proved it by topological arguments. It involves the construction of another mapping defined on polynomials $P(Y)$ of odd traces of Y, by taking the average over the Gaussian measure $d\mu_\Lambda(Y)$. The result is that as a function of $t.$ this average is still polynomial:

$$\langle P(Y) \rangle = \frac{\int d\mu_\Lambda(Y)\,P(Y)}{\int d\mu_\Lambda(Y)} = Q(t.) . \qquad (8)$$

We will compute the mapping $P \to Q$ explicitly in a purely algebraic way, and be naturally led to introduce a new set of polynomials which generalize the ordinary Schur polynomials. These polynomials will be used in sect.3 to describe the Witten mapping $R \to P$. The essence of these proofs is extremely simple and relies mainly on comparisons between integrals over $N+1 \times N+1$ matrices and over $N \times N$ matrices.

THE KONTSEVICH MAPPING

In this section, we construct the Kontsevich mapping $P \to Q$ explicitly, where $Q(\Lambda^{-1}) = \langle P(Y) \rangle$.

The main tool for working with matrix models is the formula of integration over the "angular variables" U, when one diagonalizes the hermitian matrix integration variable $Y = UyU^\dagger$, with $y = \text{diag}(y_1,..,y_n)$. The Haar measure decomposes into $dY = (2\pi)^{-N^2/2} \prod dy_i\,dU\,\Delta(Y)^2$, where $\Delta(Y) = \prod_{i<j}(y_i - y_j)$ is the Vandermonde determinant of Y. The Harish-Chandra-Itzykson-Zuber formula [18] reads, for any two diagonal matrices x and y:

$$\int dU \exp\,\text{tr}(UxU^\dagger y) \propto \frac{\det[e^{x_i y_j}]_{1\leq i,j\leq N}}{\Delta(x)\Delta(y)} , \qquad (9)$$

up to an irrelevant numerical factor depending on N only (in the following we omit most of these cumbersome factors and use the symbol $\propto$ to indicate their presence). This formula happens to be a simple case of the Duistermaat-Heckman integration formula [19]: (9) expresses nothing but the fact that the semi–classical approximation to the integral is exact (the classical solutions for the potential $\mathrm{tr}(UxU^\dagger y)$ are just permutations, and the inverse Vandermonde determinants arise from the Gaussian integral)[2]. This result enables to restrict most of the interesting matrix integrals to integrals over eigenvalues. Let us use it to rewrite the Gaussian integral:

$$Z_N = \int d\mu_\Lambda(Y) \propto \int \frac{\prod dy_i}{\Delta(\Lambda)} \prod_{i<j} \frac{y_i + y_j}{y_i - y_j} e^{-\sum \lambda_i y_i^2/2}, \qquad (10)$$

where we dropped the determinant symbol by noticing the antisymmetry of the prefactor. On the other hand this integral is easily computed by direct integration:

$$Z_N \propto \det(\Lambda)^{-1/2} \frac{\Delta(\Lambda)}{\Delta(\Lambda^2)}. \qquad (11)$$

Consider now the $N+1 \times N+1$ version of (10), this amounts to introducing additional eigenvalues y and λ to Y and Λ respectively and we compute:

$$\frac{Z_{N+1}}{Z_N} \det(\lambda - \Lambda) = \lambda^{-1/2} \det \frac{1 - \lambda \Lambda^{-1}}{1 + \lambda \Lambda^{-1}} = \int \frac{dy}{(2\pi)^{1/2}} e^{-\lambda y^2/2} \left\langle \det \frac{y - Y}{y + Y} \right\rangle. \qquad (12)$$

Let us take a closer look to the integrand of (12). It involves the determinant:

$$f_Y(y) = \det \frac{y - Y}{y + Y} = \exp -2 \sum_{k=0}^{\infty} y^{-2k-1} \mathrm{tr}(Y^{2k+1})/(2k+1) = \sum_{m=0}^{\infty} y^{-m} p_m(Y), \qquad (13)$$

where p_m denote the Schur polynomials of the odd variables:

$$\theta_{2i+1}(Y) = -\frac{2}{2i+1} \mathrm{tr}(Y^{2i+1}), \qquad (14)$$

$$p_m = \sum_{\substack{\nu_{2j+1} \geq 0 \\ \sum_{j \text{ odd}} j\nu_j = m}} \prod_{j \text{ odd}} \frac{\theta_j^{\nu_j}}{\nu_j!}. \qquad (15)$$

Therefore (12) gives us some expressions for the Gaussian averages of these polynomials. But due to the divergence of the y integral, we need to perform some analytic continuation.

To avoid this difficulty, let us compute directly as a function of the formal variable y the following Gaussian average over Y:

$$\langle f_Y(y) \rangle = 1 - 2 \sum_{i=1}^{N} \left\langle \frac{y_i}{y + y_i} \prod_{j \neq i} \frac{y_i + y_j}{y_i - y_j} \right\rangle, \qquad (16)$$

where we performed a decomposition of f defined in (13) into fractions with simple poles at $y = -y_i$. Integrating over the angular variables, we are left with

$$\langle f_Y(y) \rangle = 1 + 2 \sum_{m=1}^{\infty} (-y)^{-m} \sum_{i=1}^{N} \int dy_i \left(\frac{\lambda_i}{2\pi}\right)^{1/2} (-1)^{i-1} y_i^m e^{-\lambda_i y_i^2/2} \frac{\Delta(\Lambda_i) Z_{N-1}(\Lambda_i)}{\Delta(\Lambda) Z_N(\Lambda)}, \qquad (17)$$

---

[2] I thank J.-B. Zuber for explaining this unpublished result to me.

where we denoted by $X_{\hat i} = \text{diag}(x_1,..,x_{i-1},x_{i+1},..,x_N)$ for any diagonal matrix X. Note that we singled out the integration over y_i and that the integration over the other y's just recombined to yield $Z_{N-1}(\Lambda_{\hat i})$. Using the one dimensional integral $(2\pi)^{-1/2} \int dz\, z^{2m} e^{-\lambda z^2/2} = \lambda^{-m-1/2}(2m-1)!!$, we get:

$$\langle f_Y(y) \rangle = 1 + 2 \sum_{m=1}^{\infty} (-1)^{N-1}(2m-1)!! y^{-2m} \sum_{i=1}^{N} \lambda_i^{-m} \prod_{j \neq i} \frac{\lambda_i + \lambda_j}{\lambda_i - \lambda_j}. \qquad (18)$$

We now use the decomposition of $f_{\Lambda^{-1}}(\lambda)$ into fractions with single poles at $\lambda = -\lambda_i$ to identify for $m \geq 1$:

$$p_m(\Lambda^{-1}) = 2(-1)^{N+m-1} \sum_{i=1}^{N} \lambda_i^{-m} \prod_{j \neq i} \frac{\lambda_i + \lambda_j}{\lambda_i - \lambda_j} \qquad (19)$$

and finally:

$$\langle f_Y(y) \rangle = \sum_{k=0}^{\infty} y^{-k} \langle p_k(Y) \rangle = 1 + \sum_{m=1}^{\infty} y^{-2m} (-1)^m (2m-1)!! p_m(\Lambda^{-1}). \qquad (20)$$

Hence we get the first elements of the mapping $P \to Q$, namely

$$\begin{aligned}(i) \quad & \langle p_{2m+1}(Y) \rangle = 0, \\ (ii) \quad & \langle p_{2m}(Y) \rangle = (-1)^m (2m-1)!!\, p_m(\Lambda^{-1}).\end{aligned} \qquad (21)$$

To proceed, we need to consider more complicated polynomials, like products of p_m's, known to generate the space of polynomials of odd variables θ_i's. One can think of performing an average of the form (16), but in the presence of a "spectator" insertion of $p_k(Y)$. Actually, it is easy to see that it is more useful to consider:

$$\langle f_Y(y) p_k(\theta.(Y) + \theta.(y)) \rangle. \qquad (22)$$

If one decomposes f into fractions with single poles at $y = -y_i$ as in (16), then we see that the "spectator" $p_k(\theta.(Y) + \theta.(y)) = p_k(\theta.(Y_{\hat i})) + O(y + y_i)$ can be replaced with $p_k(\theta.(Y_{\hat i}))$ if we retain only terms y^{-k}, $k > 0$ in the formal y expansion. Therefore, we can go through the previous steps, with the only modification:

$$\langle f_Y(y) p_m(\theta.(Y) + \theta.(y)) \rangle_{<0} = 2 \sum_{m=1}^{\infty} (-y)^{-m} \sum_{i=1}^{N} \int dy_i (\frac{\lambda_i}{2\pi})^{1/2} (-1)^{i-1} y_i^m e^{\lambda_i y_i^2/2}$$

$$\times \frac{\Delta(\Lambda_{\hat i}) Z_{N-1}(\Lambda_{\hat i}) \langle p_k(\theta.(Y_{\hat i})) \rangle}{\Delta(\Lambda) Z_N(\Lambda)}, \qquad (23)$$

where the subscript < 0 indicates that we truncate the y expansion to negative powers only. We use the result (21) to compute $\langle p_k(\theta.(Y_{\hat i})) \rangle = (k-1)!!(-1)^{k/2} p_{k/2}(\Lambda_{\hat i}^{-1})$, with the convention that $(k-1)!! = 0$ when k is odd. Performing the integration over y_i, we get a generalization of (18):

$$2 \sum_{m=1}^{\infty} (-1)^{N-1}(2m-1)!! y^{-2m} (k-1)!!(-1)^{k/2} \sum_{i=1}^{N} \lambda_i^{-m} p_{k/2}(\Lambda_{\hat i}^{-1}) \prod_{j \neq i} \frac{\lambda_i + \lambda_j}{\lambda_i - \lambda_j}. \qquad (24)$$

We recognize the general term of the series generated by expanding the decomposition of $f_{\Lambda^{-1}}(\lambda)p_{k/2}(\theta.(\Lambda^{-1})+\theta.(\lambda^{-1}))$ into rational fractions with single poles at $\lambda = -\lambda_i$. If we introduce a generating function (for $z > y$):

$$f_Y(z,y) = \sum_{n\in\mathbb{Z}} z^{-n} p_n(\theta.(Y) + \theta.(y)) \det\frac{y-Y}{y+Y}$$

$$= \frac{z-y}{z+y} \det\frac{(y-Y)(z-Y)}{(y+Y)(z+Y)} \qquad (25)$$

$$= \sum_{m,n\in\mathbb{Z}} z^{-n} y^{-m} \varphi_{m,n}(Y)$$

then the result takes the simple form:

(i) $\langle\varphi_{m,n}(Y)\rangle = 0$ if m or n is odd,

(ii) $\langle\varphi_{2m,2n}(Y)\rangle = (-1)^{m+n}(2m-1)!!(2n-1)!!\, \varphi_{m,n}(\Lambda^{-1})$. $\qquad (26)$

Strictly speaking, we only proved it for $m, n \geq 1$, as we concentrated on the negative power expansion in y. But from the definition (25) it is easy to see that $\varphi_{m,0} = \varphi_m = p_m$ if $m \geq 0$, 0 if $m < 0$, that $\varphi_{m,-n} = 0$ for $n > 0$, and that $\varphi_{-m,n} = 2(-1)^m \delta_{m,n}$ for $m, n > 0$. Then if we define $(-2m-1)!! \equiv (-1)^m/(2m-1)!!$ for $m > 0$, (26) holds for any $m, n \in \mathbb{Z}$.

In view of the above, it is clear that the general Kontsevich mapping will follow from an analogous treatment of the Gaussian average over Y of

$$f_Y(z_1,..,z_p) = \prod_{a>b}\frac{z_a-z_b}{z_a+z_b}\det\prod_{a=1}^{p}\frac{z_a-Y}{z_a+Y} = \sum_{m_1,..,m_p\in\mathbb{Z}}\prod_a z_a^{-m_a}\varphi_{m_1,..,m_p}(Y) \quad (27)$$

understood as a formal series of the variables z_a in a domain where, say, $z_1 < z_2 < \ldots < z_p$. The polynomials $\varphi_{m_1,...,m_p}(Y)$ generalize the Schur polynomials, and generate the whole set of polynomials of the odd variables $\theta.(Y)$. Actually a basis is formed by the φ's with ordered indices $m_1 > m_2 > \ldots > m_p \geq 1$, $p \geq 0$. After some algebra one gets the Gaussian averages of these basis elements, defining the Kontsevich map[3]:

(i) $\langle\varphi_{m_1,..,m_p}(Y)\rangle = 0$ if at least one of the m_i is odd,

(ii) $\langle\varphi_{2m_1,..,2m_p}(Y)\rangle = \prod_{i=1}^{p}(-1)^{m_i}(2m_i-1)!!\, \varphi_{m_1,..,m_p}(\Lambda^{-1})$. $\qquad (28)$

The φ's are easily computed from the definition (27), which can be recast into the following recursive formula (by convention, the φ with no index is the constant 1):

$$\varphi_{m_1,..,m_p,m} = \sum_{s\geq 0}\varphi_{m-s}\sum_{\substack{r_1,...,r_p\geq 0 \\ r_1+..+r_p=s}}\left(\prod_{i=1}^{p}\alpha_{r_i}\right)\varphi_{m_1+r_1,...,m_p+r_p}, \qquad (29)$$

---

[3] See [17] for a detailed proof. An inductive proof can also be made along the lines of the case $p = 2$ treated above, involving a "spectator" insertion of $\varphi_{m_1,...,m_p}(\theta.(Y)+\theta.(y))$ into the Gaussian average of $f_Y(y)$.

where $\alpha_r = (-1)^r(2 - \delta_{r,0})$ are the coefficients of the expansion $(1-y)/(1+y) = \sum_{r\geq 0} \alpha_r y^r$, and $\varphi_m = p_m$ if $m \geq 0$, and vanishes for $m < 0$. We list below the first few φ's with positive ordered indices.

Table I. the φ polynomials up to degree 8. The notation $\theta_{[1^{\nu_1}3^{\nu_3}\ldots(2k+1)^{\nu_{2k+1}}\ldots]}$ is a short hand for $\frac{\theta_1^{\nu_1}}{\nu_1!}\frac{\theta_3^{\nu_3}}{\nu_3!}\ldots\frac{\theta_{2k+1}^{\nu_{2k+1}}}{\nu_{2k+1}!}\ldots$

φ_1	$= \theta_{[1^1]}$
φ_2	$= \theta_{[1^2]}$
φ_3	$= \theta_{[1^3]} + \theta_{[3^1]}$
$\varphi_{2,1}$	$= \theta_{[1^3]} - 2\theta_{[3^1]}$
φ_4	$= \theta_{[1^4]} + \theta_{[1^13^1]}$
$\varphi_{3,1}$	$= 2\theta_{[1^4]} - \theta_{[1^13^1]}$
φ_5	$= \theta_{[1^5]} + \theta_{[1^23^1]} + \theta_{[5^1]}$
$\varphi_{4,1}$	$= 3\theta_{[1^5]} - 2\theta_{[5^1]}$
$\varphi_{3,2}$	$= 2\theta_{[1^5]} - \theta_{[1^23^1]} + 2\theta_{[5^1]}$
φ_6	$= \theta_{[1^6]} + \theta_{[1^33^1]} + \theta_{[1^15^1]} + \theta_{[3^2]}$
$\varphi_{5,1}$	$= 4\theta_{[1^6]} + \theta_{[1^33^1]} - \theta_{[1^15^1]} - 2\theta_{[3^2]}$
$\varphi_{4,2}$	$= 5\theta_{[1^6]} - \theta_{[1^33^1]} + 2\theta_{[3^2]}$
$\varphi_{3,2,1}$	$= 2\theta_{[1^6]} - \theta_{[1^33^1]} + 2\theta_{[1^15^1]} - 4\theta_{[3^2]}$

THE WITTEN MAPPING

We want to investigate the mapping $R \to P$, where, hopefully for any polynomial $R(\partial_{t_\cdot})$ one can find a polynomial P of odd traces of Y such that $R(\partial_{t_\cdot})\Xi(t_\cdot) = \ll P(Y) \gg$, where the double bracket denotes the weighted average (7). Starting from the partition function $\Xi(\Lambda^{-1})$ (this notation is just to recall that the $t_\cdot$'s are themselves normalised odd traces of Λ^{-1}, and indicates that Ξ is also a function of the odd variables $\theta_{2j+1}(\Lambda^{-1})$), one can generate differentiations w.r.t. odd traces of Λ^{-1} by just expanding:

$$\Xi(\lambda^{-1} \oplus \Lambda^{-1}) = \exp\left(\sum_{i=0}^{\infty} -\frac{2\lambda^{-2j-1}}{2j+1}\frac{\partial}{\partial\theta_{2j+1}}\right)\Xi(\Lambda^{-1}) = \sum_{k=0}^{\infty} \lambda^{-k} p_k(\partial_\cdot)\,\Xi \qquad (30)$$

where $\lambda^{-1} \oplus \Lambda^{-1} = \mathrm{diag}(\lambda^{-1}, \lambda_1^{-1}, \lambda_2^{-1}, \ldots)$, and the r.h.s. is a formal power series of λ^{-1}, which is the generating function of the Schur polynomials of the odd derivatives $\partial_{2j+1} = -2/(2j+1)\partial_{\theta_{2j+1}}$ acting on Ξ, as a function of the infinitely many variables θ_{2j+1}. To evaluate this action in terms of matrix averages, we will have to compare $\Xi_{N+1}(\lambda^{-1} \oplus \Lambda^{-1})$ to $\Xi_N(\Lambda^{-1})$. To get the most general action on Ξ of polynomials of derivatives w.r.t. $\theta_\cdot$'s, we can just add p eigenvalues $\lambda_1, \ldots, \lambda_p$ to Λ, and expand

Ξ as a formal series of λ_i^{-1}. By analogy with the situation of previous section, it is very natural to consider the generating function:

$$\prod_{1\leq i<j\leq p} \frac{\lambda_i^{-1}-\lambda_j^{-1}}{\lambda_i^{-1}+\lambda_j^{-1}} \Xi(\lambda_1^{-1}\oplus..\oplus\lambda_p^{-1}\oplus\Lambda^{-1}) = \sum_{m_k\in\mathbb{Z}} \prod_{k=1}^{p} \lambda_k^{-m_k} \varphi_{m_1,..,m_p}(\partial.) \Xi(\Lambda^{-1}) \tag{31}$$

where the polynomials φ are now considered as functions of the odd differentials ∂_{2j+1} (substituted for the odd traces θ_{2j+1} of (27)), and the function is expanded in the domain $\lambda_1^{-1} > \ldots > \lambda_p^{-1}$. The reexpression of (31) as matrix averages is achieved through the following formula, where we decompose $\Lambda^{-1} = \Lambda_1^{-1}\oplus\Lambda_2^{-1}$, with $\Lambda_1 = \mathrm{diag}(\lambda_1,\ldots,\lambda_p)$ and $\Lambda_2 = \mathrm{diag}(\lambda_{p+1},\ldots,\lambda_{N+p})$:

$$\prod_{1\leq i<j\leq p} \frac{\lambda_i-\lambda_j}{\lambda_i+\lambda_j} \Xi_{p+N}(\Lambda_1^{-1}\oplus\Lambda_2^{-1}) = \int \prod_{k=1}^{p} d\nu_{\lambda_k}(y_k)$$

$$\times \prod_{1\leq m<n\leq p} \frac{2i(\lambda_m-\lambda_n)+y_m-y_n}{2i(\lambda_m+\lambda_n)+y_m+y_n} \ll \prod_{l=1}^{p} \det\left(\frac{2i\lambda_l+y_l-Y_2}{2i\lambda_l+y_l+Y_2}\right) \gg (\Lambda_2^{-1}) \tag{32}$$

where $d\nu_\lambda(y) = (\lambda/2\pi)^{\frac{1}{2}} \exp(iy^3/6 - \lambda y^2/2)dy$ is the measure of integration over the eigenvalues y adapted to our problem, and the double bracket denotes the integral over the $N\times N$ matrix Y_2 as defined in (7). The expansion of both sides of (32) in series of $\lambda_1^{-1} > \ldots > \lambda_p^{-1}$ will characterize the Witten mapping completely. Let us turn to the proof of formula (32).

At first the matrices $\Lambda, \Lambda_1, \Lambda_2$ involve diagonal real positive elements, but if we introduce a cut in the complex plane along the negative real axis, the integrals make sense for each eigenvalue having a positive real part – as absolutely convergent integrals; as semi-convergent ones we can even extend them to the imaginary axis except the origin. To give a meaning to the following operations we will first continue analytically the λ_j to imaginary non vanishing values. Similar techniques were implicit in both [11] and [13]. We consider:

$$\Xi_{p+N}(\Lambda^{-1}) = \frac{1}{Z_{p+N}(\Lambda)} \int dY e^{\frac{i}{6}\mathrm{tr}(Y^3)-\frac{1}{2}\mathrm{tr}(\Lambda Y^2)}, \tag{33}$$

where $Z_{p+N}(\Lambda)$ is defined in (10). We perform the change of variables $Z = Y + 2i(\Lambda_1 \oplus 0)$, with the obvious definition for the $(p+N)\times(p+N)$ matrix $\Lambda_1 \oplus 0 = \mathrm{diag}(\lambda_1,\ldots,\lambda_p,0,\ldots,0)$. Due to the relation

$$(\Lambda_1 \oplus 0)(0 \oplus \Lambda_2) = (0 \oplus \Lambda_2)(\Lambda_1 \oplus 0) = 0$$

the trace in the exponential becomes

$$\frac{i}{6}\mathrm{tr}(Y^3) - \frac{1}{2}\mathrm{tr}(\Lambda Y^2) = \frac{i}{6}\mathrm{tr}(Z^3) - \frac{1}{2}\mathrm{tr}([(0\oplus\Lambda_2) - (\Lambda_1\oplus 0)]Z^2) + \frac{2}{3}\mathrm{tr}(\Lambda_1^3). \tag{34}$$

We see that except for a constant term, the form of the exponential term is conserved, up to the substitution $\Lambda = \Lambda_1 \oplus \Lambda_2 \to \tilde{\Lambda} = (0\oplus\Lambda_2) - (\Lambda_1\oplus 0)$. Let us now perform the angular average over Z using (9), which results in

$$\Xi_{p+N}(\Lambda^{-1}) = \prod_{1\leq i<j\leq N+p} \frac{\tilde\lambda_j+\tilde\lambda_i}{\tilde\lambda_j-\tilde\lambda_i} \int \prod_{k=1}^{p+N} d\nu_{\tilde\lambda_k}(z_k) \prod_{n=1}^{p} e^{\frac{2}{3}\lambda_n^3} \prod_{1\leq l<m\leq N+p} \frac{z_l-z_m}{z_l+z_m} \tag{35}$$

where the $\tilde{\lambda}$'s are the diagonal elements of $\tilde{\Lambda}$, i.e. $\tilde{\lambda}_k = -\lambda_k$ for $1 \le k \le p$, $\tilde{\lambda}_k = \lambda_k$ for $p+1 \le k \le p+N$ (recall that the λ's are purely imaginary, so that the minus sign causes no harm in the integral). The antisymmetry of the integrand in z's in (35) automatically takes care of the denominators $z_l + z_m$, by antisymmetrizing the measure. We proceed and perform the opposite change of variables, but this time on the eigenvalues z by setting

$$z_k = y_k + 2i\lambda_k, \quad 1 \le k \le p,$$
$$z_k = y_k, \quad p+1 \le k \le p+N, \tag{36}$$

which leads to

$$\Xi_{p+N}(\Lambda_1^{-1} \oplus \Lambda_2^{-1}) = \prod_{1 \le i < j \le p} \frac{\lambda_i + \lambda_j}{\lambda_i - \lambda_j} \prod_{p+1 \le l < m \le p+N} \frac{\lambda_l^{-1} + \lambda_m^{-1}}{\lambda_l^{-1} - \lambda_m^{-1}} \int \prod_{k=1}^{N+p} d\nu_{\lambda_k}(y_k)$$
$$\times \prod_{1 \le l < m \le p} \frac{y_l - y_m + 2i(\lambda_l - \lambda_m)}{y_l + y_m + 2i(\lambda_l + \lambda_m)} \prod_{p+1 \le i < j \le p+N} \frac{y_i - y_j}{y_i + y_j} \prod_{\substack{1 \le l \le p \\ p+1 \le j \le p+N}} \frac{y_l + 2i\lambda_l - y_j}{y_l + 2i\lambda_l + y_j} \tag{37}$$

and amounts to (32) since

$$\prod_{p+1 \le l < m \le p+N} \frac{\lambda_l^{-1} + \lambda_m^{-1}}{\lambda_l^{-1} - \lambda_m^{-1}} \int \prod_{k=p+1}^{p+N} d\nu_{\lambda_k}(y_k) \prod_{p+1 \le i < j \le p+N} \frac{y_i - y_j}{y_i + y_j}$$
$$\times \prod_{\substack{1 \le l \le p \\ p+1 \le j \le p+N}} \frac{y_l + 2i\lambda_l - y_j}{y_l + 2i\lambda_l + y_j} = \ll \prod_{1 \le l \le p} \det\left(\frac{y_l + 2i\lambda_l - Y_2}{y_l + 2i\lambda_l + Y_2}\right) \gg (\Lambda_2^{-1}). \tag{38}$$

Let us rewrite the content of (32) for $p = 1$:

$$\Xi_{N+1}(\lambda^{-1} \oplus \Lambda^{-1}) = \int d\nu_\lambda(y) \ll \det \frac{y + 2i\lambda - Y}{y + 2i\lambda + Y} \gg = \int d\nu_\lambda(y) \ll f_{Y/2i}(\lambda + (y/2i)) \gg \tag{39}$$

where we identified the generating function f for the odd Schur polynomials (13). In the sense of asymptotic series of λ^{-1}, we are allowed to expand f and integrate term by term over y to get:

$$\Xi_{N+1}(\lambda^{-1} \oplus \Lambda^{-1}) = \sum_{m=0}^{\infty} \ll p_m(Y/2i) \gg \int d\nu_\lambda(y) \left(\lambda - \frac{iy}{2}\right)^{-m}, \tag{40}$$

so that comparing with (30), we find the first elements of the Witten mapping:

$$p_k(\partial.) \, \Xi(\theta.) = \sum_{0 \le s \le [k/3]} (-1)^s c_{s,k} \ll p_{k-3s}(Y/2i) \gg \tag{41}$$

where

$$c_{s,k} = \sum_{l=0}^{2s} \frac{1}{2^l} \frac{(k - 3s + l - 1)!}{l!(k - 3s - 1)!} \frac{(6s - 2l - 1)!!}{6^{2s-l}(2s - l)!} \tag{42}$$

and $[x]$ denotes the integral part of x.

For p generic, we are left with the easy task of expanding the r.h.s. of (32) in $\lambda_1^{-1} > \ldots > \lambda_p^{-1}$ and identifying term by term with (31). Noting that the integrand in (32) is again the generating function for φ's (27), but with the identification $Y \to Y_2/2i$ and $y_k \to \lambda_k + (y_k/2i)$, we can integrate over $y_1, \ldots, y_p$ to get the general Witten mapping in the form:

$$\varphi_{m_1,\ldots,m_p}(\partial.)\Xi(\theta.) = \sum_{\substack{s_1,\ldots,s_p \geq 0 \\ 3s_j \leq m_j + m_{j+1} + \ldots + m_p}} \prod_{i=1}^{n}(-1)^{s_i} c_{s_i, m_i} \ll \varphi_{m_1 - 3s_1, \ldots, m_p - 3s_p}(Y/2i) \gg . \tag{43}$$

We list the first few images of φ's below.

Table II. the derivatives of the Kontsevich partition function with respect to the θ.'s expressed as averages over polynomials in odd traces. The notation $\partial.$ stands for $\{-\frac{2}{2k+1}\frac{\partial}{\partial \theta_{2k+1}}\}$, $\theta. \equiv \theta(\Lambda^{-1})$, while on the r.h.s. the matrix argument of the $\varphi.$'s is $Y/2i$.

$\varphi_1(\partial.)\Xi$	$= \ll \varphi_1 \gg$
$\varphi_2(\partial.)\Xi$	$= \ll \varphi_2 \gg$
$\varphi_3(\partial.)\Xi$	$= \ll \varphi_3 - \frac{5}{24} \gg$
$\varphi_{2,1}(\partial.)\Xi$	$= \ll \varphi_{2,1} - \frac{1}{12} \gg$
$\varphi_4(\partial.)\Xi$	$= \ll \varphi_4 - \frac{17}{24}\varphi_1 \gg$
$\varphi_{3,1}(\partial.)\Xi$	$= \ll \varphi_{3,1} + \frac{5}{24}\varphi_1 \gg$
$\varphi_5(\partial.)\Xi$	$= \ll \varphi_5 - \frac{35}{24}\varphi_2 \gg$
$\varphi_{4,1}(\partial.)\Xi$	$= \ll \varphi_{4,1} \gg$
$\varphi_{3,2}(\partial.)\Xi$	$= \ll \varphi_{3,2} + \frac{5}{24}\varphi_2 \gg$
$\varphi_6(\partial.)\Xi$	$= \ll \varphi_6 - \frac{59}{24}\varphi_3 + \frac{385}{1152} \gg$
$\varphi_{5,1}(\partial.)\Xi$	$= \ll \varphi_{5,1} - \frac{35}{24}\varphi_{2,1} + \frac{35}{576} \gg$
$\varphi_{4,2}(\partial.)\Xi$	$= \ll \varphi_{4,2} + \frac{17}{24}\varphi_{2,1} - \frac{35}{576} \gg$
$\varphi_{3,2,1}(\partial.)\Xi$	$= \ll \varphi_{3,2,1} - \frac{5}{24}\varphi_{2,1} - \frac{1}{12}\varphi_3 + \frac{5}{288} \gg$

REFERENCES

[1] E. Brézin and V. Kazakov, Phys. Lett. **B236** (1990) 144; M. Douglas and S. Shenker, Nucl. Phys. **B335** (1990) 127; D. Gross and A. Migdal, Phys. Rev. Lett. **64** (1990) 127.
[2] E. Brézin, C. Itzykson, G. Parisi and J.-B. Zuber, Comm. Math. Phys. **69** (1979) 147.
[3] M. Douglas, Phys. Lett. **B238** (1990) 176.
[4] P. Di Francesco and D. Kutasov, Nucl. Phys. **B342** (1990) 589.
[5] V. Knizhnik, A. Polyakov and A. Zamolodchikov, Mod. Phys. Lett. **A3** (1988) 819; F. David, Mod. Phys. Lett. **A3** (1988) 1651; J. Distler and H. Kawai, Nucl. Phys. **B321** (1989) 509.
[6] P. Di Francesco and D. Kutasov, Phys. Lett. **B261** (1991) 385, and Nucl. Phys. **B375** (1992) 119.
[7] T. Banks, M. Douglas, N. Seiberg and S. Shenker, Phys. Lett. **B238** (1990) 279; R. Dijkgraaf, E. Verlinde and H. Verlinde, Nucl. Phys. **B348** (1991) 435.
[8] E. Witten, Comm. Math. Phys. **117** (1988) 353 and Phys. Lett. **B206** (1988) 601.

[9] E. Witten, Nucl. Phys. **B240** (1990) 281.
[10] E. Witten, Surv. Diff. Geom. **1** (1991) 243.
[11] M. Kontsevich, *Intersection theory on the moduli space of curves*, Funk. Anal. & Prilozh., **25** (1991) 50-57;
Intersection theory on the moduli space of curves and the matrix Airy function, lecture at the Arbeitstagung, Bonn, June 1991 and Comm. Math. Phys. **147** (1992) 1.
[12] E. Witten, *On the Kontsevich model and other models of two dimensional gravity*, preprint IASSNS-HEP-91/24
[13] C. Itzykson and J.-B. Zuber, Int. J. Mod. Phys. **A7** (1992) 5661.
[14] S. Kharchev, A. Marshakov, A. Mironov, A. Morozov and A. Zabrodin, Phys. Lett. **B275** (1992) 311.
[15] M. Adler and P. van Moerbeke, *The W_p-gravity version of the Witten–Kontsevich model*, Brandeis preprint, September 1991.
[16] E. Witten, *Algebraic Geometry associated with matrix models of two dimensional gravity*, preprint IASSNS-HEP-91/74.
[17] P. Di Francesco, C. Itzykson and J.-B. Zuber, Comm. Math. Phys. **151** (1993) 193.
[18] Harish-Chandra, Amer. J. Math. **79** (1957) 87-120;
C. Itzykson and J.-B. Zuber, J. Math. Phys. **21** (1980) 411-421.
[19] J. Duistermaat and G. Heckman, Invent. Math. **69** (1982) 259.

MATRIX MODELS IN STATISTICAL MECHANICS AND QUANTUM FIELD THEORY, RECENT EXAMPLES AND PROBLEMS

Giovanni M. Cicuta

Dipartimento di Fisica, Università di Bari
and INFN, Sezione di Bari
via Amendola 173, 70126 Bari, Italy

ABSTRACT

We briefly describe the possibility of studying some simple vertex models on planar random graphs and a special model, related to famous colouring problems, which seems to allow a complete analytic solution. Next the analysis done few years ago of models with rectangular random matrices will be recalled as the approach seems to provide a method for an accurate evaluation of planar Green's functions in quantum field theory in any dimension of space-time. The approach of random rectangular matrices makes use of the vast knowledge on vector models developed in the past two decades.

Before describing these two items, some basic steps in the traditional methods will be recalled, with the hope that mathematicians might be interested in solving the problems still unanswered.

1. AN OUTLINE OF SOME BASIC STEPS

Several reviews and lecture notes[1-5] are useful to learn the techniques of the random matrix models and the developments of the past few years. They led to the solution of interesting statistical mechanics models on bidimensional random lattices, the formulation of a quantum theory of gravity in two dimensions, the agreement of critical coefficients from the "continuum limit" of random matrix models with the KPZ formula from conformal field theory, suggestions for a non-perturbative theory of strings and a variety of mathematical results. The recent results and the interest in the Kontsevich model were described in this workshop by Di Francesco.

In this Section some of the most preliminary steps of the theory will be mentioned, to provide a reference for the comments in Sect. 2 on still unsolved problems.

The interest for quantum field theory in the large order limit of matrix models begins with the works by G.'t Hooft[6] and G. Veneziano[7] on the topological expansion. The nonabelian theory of the Lorentz-vector, matrix valued field $A_\mu(x) = \sum_{a=1}^{N^2-1} A_\mu^a(x)\lambda^a$, with an action invariant under the local (gauge) group $SU(N)$ is studied in the $N \to \infty$ limit, with a proper (unique) rescaling for the gluon coupling constants. The infinite number of Feynman graphs contributing to any correlation function $G_p(x_1,\ldots x_p) = \langle\Omega|TA_{\mu_1}(x_1)\ldots A_{\mu_p}(x_p)|\Omega\rangle$ fall into distinct classes characterized by increasing powers $(\frac{1}{N^2})^k$, the power k being linearly related to the genus of the orientable surface on which the Feynman graph may be properly embedded. This result stems from the group factor associated to each Feynman graph (a convolution of f_{ijk}) and it holds true also for the globally $SU(N)$ invariant model of Lorentz-scalar, matrix valued fields $A(x) = \sum_{a=1}^{N^2-1} A^a(x)\lambda^a$ with euclidean action

$$S = \int d^d x \text{ tr}\left[\frac{1}{2}(\partial_\mu A(x))^2 + g_2 A^2(x) + \frac{g_4}{N}A^4(x)\right], \qquad (1.1)$$

where the space-time dimension d is arbitrary, $\mu = 1, 2\ldots d$, $A(x)$ is a Hermitian matrix of order N (here a convolution of symmetric tensors d_{ijk} replaces the previous convolution of f_{ijk}). It does not seem possible to evaluate analytically the sum of the leading contributions in the $N \to \infty$ limit, that is the series associated to the Feynman planar graphs. Only for the small dimensions $d = 0$ and $d = 1$ the matrix model of type eq.(1) were "solved" by the two methods of frequent use today: saddle point analysis and orthogonal polynomials[8].

If $d = 0$ the first step of the analysis, for the simplest model in eq.(1.1), is to rewrite the generating functional of vacuum Feynman graphs $Z(g_2, g_4) = \int \mathcal{D}A \, e^{-S(A)}$ in terms of the eigenvalues of the matrix. Since the integrand is invariant under the unitary transformation $A = UA_dU^\dagger$ and

$$\mathcal{D}A = \prod_{i=1}^N dA_{ii} \prod_{1\le i<j\le N} d\text{Re}A_{ij}\, d\text{Im}A_{ij} = \prod_{i=1}^N d\lambda_i\, \Delta^2(\lambda)\, \mathcal{D}U$$

one obtains (by neglecting an irrelevant constant factor)

$$Z(g_2, g_4) = \int \prod_{i=1}^N d\lambda_i\, e^{-V(\lambda_i)} \Delta^2(\lambda_i),$$

$$\Delta(\lambda_i) = \prod_{1\le i<j\le N}(\lambda_i - \lambda_j), \quad V(\lambda_i) = \sum_{i=1}^N \left[g_2\lambda_i^2 + \frac{g_4}{N}\lambda_i^4\right]. \qquad (1.2)$$

Here the two methods depart. The saddle point analysis proceeds by exponentiating the Jacobian

$$\Delta(\lambda_i) = e^{\sum_{1\le i<j\le N} \log|\lambda_i - \lambda_j|} \qquad (1.3)$$

by replacing, in the $N \to \infty$ limit, the sums with integrals:

$$\lambda_i \to \sqrt{N}\lambda(i/N), \qquad \sum_i (\lambda_i)^k \to N\int_0^1 dx [\sqrt{N}\lambda(x)]^k,$$

$$2\sum_{1\le i<j\le N} \log|\lambda_i - \lambda_j| \to N^2 \int_0^1\int_0^1 dx\,dy\, \log|\lambda(x) - \lambda(y)|. \qquad (1.4)$$

The saddle point corresponds to a singular integral equation for the function $\lambda(x)$. By introducing the non-negative density of eigenvalues $u(\lambda) = \frac{dx}{d\lambda} \geq 0$, $\int_L d\lambda\, u(\lambda) = 1$, L being the (unknown) support of the eigenvalues, the singular integral equation is

$$\frac{1}{2}\frac{dV(\lambda)}{d\lambda} = g_2 \lambda + 2g_4 \lambda^3 = \mathcal{P}\int_L d\lambda' \frac{u(\lambda')}{\lambda - \lambda'}. \tag{1.5}$$

Alternatively, with the orthogonal polynomials technique, one uses the invariance of the Jacobian $\Delta(\lambda_i)$, being a Vandermonde determinant, with the replacement $\lambda^n \to P_n(\lambda)$, $\Delta(\lambda_i) = \det[(\lambda_i)^{j-1}] = \det[P_{(j-1)}(\lambda_i)]$, where $P_n(\lambda)$ are monic polynomials ($P_n(\lambda) = \lambda^n +$ lower order polynomial) which are chosen orthogonal with respect to the measure

$$\int_{-\infty}^{+\infty} d\lambda\, e^{-V(\lambda)} P_n(\lambda) P_m(\lambda) = h_n \delta_{nm}.$$

The orthogonal polynomials $P_n(\lambda)$ satisfy a recurrence relation

$$\lambda P_n(\lambda) = P_{n+1}(\lambda) + S_n P_n(\lambda) + R_n P_{n-1}(\lambda), \text{ with } R_n = \frac{h_n}{h_{n-1}}. \tag{1.6}$$

If $V(\lambda)$ has only even powers of λ (as in this example) and a solution without spontaneous symmetry breaking is expected, $S_n = 0$, and the recurrence relation

$$n = R_n \left[2g_2 + 4\frac{g_4}{N}(R_{n-1} + R_n + R_{n+1}) \right] \tag{1.7}$$

is found, sometimes called the pre-string equation.

A smooth ansatz for R_n as $N \to \infty$ reproduces the results of the planar theory, whereas a different smooth ansatz, in the double-scaling limit $N \to \infty$ and $g_i \to g_{i,\text{crit}}$, resums the contributions from any topology.

2. SOME PROBLEMS STILL UNANSWERED

a) Problems with the first step.

The first step of the analysis, to rewrite the partition function in terms of eigenvalues, is so far possible only for few cases in models with two or more matrices. This is a severe limitation since several interesting statistical mechanics models on random planar graphs are easily formulated in terms of multi-matrix models.

Let us consider a two-matrix integral

$$Z = \int \mathcal{D}A \mathcal{D}B\, e^{-V_1(A) - V_2(B) - V_3(A,B)}, \tag{2.1}$$

where $V_3(A,B) = c\,\text{tr}\,[AB]$. After the unitary transformations $A = U_1 A_d U_1^\dagger$, $B = U_2 B_d U_2^\dagger$ which leave invariant $V_1(A) = V_1(A_d)$, and $V_2(B) = V_2(B_d)$ one is led to the angular integral

$$\int \mathcal{D}U\, e^{-c\,\text{tr}\,[A_d U B_d U^\dagger]} = \frac{(\text{const})}{\Delta(\lambda)\Delta(\mu)} \det[e^{\lambda_i \mu_j}], \tag{2.2}$$

as was shown in ref.[9]. This allowed the solution of this class of two-matrix models by the orthogonal polynomial technique[10].

Other choices of the coupling potential $V_3(A, B)$, for instance $\text{tr}\,[ABAB]$, would lead to an angular integration like

$$\int \mathcal{D}U\, e^{-c\,\text{tr}\,[AUBU^\dagger AUBU^\dagger]},$$

which is only known as a series expansion of the exponential.

b) Problems with other phases.

It might seem that both eq.(1.5) and eq.(1.7) have a unique solution. Singular integral equations of type (1.5) are easily inverted, if the support L is known (one may use the Poincaré-Bertrand formula). In a generic model, however (except for the pure gaussian model of random matrices), several choices for the support L are possible that lead to multiple solutions of the model in the planar limit. The support may be one segment or the union of many segments, and, even if $V(\lambda)$ is an even function of λ, L may be symmetrical with respect to the origin or not (thus obtaining a spontaneous symmetry breaking solution)[11-13]. Each solution has a domain of existence in the space of parameters g_i of the model. If, in a given region in such space, more than one solution exist, the solution of the model (in the planar limit) corresponds to the solution of the eq.(1.5) that provides the lowest free energy. Then the boundaries, in the space of parameters, for the existence of solutions not necessarily coincide with phase transition surfaces[14,15]. However, in order to obtain a non-linear equation in the double-scaling limit, the "string equation", it is necessary to approach the boundary of existence of a given solution (obviously possible from one side only), so, for that purpose, the former boundaries are relevant.

It was shown, mainly by Molinari[16], that a period-two ansatz for R_n in eq.(1.7), that is R_{2n} and R_{2n+1} approach two different smooth functions, reproduces, in the planar limit, the results of the symmetric, two-segment support, solution of eq.(1.5).

Generalizing the Molinari ansatz to solutions with support on more than two segments seems problematic: the conjecture[15] that the proper ansatz for the continuum limit of R_n should have k-periodicity, to reproduce a k-segment saddle point solution, seems to fail[17].

Even more disturbing and relevant is the chaotic behaviour[18] exhibited by R_n. As generally happens with non-linear recurrence relations, R_n are very unstable with respect to the initial conditions (small n values) and exhibit a chaotic behaviour for large values of n. This seems to prevent a smooth ansatz to reproduce the saddle point results in the planar limit and it casts doubts on the validity of the smooth ansatz used in the double-scaling limit.

To summarize this comment, it seems obvious that each of the two methods here considered has advantages and weaknesses: the saddle point analysis easily produces the multiple solutions and the phase diagram of the model in the planar limit, but it is forbiddingly impractical to analyze, in a systematic way, the topological expansion. The orthogonal polynomial technique is the basic way to perform the double-scaling limit but, so far, it is still an art to guess **proper** ansatzes. Certainly, an understanding of the non-linear recurrence relations for R_n (pre-string equation) to know the regions in the space of the parameters g_i where regular/periodic/chaotic behaviours arise, would be very welcome.

3. VERTEX MODELS

It is well known the important role in statistical mechanics that was provided by the solution of a variety of vertex models on regular bidimensional lattices. One may recall the 6-vertex model on a square lattice, arrows are drawn on all edges of the lattice in accordance with the "ice rule": arrows on the four edges incident in each vertex are drawn so that two are incoming and two outgoing[19]. A different weight may be given to each of the six configurations for each vertex (four configurations have the two incoming arrows on adjacent edges, two have the incoming arrows on opposite edges). The weight of a configuration on the lattice is the product of the weights of each vertex.

In the formulation of a similar model on random planar graphs, such as by the set of (vacuum) planar Feynman graphs with quartic coupling, it is natural to assign equal weight, say c_1, to the 4 vertex configurations where the two incoming arrows are adjacent and a second weight, c_2, for the two remaining configurations. (If the lattice is the square lattice, this choice of two weights corresponds to the F-model[19]).

It would be an important test of the universality hypothesis for statistical models on planar random graphs, to check that the critical coefficients are independent on the specific values of the couplings c_1, c_2 (so far universality was found to hold by changing the discretization, triangular or quadrangular, for the Ising spins of a planar random graph[20,21]).

One is led to consider the partition function

$$Z(c_1, c_2) = \int \mathcal{D}\phi^\dagger \mathcal{D}\phi \; e^{-\frac{1}{2}\text{tr}[\phi^\dagger \phi] + c_1 \text{tr}[(\phi^\dagger)^2 \phi^2] + c_2 \text{tr}[\phi^\dagger \phi \phi^\dagger \phi]} , \qquad (3.1)$$

where ϕ is a square $N \times N$ matrix with complex entries. It may be useful to express $Z(c_1, c_2)$ in terms of two hermitian matrices A, B, with $\phi = A + iB$. If all the weights are equal, $c_1 = c_2$, the Boltzmann weight is just a function of $(A^2 + B^2)$ or, equivalently, of a rectangular matrix. The model is easily solved in the planar limit[22] and in the double scaling limit[23]. In the more general case where $c_1 \neq c_2$

$$Z(c_1, c_2) = \int \mathcal{D}A \mathcal{D}B \; e^{-\frac{1}{2}\text{tr}[A^2+B^2]+(c_1+c_2)\text{tr}[A^4+B^4]+2(c_1-c_2)\text{tr}[ABAB]+4c_2\text{tr}[A^2B^2]} \qquad (3.2)$$

the first step of the usual analysis, the rewriting of the two-matrix integrand in terms of the distribution of eigenvalues, has so far been impossible for the unsolved difficulty of performing the angular integration, mentioned in Sect. 2a.

We shall now describe a simpler vertex model, which seems analytically solvable, that corresponds to interesting colouring problems[24].

Let us consider the simpler two-matrix problem corresponding to the partition function

$$Z(g_1, g_2) = \int \mathcal{D}A \mathcal{D}B \; e^{-\frac{1}{2}\text{tr}[A^2+B^2]+g_1\text{tr}[ABAB]+g_2\text{tr}[A^2B^2]} . \qquad (3.3)$$

The two hermitian matrices may be written in the usual way in terms of eigenvalues and unitary transformation

$$A = U_1 A_d U_1^\dagger, \quad \int \mathcal{D}A = \int \prod_1^N d\lambda_i \, \Delta^2(\lambda) \int dU_1 ,$$

$$B = U_2 B_d U_2^\dagger, \quad \int \mathcal{D}B = \int \prod_1^N d\mu_i \, \Delta^2(\mu) \int dU_2 ,$$

$$g_1 \text{tr}[ABAB] + g_2 \text{tr}[A^2 B^2]$$
$$= g_1 \text{tr}[U_1 A_d U_1^\dagger U_2 B_d U_2^\dagger U_1 A_d U_1^\dagger U_2 B_d U_2^\dagger] + g_2 \text{tr}[U_1 A_d^2 U_1^\dagger U_2 B_d^2 U_2^\dagger] .$$

The integration over the angular variables U_2 may be replaced by the angular variables $U_3 = U_1^\dagger U_2$, the integrand then does not depend on U_1 variables, and its integration yields an irrelevant multiplicative constant. The diagonalization over matrices B may be undone, and one obtains

$$Z(g_1, g_2) \propto \int \prod_1^N d\lambda_i \, \Delta^2(\lambda) e^{-\frac{1}{2}\sum \lambda_i^2} \int \mathcal{D}B e^{-\frac{1}{2}\text{tr}[B^2] + g_1 \text{tr}[A_d B A_d B] + g_2 \text{tr}[A_d^2 B_d^2]}. \quad (3.4)$$

The integration over the matrix elements of the matrix B is gaussian and one obtains the partition function just in terms of the eigenvalues of the matrix A

$$Z(g_1, g_2) = (\text{const}) \int \prod_1^N d\lambda_i \, \Delta^2(\lambda) e^{-\frac{1}{2}\sum \lambda_i^2} \prod_{i,j} [1 - 2g_1 \lambda_i \lambda_j - g_2(\lambda_i^2 + \lambda_j^2)]^{-\frac{1}{2}}. \quad (3.5)$$

One may proceed and replace the discrete eigenvalues λ_i with the distribution $\lambda(x)$ in the large N limit and obtain, by the saddle point analysis, a singular integral equation for the density of eigenvalues $u(\lambda)$.

Before doing it, we shall show that an interesting colouring problem corresponds to the two-matrix partition function in eq.(3.3) with equal couplings.

Let us consider the partition function

$$Z(g) = \int \mathcal{D}A \mathcal{D}B \mathcal{D}C \, e^{-\frac{1}{2}\text{tr}[A^2 + B^2 + C^2] + \frac{g}{\sqrt{N}}\text{tr}[ABC + CBA]} \quad (3.6)$$

in terms of three hermitian matrices A, B, C. In the large N limit it is the generating functional of the planar graphs, with valence three, such that the three edges incident in each vertex are coloured with the three distinct colours A, B, C. Equal weight is given to a proper colouring of the edges, A, B, C, being oriented clockwise or anti-clockwise.

By standard methods of quantum field theory one may write a related partition function where some unwanted classes of graphs are eliminated: the "tadpole" graphs ("loops" in the terminology of multigraph theory), the graphs where two edges connect the same couple of vertices ("parallel lines") and the one-particle reducible graphs (to remain only with the "strongly connected" graphs).

Let us imagine all this being performed, one would then have, in the large N limit, the functional generator of planar graphs, each multiplied by the number of inequivalent Tait colourings. It is well known that they correspond to proper colourings, with four colours, of the regions of the same map (proper relates to the requirement that adjacent regions are given different colours)[25].

The correspondence may be seen by using, for the regions, the four colours 1, A, B, C, seen as the elements of the Klein four group, and by labelling the edges with the three colours A, B, C, where the colour of an edge is the (commutative) product, under the group rule, of the colours of the two regions separated by the edge.

The hermitian matrix C in eq.(3.6) may be integrated and one obtains

$$Z(g) = (\text{const}) \int \mathcal{D}A \mathcal{D}B \, e^{-\frac{1}{2}\text{tr}[A^2 + B^2] + \frac{g^2}{N}\text{tr}[ABAB + A^2 B^2]}$$

$$= (\text{const}) \int \prod_1^N d\lambda_i \, \Delta^2(\lambda) \, e^{-\frac{1}{2}\sum \lambda_i^2} \prod_{i,j} \left[1 - \frac{g^2}{N}(\lambda_i + \lambda_j)^2\right]^{-\frac{1}{2}}. \quad (3.7)$$

If one were to consider the edge-colourings where the three colours appear with definite orientation (clockwise or anti-clockwise), after integration over the matrix C, one would obtain eqs.(3.4) and (3.5) with $g_2 = 0$.

One may proceed with the usual replacements, in the large N limit, $x = \frac{i}{N}$, $\lambda_i = \sqrt{N}\lambda(x)$, $\sum_{i=1}^{N} = N \int_0^1 dx$ to obtain

$$\frac{Z(g)}{Z(0)} = \int \mathcal{D}\lambda(x)\, e^{-N^2 S(\lambda)},$$

$$S(\lambda) = \frac{1}{2} \int dx\, \lambda^2(x) - \int\int dx dy\, \log[\lambda(x) - \lambda(y)] \qquad (3.8)$$

$$+ \frac{1}{2} \int\int dx dy\, \log[1 - g^2(\lambda(x) + \lambda(y))^2],$$

and the saddle point equation for the density of eigenvalues $u(\lambda) = \frac{dx}{d\lambda}$

$$\frac{\lambda}{2} - \mathcal{P} \int_L d\mu \frac{u(\mu)}{\lambda - \mu} = g^2 \int_L d\mu \frac{u(\mu)[\lambda + \mu]}{1 - g^2(\lambda + \mu)^2} \qquad (3.9),$$

for $\lambda \in L$, L being the support of $u(\lambda)$. It seems possible to obtain an analytic solution for the singular integral equation (3.9), which is presently being done[26].

4. SYSTEMATIC EVALUATION IN ARBITRARY DIMENSION AND RECTANGULAR MATRICES

Models with rectangular matrices are relevant for several reasons:
a) to evaluate in a reliable way planar Green's functions in any dimension d of space-time
b) to interpolate between vector models and square matrix models and use the vast knowledge available on vector models (for any d)
c) to study topological micro-fluctuations of random surfaces by a triple-scaling limit. In this contribution only points a) and b) will be introduced, which are described in the papers in ref.22, while point c) is described in the preprint in ref.23.

Consider a rectangular matrix M with N_1 rows and N_2 columns with complex entries and the action S

$$S = \int d^d x\, \text{tr}\left[\partial_\mu M^\dagger(x)\partial_\mu M(x) + g_2 M^\dagger(x)M(x) + \frac{2g_4}{N} M^\dagger(x)M(x)M^\dagger(x)M(x)\right],$$

$$Z(M^\dagger, M) = \int \mathcal{D}M^\dagger \mathcal{D}M\, e^{-S(M^\dagger,M)}. \qquad (4.1)$$

Let $L = \frac{N_1}{N_2}$ be fixed as $N_1, N_2 \to \infty$. Again the Feynman graphs are arranged into a topological expansion. The sum of the infinite Feynman graphs of fixed genus contributing to the free energy or to a correlation function in the model, may further be expanded in a Taylor series in the parameter L. Well defined classes of Feynman graphs contribute to a given term in the topological expansion and to a given power in L. They are easily identified after the introduction of an hermitian $N_2 \times N_2$ matrix $H(x)$

$$e^{\frac{2g_4}{N}\text{tr}[M^\dagger(x)M(x)M^\dagger(x)M(x)]} = \int \mathcal{D}H(x)\, e^{-\frac{1}{2}\text{tr}[H^2(x) - 2i\sqrt{\frac{g_4}{N}}M^\dagger(x)M(x)H(x)]}. \qquad (4.2)$$

Since the action $S(M^\dagger, M)$ is invariant under the replacement of L with $\frac{1}{L}$, that is $M^\dagger(x)$ with $M(x)$, the coefficients of the expansion of planar Green's functions in powers of L (the charged-loop expansion for the action $S(M^\dagger, M, H)$) are equal, after suitable rescaling of the coupling, to the coefficients of the expansion in inverse powers of L (which are the familiar terms occurring in the $\frac{1}{n}$ expansion of vector models[22,27]).

The method provides a systematic way to evaluate in approximate form the sum of the planar graphs in any theory with large global symmetry group (it is not known how to extend it to non abelian gauge symmetry). For those models where exact planar results for the free energy or the propagator are known, like the models in $d = 0$ or $d = 1$, it was checked that the first few orders in the Taylor expansion in L provide an approximation accurate in the whole range $-\infty < g_2 < \infty$, $g_4 > 0$, for $L = 1$ (that is for the planar theory).

This approach does not depend on the dimension d of space-time (except for the different renormalizations needed in different d) and it seems very promising to study non perturbative features of quantum field theory in the continuum (bound states, the strongly interacting sector of the Higgs meson,...) for $d = 2, 3$ and 4.

REFERENCES

[1] V.A. Kazakov, Lecture at Les Houches, Session XLIX, edited by E. Brezin and J. Zinn-Justin, North-Holland 1990 and Lecture at Cargese 1990, LPTENS 90/30, to appear in Random Surfaces and Quantum Gravity, ed. by O. Alvarez *et al.*

[2] Two dimensional quantum gravity and random surfaces, ed. by D. Gross, T. Piran, S. Weinberg, World Scientific, 1992.

[3] F. David, Lecture at Trieste 1990, S.Ph-T/90-127.
P. Ginsparg, Lectures at Trieste 1991, LA-UR-91-4101.

[4] Y. Lozano, J.L. Manes, Introduction to Nonperturbative 2D Quantum Gravity, UB-ECM-PF 3/91, Barcelona 1991.

[5] T. Tada, Ph.D.Thesis at Univ. of Tokyo, Komaba, 1990.

[6] G.'t Hooft, Nucl. Phys. **B72** (1974) 461 and **B75** (1974) 461.

[7] G. Veneziano, review talk at the 1978 Tokyo meeting.

[8] E. Brezin, C. Itzykson, G. Parisi, J.B. Zuber, Comm. Math. Phys. **59** (1978) 35. D. Bessis, Comm. Math. Phys. **69** (1979) 147. D. Bessis, C. Itzykson, J.B. Zuber, Adv. Appl. Math. **1** (1980) 109.

[9] Harish-Chandra, Amer. J. Math. **79** (1957) 87. C. Itzykson, J.B. Zuber, J. Math. Phys. **21** (1980) 411.

[10] M.L. Mehta, Comm. Math. Phys. **79** (1981) 327.

[11] Y. Shimamune, Phys. Lett. **108B** (1982) 407. I.Ya Arefieva, 3rd Int. Symposium Dubna 1984.

[12] Similar multiple solutions, in unitary matrix models, were found by J. Jurkiewicz and K. Zalewski, Nucl. Phys. **B220** (1983) 167 and T.L. Chen, C-I. Tan, Z.T. Zheng, Phys. Lett. **123B** (1983) 423.

[13] Unaware of refs.[11,12], the multiple solutions in hermitian models in dimension $d = 0$ and $d = 1$ were discussed in G.M. Cicuta, L. Molinari, E. Montaldi, Mod. Phys. Lett. **A1** (1986) 125 and Jour. of Phys. **A20** (1987) L67.

[14] J. Jurkiewicz, Phys. Lett. **245B** (1990) 178. G.M. Cicuta, L. Molinari, E. Montaldi, Jour. of Phys. **A23** (1990) L421.

[15] K. Demeterfi, N. Deo, S. Jain, C-I. Tan, Phys. Rev. **D42** (1990) 4105.

[16] L. Molinari, J.of Phys. A21 (1988) 1. G.M. Cicuta, L. Molinari, E. Montaldi, Nucl. Phys. **B300** (1988) 197. L. Molinari, E. Montaldi, J. of Phys. **A23** (1990) 4995.

[17] G. Bhanot, G. Mandal, O. Narayan, Phys. Lett. **B251** (1990) 388. O. Lechtenfeld, R. Ray, A. Ray, Int. Jour. Mod. Phys. **A6** (1991) 4491. M. Sasaki, H. Suzuki, Phys. Rev. **D43** (1991) 4015.

[18] O. Lechtenfeld, Int. Jour. Mod. Phys. **A7** (1992) 2335. J. Jurkiewicz, Phys. Lett. **261B** (1991) 260. D. Senechal, LAVAL-PHY-22/91 (1991) to appear in Int. Jour. Mod. Phys.

[19] See, for instance, chapter 8 in R.J. Baxter, Exactly solved models in statistical mechanics, Academic Press 1982.

[20] V.A. Kazakov, Phys. Lett. **A119** (1986) 140.

[21] D.V. Boulatov, V.A. Kazakov, Phys. Lett. **B186** (1987) 379.

[22] A. Barbieri, G.M. Cicuta, E. Montaldi, Nuovo Cim. **84A** (1984) 173. C.M. Canali, G.M. Cicuta, L. Molinari, E. Montaldi, Nucl. Phys. **B265** (1986) 485. G.M. Cicuta, L. Molinari, E. Montaldi, F. Riva, J. Math. Phys. **28** (1987) 1716.

[23] A. Anderson, R.C. Myers, V. Periwal, Phys. Lett. **254B** (1991) 89 and Nucl. Phys. **B360** (1991) 463. R.C. Myers, V. Periwal, NSF-ITP-91-01, Princeton 1991.

[24] An early attempt to evaluate the number of proper colourings with the methods of quantum field theory (without using random matrices) is in N. Nakanishi, Comm. Math. Phys. **32** (1973) 167.

[25] See, for instance, B. Bollobas, Graph Theory, Springer-Verlag, New York, 1979, exerc. 27, page 101, or the Tait theorem in S. Fiorini, R.J. Wilson, Edge-colourings of graphs, Pitman, London 1977, page 26, or page 103 in T.L. Saaty, P.C. Kainen, The Four Color Problem, McGraw-Hill, New York 1977.

[26] G.M. Cicuta, E. Montaldi, L. Molinari, preprint BARI-Th.92/130, to appear in Phys. Lett.

[27] A similar expansion is described in A.A. Slavnov, Lecture at Schladming 1983 and in F. Green, S. Samuel, Nucl. Phys. **B190** (1981) 113.

DILOGARITHMS AND W-ALGEBRAS

Werner Nahm

Physikalisches Institut der Universität Bonn
Nussallee 12
D-5300 Bonn 1, Germany

The Rogers dilogarithm is defined by

$$L(z) = \sum_{1}^{\infty} \frac{z^n}{n^2} + \frac{1}{2} \log z \log(1-z) . \tag{1}$$

We always restrict the argument to the interval $[0, 1]$. The logarithmic correction term is chosen such that $L(z) + L(1 - z) = L(1)$. Since Euler it is known that $L(1) = \pi^2/6$. Around 1985 the Leningrad school discovered from a calculation of the magnetic susceptibility in the XXZ model that special values of the dilogarithm are related to the central charges of conformal theories. Further work showed close relations of this function to other integrable models in two dimensions, with arguments of L given by the Bethe ansatz (see [1,2,3], which contain further references). I'll argue that the basic phenomena can be found in conformal quantum field theory itself and will make no reference to integrable perturbations. Many ideas are still tentative, but the following conjectures at least define a research program.

The central charge c of unitary conformal field theories yields both the central extension of the diffeomorphism group and the boundary contribution to the free energy. The latter is a Casimir effect and reflects the behaviour of the spectrum at high energy. For rational non-unitary theories, this behaviour is determined not by the central extension c but by $c_{\text{eff}} = c - 24 h_{\min}$, where $h_{\min}$ it the minimal eigenvalue of L_0 occurring in any representation. Obviously, c_{eff} has to be positive, except for the trivial theory, where $c = c_{\text{eff}} = 0$.

Since rational conformal field theories can be tensored, the set $\mathcal{C}$ of possible values of c_{eff} is additive. It is supposed to be well-ordered, i.e. between each value and the next highest one there is a finite gap.

I conjecture that this set is identical to the union $\mathcal{D}$ of the sets $\mathcal{D}^N$ of those rational numbers which can be written in the form $\sum_{i=1}^{N} L(x_i)/L(1)$, where the x_i are algebraic numbers in the interval $(0, 1)$. Any rational conformal theory should produce a number in $\mathcal{D}$ such that the x_i lie in the field generated by its quantum dimensions. Very similar formulas should exist for $c - 24h_i$, where h_i runs over the dimensions of all primary fields, but this will be discussed elsewhere.

The formulas given by Kirillov yield $c \in \mathcal{D}$ for all Kac-Moody current algebra models at integral level, though his proofs so far only cover the A- and D-series. The formulas also extend to the cosets G/T, where T is a maximal torus of G.

Conversely, consider $\mathcal{D}^1$. This is believed to be the set $\{1/2, 2/5, 3/5\}$. Here $L(1/2) = L(1)/2$ should correspond to the free Majorana fermion. The other two values should correspond to the $(3,5)$ and $(2,5)$ minimal models. In the latter case, the only non-trivial quantum dimension τ^{-1} is given by the golden ratio equation

$$\tau + \tau^2 = 1 . \tag{2}$$

We have $L(\tau)/L(1) = 3/5$ and $L(\tau^2)/L(1) = 2/5$. Minimal models with $c \in \mathcal{D}^2$ will be discussed below.

An algorithm invented in the context of pure combinatorics and number theory can be interpreted as relating quantum dimensions and central charge in conformal field theory [4]. Let us start with an example.

The Virasoro algebra with central charge $c = -22/5$ has two irreducible representations: The basic one with $h = 0$ and another one with $h = -1/5$. The effective central charge, which yields the asymptotic behaviour of the partition function and thus the boundary contribution to the free energy is thus $c_{\text{eff}} = c - 24h_{\min} = 2/5$. The partition functions have the form

$$Z_0 = \sum_n \frac{q^{n(n+1)}}{(q)_n} , \tag{3}$$

$$Z_1 = \sum_n \frac{q^{n^2}}{(q)_n} , \tag{4}$$

where

$$(q)_n = (1-q)\ldots(1-q^n) . \tag{5}$$

These relations can be interpreted as consequences of the fact that all normal ordered product relations in the W-algebra arise from the fact that the normal ordered product of two energy-momentum fields is proportional to the second derivative of the same field [5]. For general (p,q), p,q coprime, minimal models $c = 1 - 6(p-q)^2/pq$ and, since $\min_{0<m<q, 0<n<p}(mp-nq)^2 = 1$, choosing m,n so that $pm = 1 \mod q$, $-qn = 1 \mod p$, then from the Kac formula $h_{\min} = (1-(p-q)^2)/4pq$ and hence $c_{\text{eff}} = 1 - 6/pq$.

If one writes

$$Z = \sum a_n q^n , \tag{6}$$

it is well known that

$$\log a_n \sim 2\pi\sqrt{c_{\text{eff}} n/6} . \tag{7}$$

We will consider the slightly more general case

$$Z = \sum_n \frac{q^{\alpha n^2 + \beta n}}{(q)_n} . \tag{8}$$

One can calculate a_{N-1} by a contour integral over $Zq^{-N}/2\pi i$ around the origin, using the residue theorem. The integral can be estimated by a stationary phase approximation, if we replace the sum over n by an integral and $\log(q)_n$ by $\int_0^n \log(1-q^x)dx$.

This procedure can be justified by the Euler-McLaurin formula. Stationarity with respect to n yields
$$\xi^{2\alpha} = 1 - \xi + O(1/n), \tag{9}$$
where $\xi = q^n$, and stationarity with respect to q yields
$$\alpha n^2 - N + (\log \xi)^{-2} Li_2(1-\xi) n^2 = O(n), \tag{10}$$
where
$$Li_2(z) = -\int_0^z \frac{\log(1-t)}{t} dt = \sum_n z^n/n^2. \tag{11}$$
Using the Rogers dilogarithm (1)
$$L(z) = Li_2(z) + \tfrac{1}{2} \log(z) \log(1-z) \tag{12}$$
we obtain from (9) and (10)
$$n^2 = N(\log \xi)^2 / L(\xi^{2\alpha}) + O(n). \tag{13}$$
From the integral for a_N at the stationary point
$$\log a_N \sim (\alpha n^2 - N) \log q - \log(q) n \sim 2(NL(\xi^{2\alpha}))^{\frac{1}{2}} \tag{14}$$
using (9) and (13) as well as
$$\log(q) n = n\big(L(1-\xi)/\log \xi + \tfrac{1}{2} \log(1-\xi)\big) + O(1). \tag{15}$$
Hence
$$c_{\text{eff}} = L(\xi^{2\alpha})/L(1). \tag{16}$$
The quantum dimensions are given by the ratios of the a_n at large n, i.e. by the values of ξ^β.

The $(2, 5)$ minimal model yields $\alpha = 1$. For the $(3, 4)$ model, i.e. for free fermions, one has $\alpha = 1/2$, and for the $(3, 5)$ model a sum of two characters yields $\alpha = 1/4$. In all cases, β takes the values $0, 1$. Thus in these cases a relation between central extensions, dilogarithms and quantum dimensions is manifest. The models yields exactly the elements of the conjectured $\mathcal{D}^1$, namely $L(\tau^2)/L(1) = 2/5$, $L(1/2)/L(1) = 1/2$, $L(\tau)/L(1) = 3/5$.

The fact that the examples yielded elements of $\mathcal{D}^1$ comes from the fact that all of the states contributing to the partition function come from one field with a single normal ordered product relation, namely the energy momentum tensor, the free fermion field, and the primary field of conformal dimension $3/4$, respectively. As long as the numbers of generators and relations remain finite, there is hope to generalize the preceding calculation.

Some dilogarithm formulas have moduli, i.e. they allow the insertion of arbitrary chosen algebraic numbers. In such case the corresponding values of c_{eff} seem to describe conformal theories with a related number of moduli. In particular, one has the standard formulas
$$L(x) + L(1-x) = L(1),$$
$$L(1-x) + L(1-y) + L(xy) + L\left(\frac{x(1-y)}{1-xy}\right) + L\left(\frac{y(1-x)}{1-xy}\right) = 2L(1). \tag{17}$$

At the values 1,2 of c one has free boson theories, which of course have moduli.

The former equation yields $\mathcal{D}^N = N - \mathcal{D}^N$. The corresponding statement for central charges should generalize an aspect of the level-rank duality.

It is very likely that the x_i satisfy stronger number theoretic constraints, like the quantum dimensions, which are particular algebraic numbers. In the examples above, τ and τ^2 satisfying (2) are units in the field of algebraic numbers, and an analogous situation occurs for the $(2,7)$, $(3,7)$ and $(6,7)$ minimal models. Here the quantum dimension of the primary field ϕ_{12} is the unit $1 + a = 2\cos(\pi/7)$, such that

$$a^3 + 2a^2 - a - 1 = 0 . \tag{18}$$

Other units are $1 - a$, $1 - a^2$ and $b = (1+a)^{-1}$. We find

$$L(1-a) + L(b^2) = \frac{4}{7}L(1) ,$$
$$L(1-a^2) + L(b^2) = \frac{5}{7}L(1) , \tag{19}$$
$$L(1-a) + L(a^2) = \frac{6}{7}L(1) .$$

On the other hand, for the free fermion we needed $x = 1/2$. The situation is still a bit too complex to make further conjectures.

ACKNOWLEDGEMENTS

Most of this work was done during a visit to the Isaac Newton Institute for Mathematical Sciences. I thank A. Goncharev, A. Kirillov, A. Recknagel, M. Terhoeven and D. Zagier for very helpful discussions.

REFERENCES

[1] A.N. Kirillov, N.Yu. Reshetikhin, Zap. Nauch. Semin. LOMI 160 (1987) 211.
[2] A.N. Kirillov, Zap. Nauch. Semin. LOMI 164 (1987) 121.
[3] A. Kuniba, T. Nakanishi, Mod. Phys. Lett. **A7** (1992) 3487.
[4] L. Lewin, Polylogarithms and associated functions, North-Holland 1981.
[5] B.L. Feigin et al., 'The Annihilating Ideals of Minimal Models', in: Infinite Analysis, Proceedings of the RIMS Research Project 1991, A. Tsuchiya et al. eds., (World Scientific 1992).

DILOGARITHM IDENTITIES AND SPECTRA IN CONFORMAL FIELD THEORY

Anatol N. Kirillov

Isaac Newton Institute for Mathematical Sciences,
20 Clarkson Road, Cambridge, CB3 OEH, U.K.
and
Steklov Mathematical Institute,
Fontanka 27, St.Petersburg, 191011, Russia

ABSTRACT

We prove new identities between the values of the Rogers dilogarithm function and describe a connection between these identities and spectra in conformal field theory.

INTRODUCTION

The dilogarithm function defined for $0 \le z \le 1$ by

$$Li_2(z) = \sum_{n=1}^{\infty} \frac{z^n}{n^2} = -\int_0^z \frac{\log(1-t)}{t} dt,$$

is one of the lesser transcendental functions. Nonetheless, it has many intriguing properties and has appeared in various branches of mathematics and physics. Here we mention only the appearances of dilogarithm function in number theory (the study of asymptotic behaviour of partitions [RS]; the values of ζ-functions in some special points [Za]) in algebraic K-theory (A. Beilinson, S. Bloch, A. Groncharov,...).

In physics, the dilogarithm appears at first from a calculation of magnetic susceptibility in the XXZ model at small magnetic field [KR]. More recently [Z], the dilogarithm identities (through the Thermodynamic Bethe Ansatz (TBA)) through appear in the context of investigation of UV limit or the critical behaviour of integrable 2-dimensional quantum field theories and lattice models [Z], [DR], [KP],...). One aim of this paper is to prove some new identities between the values of Rogers dilogarithm function and to show that any rational number may be obtained as the value of some dilogarithm sum. More details and further results will appear in [Ki2].

1. DEFINITION AND THE BASIC PROPERTIES OF ROGERS DILOGARITHM

Let us recall the definition of the Rogers dilogarithm function $L(x)$ for $x \in (0,1)$

$$L(x) = -\frac{1}{2}\int_0^x \left[\frac{\log(1-t)}{t} + \frac{\log t}{1-t}\right] dt = \sum_{n=1}^{\infty} \frac{x^n}{n^2} + \frac{1}{2}\log x \cdot \log(1-x). \quad (1.1)$$

The following two classical results (see e.g. [Le], [GM], [Ki]) contain the basic properties of the function $L(x)$.

Theorem A. The function $L(x) \in C^{\infty}((0,1))$ and satisfies the following functional equations

1. $L(x) + L(1-x) = \dfrac{\pi^2}{6}, \quad 0 < x < 1,$ \hfill (1.2)

2. $L(x) + L(y) = L(xy) + L\left(\dfrac{x(1-y)}{1-xy}\right) + L\left(\dfrac{y(1-x)}{1-xy}\right),$ \hfill (1.3)

where $0 < x, y < 1$.

Theorem B. Let $f(x)$ be a function of class $C^3((0,1))$ and satisfy the relations (1.2) and (1.3). Then we have

$$f(x) = \text{const} \cdot L(x).$$

We continue the function $L(x)$ on all real axis $\mathbf{R} = \mathbf{R}^1 \cup \{\pm\infty\}$ by the following rules

$$L(x) = \frac{\pi^2}{3} - L(x^{-1}), \quad \text{if } x > 1, \quad (1.4)$$

$$L(x) = L\left(\frac{1}{1-x}\right) - \frac{\pi^2}{6}, \quad \text{if } x < 0, \quad (1.5)$$

$$L(0) = 0, \quad L(1) = \frac{\pi^2}{6}, \quad L(+\infty) = \frac{\pi^2}{3}, \quad L(-\infty) = -\frac{\pi^2}{6}.$$

The present work will concern with relations between the values of the Rogers dilogarithm function at certain algebraic numbers. More exactly, let us consider an abelian subgroup W in the field of rational numbers $\mathbf{Q}$

$$W = \left\{\sum_i \frac{n_i L(\alpha_i)}{L(1)} \mid n_i \in \mathbf{Z}, \; \alpha_i \in \overline{\mathbf{Q}} \cap \mathbf{R} \text{ for all } i\right\} \cap \mathbf{Q}. \quad (1.6)$$

According to a conjecture of W. Nahm [Na] the abelian group W "coincides" with the spectra in rational conformal field theory. Thus it seems very interesting task to obtain more explicit description of the group W (e.g. to find a system of generators for W) and also to connect already known results about the spectra in conformal field theory (see e.g. [BPZ], [GKO], [FF], [FQS], [Ka], [KW], [KP]) with suitable elements in W. One of our main results of the present paper allows to describe some part of a system of generators for abelian group W. As a corollary, we will show that the spectra of unitary minimal models [BPZ] and some others are really contained in W.

2. BASIC IDENTITIES AND CONFORMAL WEIGHTS

In this section we present our main results dealing with a computation of the following dilogarithm sum

$$\sum_{k=1}^{n-1}\sum_{m=1}^{r} L\left(\frac{\sin k\varphi \cdot \sin(n-k)\varphi}{\sin(m+k)\varphi \cdot \sin(m+n-k)\varphi}\right) := \frac{\pi^2}{6} s(j,n,r), \qquad (2.1)$$

where $\varphi = \dfrac{(j+1)\pi}{n+r}$, $0 \leq j \leq n+r-2$.

It is clear that $s(j,n,r) = s(n+r-2-j,n,r)$, so we will assume in sequel that $0 \leq 2j \leq n+r-2$.

The dilogarithm sum (2.1) corresponds to the Lie algebra of type A_{n-1}. The case $j = 0$ was considered in our previous paper [Ki], where it was proved that $s(0,n,r) = \dfrac{(n^2-1)r}{n+r}$. It was stated in [Ki] that this number coincides with the central charge of the $SU(n)$ level r WZNW model. Before formulating our result about computation of the sum $s(j,n,r)$ let us remind the definition of Bernoulli polynomials. They are defined by the generating function

$$\frac{te^{xt}}{e^t - 1} = \sum_{n=0}^{\infty} B_n(x) \frac{t^n}{n!}, \quad |t| < 2\pi.$$

We use also modified Bernoulli polynomials

$$\overline{B}_n(x) = B_n(\{x\}), \text{ where } \{x\} = x - [x]$$

is the fractional part of $x \in \mathbf{R}$. It is well-known that

$$\overline{B}_{2n}(x) = (-1)^n \frac{2(2n)!}{(2\pi)^{2n}} \sum_{k=1}^{\infty} \frac{\cos 2k\pi x}{k^{2n}}, \qquad (2.2)$$

$$\overline{B}_{2n+1}(x) = (-1)^{n-1} \frac{2(2n+1)!}{(2\pi)^{2n+1}} \sum_{k=1}^{\infty} \frac{\sin 2k\pi x}{k^{2n+1}}.$$

Theorem 2.1. We have

$$s(j,n,r) = 6(r+n) \sum_{k=0}^{[\frac{n-1}{2}]} \left\{\frac{1}{6} - \overline{B}_2\big((n-1-2k)\theta\big)\right\} - \frac{1}{4}\left\{2n^2 + 1 + 3(-1)^n\right\}, \quad (2.3)$$

where $\theta = \dfrac{j+1}{r+n}$ and g.c.d.$(j+1, r+n) = 1$.

Theorem 2.2. (level-rank duality)

$$s(j,n,r) + s(j,r,n) = nr - 1. \qquad (2.4)$$

Corollary 2.3. We have
$$s(j,n,r) = c_r^{(n)} - 24 h_j^{(r,n)} + 6 \cdot \mathbb{Z}_+, \qquad (2.5)$$

where
$$c_r^{(n)} = \frac{(n^2-1)r}{n+r}, \quad h_j^{(r,n)} = \frac{n(n^2-1)}{24} \cdot \frac{j(j+2)}{r+n}, \quad 0 \le j \le r+n-2, \qquad (2.6)$$

are the central charge and conformal dimensions of the $SU(n)$ level r WZNW primary fields, respectively.

Proof. Let us remember that
$$B_2(x) = x^2 - x + \frac{1}{6} \quad \text{and} \quad \overline{B}_2(x) = B_2(\{x\}).$$

Thus,
$$6(r+n) \sum_{k=0}^{[\frac{n-1}{2}]} \left(\frac{1}{6} - \overline{B}_2\Big((n-1-2k)\theta\Big) \right)$$

$$= 6(r+n) \sum_{k=0}^{[\frac{n-1}{2}]} (n-1-2k)\theta - 6(r+n) \sum_{k=0}^{[\frac{n-1}{2}]} (n-1-2k)^2 \theta^2$$

$$+ 6(r+n) \sum_{k=0}^{[\frac{n-1}{2}]} \Big[(n-1-2k)\theta\Big]\left((n-1-2k)\theta - 1 + \Big\{(n-1-2k)\theta\Big\} \right)$$

$$:= 6\Sigma_1 - 6\Sigma_2 + 6\Sigma_3.$$

Now if we take $\theta = \dfrac{j+1}{r+n}$, then it is clear that $\Sigma_3 \in \mathbb{Z}_+$. In order to compute Σ_1 and Σ_2 we use the following summation formulae

$$\sum_{k=0}^{[\frac{n-1}{2}]} (n-1-2k) = \frac{2n^2 - 1 + (-1)^n}{8} = \left[\frac{n}{2}\right] \cdot \left[\frac{n+1}{2}\right],$$

$$\sum_{k=0}^{[\frac{n-1}{2}]} (n-1-2k)^2 = \frac{n(n^2-1)}{6}.$$

Consequently the sum becomes

$$s(j,n,r)$$
$$= \frac{3(2n^2 - 1 + (-1)^n)(j+1)}{4} - \frac{n(n^2-1)(j+1)^2}{r+n} - \frac{2n^2 + 1 + 3(-1)^n}{4} + 6\Sigma_3$$
$$= \frac{(n^2-1)r}{r+n} - \frac{n(n^2-1)j(j+2)}{r+n} + 6j \left[\frac{n}{2}\right] \cdot \left[\frac{n+1}{2}\right] + 6\Sigma_3. \qquad \blacksquare$$

For small values of j we may compute the sum in (2.3) and thus to find corresponding positive integer in (2.5).

Corollary 2.4.
i) if $j \le r$, then

$$s(j,2,r) = c_r^{(2)} - 24h_j^{(r,2)} + 6j = \frac{3r}{r+2} + 6\frac{j(r-j)}{r+2}. \tag{2.7}$$

ii) if $2j \le r+1$, then

$$s(j,3,r) = \frac{8r}{r+3} - 24\frac{j(j+2)}{r+3} + 12j. \tag{2.8}$$

iii) if $(n-1)j < r+1$, then

$$s(j,n,r) = c_r^{(n)} - 24h_j^{(r,n)} + 6j \cdot \left[\frac{n}{2}\right] \cdot \left[\frac{n+1}{2}\right]. \tag{2.9}$$

Proof. An assumption $(n-1)j < r+1$ is equivalent to a condition $\frac{(n-1)(j+1)}{n+r} < 1$. So the term Σ_3 (see a proof of Corollary 2.3) is equal to zero. ∎

It seems interesting to find a meaning of the positive integer in (2.5). Now we want to find a "dilogarithm interpretation" of the central charges and conformal dimensions for some well-known conformal models.

Corollary 2.5. We have

$$s(j_1,2,k)+s(0,2,1)-s(j_2,2,k+1) = c_k - 24h_{j_1+1,j_2+1} + 6(j_1-j_2)(j_1-j_2+1), \tag{2.10}$$

where

$$c_k = 1 - \frac{6}{(k+2)(k+3)}, \quad h_{r,s}^{(k)} = \frac{[(k+3)r - (k+2)s]^2 - 1}{4(k+2)(k+3)} \tag{2.11}$$

are the central charge and conformal dimensions of the primary fields for unitary minimal conformal models [BPZ], [Ka], [FQS].

Corollary 2.6.

$$s(j_1,n,k) + s(0,n,1) - s(j_2,n,k+1) = c_{k,n} - 24h_{j_1+1,j_2+1}^{(n)} + 12Z_+, \tag{2.12}$$

where

$$c_{k,n} = (n-1)\left\{1 - \frac{n(n+1)}{(k+n)(k+n+1)}\right\}, \tag{2.13}$$

$$h_{r,s}^{(n)}(k) = \frac{n(n^2-1)}{24} \cdot \frac{[(k+n+1)r - (k+n)s]^2 - 1}{(k+n)(k+n+1)}$$

are the central charge and conformal dimensions of the primary fields for W_n models [Bi].

Finally we give a "dilogarithm interpretation" for the central charge and conformal weights of restricted solid-on-solid (RSOS) lattice models and their fusion hierarchies [KP].

Corollary 2.7. We have
$$s(l,2,N) + s(N-1,2,N-2) - s(m-1,2,N-2) = \qquad (2.14)$$
$$= c + 1 - 24\Delta + 6(l - |m|), \quad m \in \mathbf{Z}, \quad 0 \leq l \leq N,$$

where
$$c = \frac{2(N-1)}{N+2} \quad \text{and} \quad \Delta = \frac{l(l+2)}{4(N+2)} - \frac{m^2}{4N} \qquad (2.15)$$

are the central charge and conformal weights of $\mathbf{Z}_N$ parafermion theories [FZ]. The members of (2.15) may be also realised as the central charge and conformal weights of fusion $N+1$-state RSOS(p,p) lattice models [DJKMO], [BR] on the regime I/II critical line. Note that physical constraints

$$|m| \leq l, \quad m \equiv l \pmod{2}$$

for values of m in (2.15) are equivalent to a condition that "remainder term" in (2.14), namely, $6(l - |m|)$, must belong to $12\mathbf{Z}_+$.

Corollary 2.8. Let us fix the positive integers k, $p = 1, 2, \ldots$ (the fusion level), j_1 and j_2 such that $0 \leq j_1 \leq k$, $0 \leq j_2 \leq k + l$.
Let us put $r_0 = p \left\{ \dfrac{j_1 - j_2}{p} \right\}$. Then we have

$$s(j_1, 2, k) + s(r_0, 2, p) - s(j_2, 2, k + p)$$
$$= c - 24\Delta + 12 \, \frac{(j_1 - j_2)(p + j_1 - j_2) + r_0(p - r_0)}{2p}, \qquad (2.16)$$

where
$$c = \frac{3p}{p+2}\left(1 - \frac{2(p+2)}{(k+2)(k+p+2)}\right), \qquad (2.17)$$

$$\Delta = \frac{[(k+p+2)(j_1+1) - (k+2)(j_2+1)]^2 - p^2}{4p(k+2)(k+p+2)} + \frac{r_0(p-r_0)}{2p(p+2)}$$

are the central charge and conformal weights of the fusion $(k+p+1)$-state RSOS(p,p) lattice models [KP] on the regime III/IV critical line. It is easy to see that "remainder term" in (2.16) belongs to $12\mathbf{Z}_+$. Note also that the fusion RSOS(p,q) lattice models, obtained by fusing $p \times q$ blocks of face weights together, are related to coset conformal fields theories obtained by the Goddard-Kent-Olive (GKO) construction [GKO]. Namely, c and Δ in (2.17) are the central charge and conformal dimensions of conformal field theory, which corresponds to the coset pair [GKO]

$$\begin{array}{cccc} & A_1 & \oplus \; A_1 & \supset \; A_1 \\ \text{levels} & k & p & k+p \, . \end{array}$$

3. A_1-type DILOGARITHM IDENTITIES

As is well-known [Le], the Rogers dilogarithm function $L(x)$ admits a continuation to all the complex plane $\mathbf{C}$. Following [Le], [KR] we define a function

$$L(x,\theta) := -\frac{1}{2}\int_0^x \frac{\log(1 - 2x\cos\theta + x^2)}{x}dx + \frac{1}{4}\log|x| \cdot \log(1 - 2x\cos\theta + x^2)$$
$$= \mathrm{Re}L(xe^{i\theta}), \quad x, \theta \in \mathbf{R} \, . \qquad (3.1)$$

Our proof of Theorem 2.1 is based on a study of properties of the function $L(x,\theta)$.

Proposition 3.1. For all real φ, θ we have

$$L\left(\left(\frac{\sin\theta}{\sin\varphi}\right)^2\right) = \pi^2\left\{\overline{B}_2\left(\frac{\theta+\varphi}{\pi}\right) - \overline{B}_2\left(\frac{\varphi}{\pi}\right) - \overline{B}_2\left(\frac{\theta}{\pi}\right) + \frac{1}{6}\right\}$$
$$+ 2L\left(-\frac{\sin(\varphi-\theta)}{\sin\theta},\varphi\right) - 2L\left(-\frac{\sin\varphi}{\sin\theta},\varphi+\theta\right). \qquad (3.2)$$

Before proving a Proposition 3.1 let us give the others useful properties of function (3.1) (compare with [Le]).

Lemma 3.2.

(i) $\quad L(x,0) = L(x), \quad L(-x,\varphi) = L(-x,\pi-\varphi),$ \hfill (3.3)

(ii) $\quad L(x,\varphi) = L(x, 2\pi k \pm \varphi), \quad k \in \mathbf{Z},$ \hfill (3.4)

(iii) $\quad L(-1,\varphi) = \pi^2 \overline{B}_2\left(\frac{\varphi}{2\pi} + \frac{1}{2}\right),$

$\qquad L(1,\varphi) = \pi^2 \overline{B}_2\left(\frac{\varphi}{2\pi}\right),$ \hfill (3.5)

(iv) $\quad L(x,\varphi) + L(x^{-1},\varphi) = 2\pi^2 \overline{B}_2\left(\frac{\varphi}{2\pi}\right), \quad x > 0,$

$\qquad L(-x,\varphi) + L(-x^{-1},\varphi) = 2\pi^2 \overline{B}_2\left(\frac{\varphi}{2\pi}+\frac{1}{2}\right), \quad x < 0,$ \hfill (3.6)

(v) $\quad L(0,\varphi) = 0, \quad L(+\infty,\varphi) = 2\pi^2 \overline{B}_2\left(\frac{\varphi}{2\pi}\right),$

$\qquad L(-\infty,\varphi) = 2\pi^2 \overline{B}_2\left(\frac{\varphi+\pi}{2\pi}\right)$ \hfill (3.7)

(vi) $\quad L(2\cos\varphi,\varphi) = \pi^2\left\{\overline{B}_2\left(\frac{\varphi}{\pi}\right) + \frac{1}{12}\right\},$ \hfill (3.8)

(vii) $\quad L(x^n, n\varphi) = n \sum_{k=0}^{n-1} L\left(x, \varphi + \frac{2k\pi}{n}\right), \quad x \in \mathbf{R}_+,$

$\qquad L(x^n) = n \sum_{k=0}^{n-1} L\left(x \cdot \exp\frac{2k\pi i}{n}\right), \quad x \in (0,1).$ \hfill (3.9)

More generally (Rogers' identity [Ro])

$$L(1-y^n) = \sum_{k=1}^{n}\sum_{l=1}^{n}\left[L(\lambda_k/\lambda_l) - L(x_k\lambda_l)\right],$$

where $\{x_k\}_{k=1}^n$ are the roots of the equation

$$1 - y^n = \prod_{k=1}^{n}(1 - \lambda_k x).$$

Proof of the Theorem 2.1 for the case $n = 2$. If we substitute $\theta = m\varphi$ in (3.2) we then obtain

$$L\left(\left(\frac{\sin m\varphi}{\sin\varphi}\right)^2\right) = \pi^2\left\{\overline{B}_2\left(\frac{(m+1)\varphi}{\pi}\right) - \overline{B}_2\left(\frac{m\varphi}{\pi}\right) - \overline{B}_2\left(\frac{\varphi}{\pi}\right) + \frac{1}{6}\right\}$$
$$+ 2L\left(-\frac{\sin((m-1)\varphi)}{\sin\varphi},m\varphi\right) - 2L\left(-\frac{\sin m\varphi}{\sin\varphi},(m+1)\varphi\right). \qquad (3.10)$$

105

Further let us introduce the notation

$$f_m(\varphi) := 1 - \frac{Q_{m-1}(\varphi)Q_{m+1}(\varphi)}{Q_m^2(\varphi)} = \frac{1}{Q_m^2(\varphi)}.$$

Then using (3.10) we find

$$\sum_{m=1}^{r} L(f_m(\varphi)) = -2\left\{L\left(-Q_r(\varphi), (r+2)\varphi\right) - \frac{\pi^2}{6}\right\} \tag{3.11}$$

$$+ \pi^2 \left\{\overline{B}_2\left(\frac{(r+2)\varphi}{\pi}\right) - \frac{1}{6}\right\} + (r+2)\pi^2 \left\{\frac{1}{6} - \overline{B}_2\left(\frac{\varphi}{\pi}\right)\right\} - \frac{\pi^2}{2}.$$

Now let us put $\varphi = \dfrac{(j+1)\pi}{r+2}$, $0 \leq j \leq r+1$. Then $Q_r(\varphi) = (-1)^j$ and it is clear from (3.3) and (3.4) that

$$L((-1)^{j+1}, (j+1)\pi) = L(1) = \frac{\pi^2}{6}. \qquad \blacksquare$$

Note that the polynomials $Q_m := Q_m(\varphi)$ satisfy the following recurrence relation

$$Q_m^2 = Q_{m-1}Q_{m+1} + 1, \quad Q_0 \equiv 1, \quad m \geq 1,$$

whereas the polynomials $y_m := y_m(\varphi) = Q_{m-1}(\varphi) \cdot Q_{m+1}(\varphi)$ satisfy the following one

$$y_m^2 = (1 + y_{m-1})(1 + y_{m+1}), \quad y_0 \equiv 0, \quad m \geq 1.$$

Now we propose a generalisation of (2.12). Given a rational number p and a decomposition of p into the continued fraction

$$p = [b_r, b_{r-1}, \ldots, b_1, b_0] = b_r + \cfrac{1}{b_{r-1} + \cfrac{1}{\cdots + \cfrac{1}{b_1 + \cfrac{1}{b_0}}}}. \tag{3.12}$$

We will assume that $b_i > 0$ if $0 \leq i < r$ and $b_r \in \mathbb{Z}$. Using the decomposition (3.12) we define the set of integers y_i and m_i:

$$y_{-1} = 0, \quad y_0 = 1, \quad y_1 = b_0, \ldots, y_{i+1} = y_{i-1} + b_i y_i, \quad 0 \leq i \leq r, \tag{3.13}$$
$$m_0 = 0, \quad m_1 = b_0, \quad m_{i+1} = |b_i| + m_i, \quad 0 \leq i \leq r.$$

The following sequences of integers were first introduced by Takahashi and Suzuki [TS]: $r(j) = i$, if $m_i \leq j < m_{i+1}$, $0 \leq i \leq r$,

$$n_j = y_{i-1} + (j - m_i)y_i, \quad \text{if} \quad m_i \leq j < m_{i+1} + \delta_{i,r}, \quad 0 \leq i \leq r.$$

Finally we define a dilogarithm sum of "fractional level p":

$$\sum_{j=1}^{m_{r+1}} (-1)^{r(j)} L\left(\left(\frac{\sin y_{r(j)}\theta}{\sin(n_j + y_{r(j)})\theta}\right)^2\right) := (-1)^r \frac{\pi^2}{6} s(k, 2, p), \tag{3.14}$$

where $\theta = \dfrac{(k+1)\pi}{y_{r+1} + 2y_r}$.

Proposition 3.3. We have

$$s(k, 2, p) = \frac{3p}{p+2} - 6\frac{k(k+2)}{p+2} + 6\mathbf{Z}. \tag{3.15}$$

Conjecture 3.4. For all positive $p \in \mathbf{Q}$ the remainder term in (3.15) lies in $6\mathbf{Z}_+$.

Corollary 3.5. Let us fix the positive integers $l = 1, 2, 3, \ldots$ (the fusion level), $p > q$, j_1 and j_2. Then

$$s\left(j_1, 2, \frac{ql}{p-q} - 2\right) + s(r_0, 2, l) - s\left(j_2, 2, \frac{pl}{p-q} - 2\right) = c - 24\Delta + 6\mathbf{Z}, \tag{3.16}$$

where $r_0 = l \cdot \left\{\frac{j_1 - j_2}{l}\right\}$ and

$$c = \frac{3l}{l+2}\left(1 - \frac{2(l+2)(p-q)^2}{l^2 pq}\right), \tag{3.17}$$

$$\Delta = \frac{[p(j_1+1) - q(j_2+1)]^2 - (p-q)^2}{4lpq} + \frac{r_0(l-r_0)}{2l(l+2)}$$

are the central charge and conformal dimensions of RCFT, which corresponds to the coset pair [GKO]

$$\begin{array}{ccccc} & A_1 & \oplus & A_1 & \supset & A_1 \\ \text{levels} & \frac{ql}{p-q} - 2 & & l & & \frac{pl}{p-q} - 2 \end{array}.$$

Acknowledgements

Most of this work was done during a visit to the Isaac Newton Institute for Mathematical Sciences. I thank W. Nahm, E.K. Sklyanin, F. Ravanini and E. Corrigan for very helpful discussions.

REFERENCES

[Bi] A. Bilal, Nucl. Phys. B330 (1990) 399.
[BPZ] A. Belavin, A. Polyakov, A. Zamolodchikov, J. Stat. Phys. 34 (1984) 763; Nucl. Phys. B241 (1984) 333.
[BR] V.V. Bazhanov, N.Yu. Reshetikhin, suppl. Prog. Theor. Phys. 102 (1990).
[CR] P. Christe, F. Ravanini, $G_N \otimes G_L/G_{N+L}$ conformal field theories and their modular invariant partition functions, Int. J. Mod. Phys. A4 (1989) 897.
[DR] P. Dorey, F. Ravanini, Staircase model from affine Toda field theory, Int. J. Mod. Phys. A8 (1993) 873.
[FL] V.A. Fateev, S. Lukyanov, Int. J. Mod. Phys. A3 (1988) 507.
[FQS] D. Friedan, Z. Qiu, S. Shenker, Phys. Rev. Lett. 52 (1984) 1575.
[GKO] P. Goddard, A. Kent, D. Olive, Phys. Lett. B152 (1985) 105.
[GM] I.M. Gelfand, R.D. MacPherson, Advan. in Math. 44 (1982) 279.

[Ka]   V.G. Kac, Infinite dimensional Lie algebras, Cambridge Univ. Press, 1990.

[Ki1] A.N. Kirillov, Zap. Nauch. Semin. LOMI, 164 (1987) 121.

[Ki2] A.N. Kirillov, Dilogarithm identities, partitions and spectra in conformal field theory, in preparation.

[KN1] A. Kuniba, T. Nakanishi, Level-rank duality in fusion RSOS models, Proc. of Int. Colloq. on Modern Quantum Field Theory, Bombay, India, 1990.

[KN2] A. Kuniba, T. Nakanishi, Spectra in conformal field theories from Rodgers dilogarithm, Mod. Phys. Lett. A7 (1992) 3487.

[KP]   A. Klumper, P. Pearce, Physica A183 (1992) 304.

[KR]   A.N. Kirillov, N.Yu. Reshetikhin, Zap. Nauch. Semin. LOMI, 160 (1987) 211.

[Ku]   A. Kuniba, Thermodynamics of the $U_q(X_r^{(1)})$ Bethe ansatz system with q a root of Unity, Nucl. Phys. B389 (1993) 209.

[KW]   V. Kac, M. Wakimoto, Adv. Math. 70 (1988) 156.

[Le]   L. Lewin, Polylogarithms and associated functions, (North-Holland, 1981).

[MSW] P. Mathieu. D. Senechal, M. Walton, Field identification in nonunitary diagonal cosets, prepr. LETH-PHY-9/91.

[Na]   W. Nahm, Dilogarithm and W-algebras, contribution to this volume.

[NRT] W. Nahm, A. Recknagel, M. Terhoeven, Dilogarithm identities in conformal field theory, prep. BONN-92-35, hepth 9211034.

[Ro]   L.J. Rogers, Proc. London Math. Soc. 4 (1907) 169.

[RS]   B. Richmond, G. Szekeres, J. Austral. Math. Soc., A31 (1981) 362.

[SA]   H. Saleur, D. Altshuler, Nucl. Phys. B354 (1991) 579.

[TS]   M. Takahashi, M. Suzuki, Prog. Theor. Phys., 48 (1972) 2187.

[Z]   Al.B. Zamolodchikov, Nucl. Phys. B342 (1990) 695.

[Za]   D. Zagier, The remarkable dilogarithm (Number theory and related topics. Papers presented at the Ramanujan colloquium, Bombay, 1988, TIFR).

PHYSICAL STATES IN TOPOLOGICAL COSET MODELS

J. Sonnenschein

School of Physics and Astronomy
Beverly and Raymond Sackler Faculty of Exact Sciences
Ramat Aviv Tel-Aviv, 69978, Israel

ABSTRACT

Recent results about topological coset models are summarized. The action of a topological $\frac{G}{H}$ coset model (rank H = rank G) is written down as a sum of "decoupled" matter, gauge and ghost sectors. The physical states are in the cohomology of a BRST-like operator that relates these sectors. The cohomology on a free field Fock space as well as on an irreducible representation of the "matter" Kac-Moody algebra are extracted. We compare the results with those of (p,q) minimal models coupled to gravity and with (p,q) W_N strings for the case of $A_1^{(1)}$ at level $k = \frac{p}{q} - 2$ and $A_1^{(N-1)}$ at level $k = \frac{p}{q} - N$ respectively.

1. FROM TQFT'S TO TOPOLOGICAL COSET MODELS

A Topological Quantum Field Theory (TQFT)[1] is a QFT in which all the "observables", namely, all correlators of "physical operators", are invariant under any arbitrary deformation of $g_{\alpha\beta}$ the metric of the underlying space-time. Given a set of physical operators $F_i[\Phi^a(x_i)]$ which are functionals of the fields $\Phi^a(x)$, $a = 1,\ldots p$ of the theory and which are invariant under the symmetries of the theory, then the theory is topological iff

$$\delta_{g_{\alpha\beta}} \langle \prod_i F_i[\Phi^a(x)] \rangle = 0 . \tag{1}$$

for any product which is a Lorentz scalar. In particular this definition implies that all correlators are independent of distances between the operators. In a theory where the energy-momentum tensor $T_{\alpha\beta}$ is exact under a BRST-like symmetry operator Q, namely, $T_{\alpha\beta} = \{Q, G_{\alpha\beta}\}$, and provided that all physical operators are in the cohomology of Q then for every product defined above property (1) is obeyed. Thus, it is a TQFT. Schematically the proof of the latter statement is the following

$$\delta_{g_{\alpha\beta}}\langle\prod_i F_i[\Phi^a(x_i)]\rangle = \delta_{g_{\alpha\beta}}\int D\Phi^a e^{iS(\Phi^a)}\prod_i F_i[\Phi^a(x_i)]$$
$$= \int D\Phi^a e^{iS(\Phi^a)}\int dx' T_{\alpha\beta}(x')\prod_i F_i[\Phi^a(x_i)] \qquad (2)$$
$$= \int D\Phi^a e^{iS(\Phi^a)}\int dx' \{Q,G_{\alpha\beta}\}(x')\prod_i F_i[\Phi^a(x_i)] = \langle\{Q,\}\rangle = 0 .$$

A two-dimensional theory which is a TQFT as well as conformal, namely, $T_{zz} = T(z)$ and $T_{\bar{z}\bar{z}} = \bar{T}(\bar{z})$ and similarly for all the symmetry currents, is a topological conformal field theory (TCFT). Any theory which has the following algebra

$$T(z) = \{Q,G(z)\} ,$$
$$J^{(\mathrm{BRST})} = \{Q,j^\#(z)\} , \qquad (3)$$

where $Q = \oint J^{(\mathrm{BRST})}(z)$ is the BRST charge is a TCFT. This algebra is referred to as the TCFT algebra[2] and in fact it is a sub-algebra of a larger algebra which is shared by the TCFT's[3]. An immediate consequence of the fact that T is exact is that the theory has a vanishing Virasoro anomaly.

Obviously every conformal field theory coupled to 2-D gravity is a TCFT after integration over all metric degrees of freedom. We now define the concept of a "topological model" which is a TQFT without the introduction and integration over the metric. In more than two dimensions examples of such models are the Chern-Simons theory and the four dimensional theory which corresponds to the Donaldson invariants[1]. In two dimensions an example is the theory of flat gauge connections[5]. We are now ready to introduce the notion of "topological coset models" which are gauged WZW models that are also topological models.

Gravitational models in two dimensions were a subject of intensive study in recent years. Various different approaches were invoked in this domain of research. Starting from continuum Liouville theory and the light-cone world sheet formulation[7] through matrix[8] models, KdV hierarchies to the Kontsevich integral[9]. Yet another possible approach is that of the topological coset models. Here we summarize some recent work[10,11,12] in extracting the space of physical states of those models and reveal some correspondence between these models and matter models like the (p,q) minimal models coupled to two dimensional gravity

2. THE QUANTUM ACTION OF A TWISTED $\frac{G}{H}$ MODEL

The classical action of the twisted $\frac{G}{H}$ model[13] is that of level k twisted supersymmetric G-WZW model coupled to gauge fields in the algebra of $H \in G$. In other words it is the usual $\frac{G}{H}$ model with an extra set of $(1,0)$ anti-commuting ghosts where the dimension one fields take their values in the positive roots of $\frac{G}{H}$ and the dimension zero fields in the negative ones. The action of the model reads

$$S_{(tKS)} = S_k(g, A, \bar{A}) + S^{\frac{G}{H}}_{(gh)} ,$$

$$S_k(g, A, \bar{A}) = S_k(g) - \frac{k}{2\pi} \int_\Sigma d^2z \, \mathrm{Tr}_G[g^{-1}\partial g \bar{A}_{\bar{z}} + g\bar{\partial}g^{-1}A - \bar{A}g^{-1}Ag + A\bar{A}] , \quad (4)$$

$$S^{\frac{G}{H}}_{gh} = \frac{i}{2\pi} \int d^2z \sum_{\alpha \in \frac{G}{H}} [\rho^{+\alpha}(\bar{D}\chi)^{-\alpha} + \bar{\rho}^{+\alpha}(D\bar{\chi})^{-\alpha}] ,$$

where $g \in G$ and $S_k(g)$ is the WZW action at level k. In the case that $H = G$ the model coincides with the $\frac{G}{G}$ model[14]. In the case that Σ is topologically trivial the gauge fields can be parametrized as follows $A = ih^{-1}\partial h, \bar{A} = i\bar{h}\bar{\partial}\bar{h}^{-1}$ where $h(z), \bar{h}(z) \in H^{c\dagger}$. For the $U(1)$ parts of H we take $A = i\partial\mathcal{H}_s$ where s goes over the $U(1)$ factors. The WZW part of the action then[15,16,17] takes the form

$$S_k(g, A) = S_k(hg\bar{h}) - S_k(h\bar{h}) . \quad (5)$$

The Jacobian of the change of variables introduces a dimension $(1,0)$ system of anti-commuting ghosts χ and ρ in the adjoint representation of H. The WZW action thus becomes

$$S_k(g, A) = S_k(hg\bar{h}) - S_k(h\bar{h}) + \frac{i}{2\pi} \int d^2z \, \mathrm{Tr}_H[\rho\bar{D}\chi + \bar{\rho}D\bar{\chi}] \quad (6)$$

where $D\chi = \partial\chi - i[A, \chi]$. One then fixes the gauge by setting $\bar{h} = 1$ which implies $\bar{A} = 0$ and redefining $hg \to g$. We shall treat in detail the case of $G = SU(N)$ and $H = SU(N_1) \times \ldots \times SU(N_n) \times U(1)^r$ with $r = N - 1 - \sum_{I=1}^n N_I + n$, the gauge fields A take the form $A = i\sum_{I=1}^n h^{(I)(-1)}\partial h^{(I)} + i\sum_{s=1}^r \partial\mathcal{H}_s$ and the twisted Kazama-Suzuki[18] action is given by

$$S_{(tKS)} = S_k(g) - \sum_{I=1}^n S_k(h^{(I)}) - \frac{k}{4\pi} \int d^2z \sum_{s=1}^r \partial\mathcal{H}_s\bar{\partial}\mathcal{H}_s$$
$$+ \frac{i}{2\pi} \int d^2z \, \mathrm{Tr}_H[\rho\bar{D}\chi + \bar{\rho}\partial\bar{\chi}] + \frac{i}{2\pi} \int d^2z \sum_{\alpha \in \frac{G}{H}} [\rho^{+\alpha}(\bar{\partial}\chi)^{-\alpha} + \bar{\rho}^{+\alpha}(D\bar{\chi})^{-\alpha}] . \quad (7)$$

To achieve in the level of the action a complete decoupling of the matter, gauge and ghost sectors, one has to perform now a chiral rotation to eliminate the coupling between the ghost and A in (7). This can be performed by an explicit computation of the change in the measure of the ghosts or by non-abelian bosonization of the latter and using the Polyakov Wiegmann formula[19] in the corresponding WZW actions. These two approaches lead to the same result[12] which for rank H = rank G is the

---

† For a comment about a possible difficulty in the case of non-compact groups see ref. 11.

following quantum action,

$$S_k = S_k(g) + \sum_{I=1}^{n} S_{-(k+C_G+C_{H^{(I)}})}(h^{(I)})$$

$$+ \frac{1}{2\pi} \int d^2z \left[\sum_{s=1}^{r} \partial \mathcal{H}_s \bar{\partial} \mathcal{H}_s + i\sqrt{\frac{2}{k+C_G}} (\vec{\rho}_G - \vec{\rho}_H) \cdot \vec{\mathcal{H}} R \right] \quad (8)$$

$$+ \frac{i}{2\pi} \int d^2z \, \text{Tr}_H[\bar{\rho}\partial\chi + \rho\bar{\partial}\bar{\chi}] + \frac{i}{2\pi} \int d^2z \sum_{\alpha \in \frac{G}{H}} [\rho^{+\alpha}(\bar{\partial}\chi)^{-\alpha} + \bar{\rho}^{+\alpha}(\partial\bar{\chi})^{-\alpha}],$$

where we have normalized the $\vec{\mathcal{H}}$ fields to be free bosons, and $\vec{\rho}_G$ and $\vec{\rho}_H$ are half the sums of the positive roots of G and H respectively. The action is composed of three decoupled sectors: the matter sector, the gauge sector and the ghost sector involving ghosts in H and $\frac{G}{H}$.

3. THE ALGEBRAIC STRUCTURE OF THE TWISTED KAZAMA-SUZUKI MODEL

The next step in analyzing the twisted $\frac{G}{H}$ models is to determine under what conditions are they TCFT's. For that purpose one needs to derive the algebraic structure of these models. We start with the Kac-Moody algebra associated with the group H. We define the currents $J^{(\text{tot})a}{}_I$ for each non-abelian group factor $H^{(I)}$, and for each $U(1)$ group, as

$$J^{(\text{tot})a}{}_I = J^a + I^a + if^a_{bc}\rho^b\chi^c + i \sum_{\alpha\beta \in \frac{G}{H}} f^a_{\alpha,-\beta}\rho^\alpha\chi^{-\beta}, \quad (9)$$

where J^a, I^a are the contributions of the g and h sectors respectively. The contribution of the ghost currents consists of both the H and $\frac{G}{H}$ parts to be denoted by J^a_H and $J^a_{\frac{G}{H}}$ respectively. The level of these currents vanishes

$$k_I^{(\text{tot})} = k - (k + C_G + C_{H_I}) + 2C_{H_I} + (C_G - C_{H_I}) = 0. \quad (10)$$

For the abelian case there is a similar expression now with $C_{H_I} = 0$.

The energy momentum tensor T can be decomposed, in a way which will be found later to be natural, into $T = T^H + T^{\frac{G}{H}}$ as follows

$$T(z) = \frac{1}{2(k+c_G)} g_{\tilde{a}\tilde{b}} : J^{\tilde{a}} J^{\tilde{b}} : - \frac{1}{2(k+c_G)} g_{ab} : I^a I^b : + g_{ab}\rho^a \partial\chi^b$$
$$- \frac{\sqrt{2}}{k+C_G}(\vec{\rho}_G - \vec{\rho}_H)\partial\vec{I} + \sum_{\alpha \in \frac{G}{H}} \rho^{+\alpha}(\partial\chi)^{-\alpha},$$

$$T^H(z) = \frac{1}{2(k+c_G)} g_{ab} : (J^a + J^a_{\frac{G}{H}})(J^b + J^b_{\frac{G}{H}}) : - \frac{1}{2(k+c_G)} g_{ab} : I^a I^b : \quad (11)$$
$$- \frac{\sqrt{2}}{k+C_G}(\vec{\rho}_G - \vec{\rho}_H) \cdot \partial(\vec{J} + \vec{I} + \vec{J}_{\frac{G}{H}}) + g_{ab}\rho^a\partial\chi^b,$$

$$T^{\frac{G}{H}}(z) = \frac{1}{2(k+c_G)} g_{\tilde{a}\tilde{b}} : J^{\tilde{a}} J^{\tilde{b}} : - \frac{1}{2(k+c_G)} g_{ab} : (J^a + J^a_{\frac{G}{H}})(J^b + J^b_{\frac{G}{H}}) :$$
$$+ \frac{\sqrt{2}}{k+C_G}(\vec{\rho}_G - \vec{\rho}_H) \cdot \partial(\vec{J} + \vec{J}_{\frac{G}{H}}) + \sum_{\alpha \in \frac{G}{H}} \rho^{+\alpha}(\partial \chi)^{-\alpha} ,$$

where $\tilde{a}$ and $\tilde{b}$ go over the adjoint of G. $\vec{J}, \vec{I}$ and $\vec{J}_{\frac{G}{H}}$ are the Cartan-subalgebra currents given in the basis in which $[J^i_n, J^j_m] = kn\delta^{ij}\delta_{m+n}$. The total Virasoro central charge is found to be

$$c = \frac{kd_G}{k+C_G} + \sum_{I=1}^{n} \frac{(k+C_G+C_{HI})d_{HI}}{k+C_G} + r - 2d_H - (d_G - d_H)$$
$$+ 6\left[\sqrt{\frac{2}{k+C_G}}(\vec{\rho}_G - \vec{\rho}_H)\right]^2 \tag{12}$$
$$= 0 ,$$

where we have used, assuming G is a simply laced group, the relations $12\rho^2_G = d_G C_G$, and $\vec{\rho}_H \cdot (\vec{\rho}_G - \vec{\rho}_H) = 0$.

Upon gauge fixing, the gauge invariance is transformed into a BRST symmetry generated by a dimension one current $J^{(\text{BRST})}$. This current has a dimension two partner G. These two anti-commuting currents are given by

$$J^{(\text{BRST})} = g_{ab}\chi^a[J^b + I^b + J^b_{\frac{G}{H}} + \frac{1}{2}J^{(gh)b}_H]$$
$$= g_{ab}\chi^a[J^b + I^b + i\frac{1}{2}f^b_{cd}\rho^c\chi^d + if^b_{\gamma,-\beta}\rho^\gamma\chi^{-\beta}] , \tag{13}$$
$$G^H = \frac{g_{ab}}{2(k+c_G)}\rho^a[J^b - I^b + if^b_{\gamma,-\beta}\rho^\gamma\chi^{-\beta}] - \frac{\sqrt{2}}{k+C_G}(\vec{\rho}_G - \vec{\rho}_H) \cdot \partial\vec{\rho} .$$

It is straightforward to realize that $T^H(z)$ is BRST exact

$$T^H(z) = \{Q^{(\text{BRST})}, G^H(z)\} \tag{14}$$

where $Q^{(\text{BRST})} = \oint dz J^{(\text{BRST})}$. As was shown in ref. [13] the addition of the coset ghosts turned the model into a twisted $N = 2$ model. The twisted $N = 2$ algebra is generated by $Q^{\frac{G}{H}}$ and $G^{\frac{G}{H}}$ given by their $N = 2$ counterparts[18] with the fermions replaced with ghosts,

$$Q^{\frac{G}{H}} = \sum_{\alpha\beta\gamma \in \frac{G}{H}} \chi^{-\alpha}(J^\alpha + i\frac{1}{2}f^\alpha_{\gamma,-\beta}\rho^\gamma\chi^{-\beta}) ,$$
$$G^{\frac{G}{H}} = \frac{1}{k+C_G} \sum_{\alpha\beta\gamma \in \frac{G}{H}} \rho^\alpha(J^{-\alpha} + i\frac{1}{2}f^{-\alpha}_{\gamma,-\beta}\rho^\gamma\chi^{-\beta}) . \tag{15}$$

$T^{\frac{G}{H}}$ defined above is exact with respect to $Q^{\frac{G}{H}}$

$$T^{\frac{G}{H}} = \{Q^{\frac{G}{H}}, G^{\frac{G}{H}}\} . \tag{16}$$

The various Q's and G's obey the following anti-commutation relations:

$$\{Q^{\frac{G}{H}}, Q^{(\text{BRST})}\} = \{Q^{\frac{G}{H}}, G^H\} = \{Q^{(\text{BRST})}, G^{\frac{G}{H}}\} = 0 ,$$
$$\{Q^{(\text{BRST})}, Q^{(\text{BRST})}\} = \{Q^{\frac{G}{H}}, Q^{\frac{G}{H}}\} = \{G^{\frac{G}{H}}, G^{\frac{G}{H}}\} = 0 , \tag{17}$$
$$\{G^H, G^H\} = \frac{1}{4(k+C_G)^2} f_{abc}\rho^a\rho^b J^{(\text{tot})c} .$$

The fact that G^H is not nilpotent is shared by several other TCFT's, in particular the $\frac{G}{G}$ models. Defining now the combined generators

$$Q = Q^{(\text{BRST})} + Q^{\frac{G}{H}}, \qquad G = G^H + G^{\frac{G}{H}}, \tag{18}$$

we find that (for rank G = rank H) the following relations hold

$$\begin{aligned} T(z) &= \{Q, G(z)\}, \\ J^{(\text{BRST})} &= \{Q, j^{\#}(z)\}, \\ J^{(\text{tot})a} &= \{Q, \rho^a\}, \end{aligned} \tag{19}$$

where $J^{(\text{tot})a}$ denotes a current in the algebra of H and

$$j^{\#}(z) = g_{ab}\rho^a \chi^b + \sum_{\alpha \in \frac{G}{H}} \rho^{+\alpha} \chi^{-\alpha}.$$

Hence, from the algebraic structure one finds that the twisted $\frac{G}{H}$ models are indeed TCFT's provided that rank G = rank H.

4. BRST COHOMOLOGY AND PHYSICAL STATES

Next we proceed to extract the space of physical states of the model. We take as our definition of a physical state a state in the cohomology of $Q = Q^{(\text{BRST})} + Q^{\frac{G}{H}}$, namely, $|\text{phys}\rangle \in H^*(Q)$. In the case that the spectral sequence[20] of the double complex of $Q^{(\text{BRST})}$ and $Q^{\frac{G}{H}}$ degenerates at the E_2 term, this is the same as taking one cohomology and then the other. The extraction of the physical states was worked out in detail in refs. 10,11,12. The procedure for the twisted $\frac{G}{H}$ is very much like the one of the $\frac{G}{G}$ models. Hence, we summarize here the analysis of the $A_1^{(1)}$ $\frac{G}{G}$ model. Expanding the currents J^a, I^a and the $(1,0)$ ghost fields ρ^a, χ^a in modes and inserting them into eqn. (13) we obtain the following BRST charge

$$Q = \sum_{n=-\infty}^{\infty} \left[g_{ab} \chi_n^a (J_{-n}^b + I_{-n}^b) - \tfrac{1}{2} f_{abc} \sum_{m=-\infty}^{\infty} : \chi_{-n}^a \chi_{-m}^b \rho_{n+m}^c : \right], \tag{20}$$

where : : denotes normal ordering namely putting modes with negative subscripts to the left of those with positive ones and ρ_0^a to the right of χ_0^a. Since both $J^{(\text{tot})a}_n$ and L_n are Q exact namely

$$\{Q, \rho_n^a\} = J^{(\text{tot})a}_n, \qquad \{Q, G_n\} = L_n, \tag{21}$$

it follows that

$$L_0|\text{phys}\rangle = 0, \qquad J^{(\text{tot})0}_0|\text{phys}\rangle = 0. \tag{22}$$

For non-vanishing eigenvalues of L_0 and $J^{(\text{tot})0}_0$ it is easy to see that $|\text{phys}\rangle$ is in the image of Q which cannot be true for a non-trivial $|\text{phys}\rangle \in H^*(Q)$.

Let us now select a sub-space $\mathcal{F}(J, I)$ of the space of physical states on which $\rho_0^0 = 0$ in addition to $J^{(\text{tot})0}_0 = L_0 = 0$. On this sub-space Q which may be written as

$$Q = \chi_0^0 J^{(tot)0}{}_0 + M\rho_0^0 + \hat{Q} ,$$
$$M = -\tfrac{1}{2}f_{0bc} \sum_{n\neq 0} :\chi_{-n}^b \chi_n^c: - \tfrac{1}{2}f_{0bc} :\chi_0^b \chi_0^c: , \quad (23)$$

equals $\hat{Q}$. We thus start by deducing $H^*(\hat{Q})$ the cohomology of $\hat{Q}$. The states which correspond to the latter are built on a highest weight state vacuum $|J,I\rangle$. A convenient way to handle the J^a and I^a currents is to invoke the following "bosonization"[21]

$$J_n^+ = \beta_n ,$$
$$J_n^0 = \sum_m :\beta_m \gamma_{n-m}: + \frac{a}{\sqrt{2}}\phi_n , \quad (24)$$
$$J_n^- = -\sum_{k,m} :\gamma_m \gamma_k \beta_{n-m-k}: -\sqrt{2}a\sum_m \phi_m\gamma_{n-m} + kn\gamma_n ,$$

where $a^2 = k+2$. The fields β and γ form a bosonic $(1,0)$ system with $[\gamma_m, \beta_n] = \delta_{m+n}$. The modes ϕ_n correspond to the dimension one operator $i\partial\phi$ and they obey $[\phi_m, \phi_n] = m\delta_{m+n}$. In the I sector we use an inverted bosonization $I_n^0 \leftrightarrow -I_n^0$, $I_n^+ \leftrightarrow I_n^-$, expressed by operators with tildes $\tilde{I}_n^- = \tilde{b}_n$ etc, and also take $k \to -k-4$. It is easy to realize that the highest weight conditions are obeyed only provided $\beta_0|J,I\rangle = \tilde{\gamma}_0|J,I\rangle = 0$. The normal ordering, however, is with respect to the usual $SL(2,R)$ invariant vacuum $\beta_0|J,I\rangle = \tilde{\beta}_0|J,I\rangle = 0$. The next step following ref. 22 is to assign a degree to the various fields. The idea is to decompose $\hat{Q}$ into terms of different degrees in such a way that there is a nilpotent operator that carries the lowest degree which is zero. An assignment that obeys this requirement is the following

$$\deg(\chi) = \deg(\gamma) = \deg(\tilde{\gamma}) = \deg(\phi^+) = 1 ,$$
$$\deg(\rho) = \deg(\beta) = \deg(\tilde{\beta}) = \deg(\phi^-) = -1 . \quad (25)$$

The spectral sequence decomposition of $\hat{Q}$ takes the form

$$\hat{Q} = Q^{(0)} + Q^{(1)} + Q^{(2)} + Q^{(3)} ,$$
$$Q^{(0)} = \sum_n \chi_{-n}^- \beta_n + 2a \sum_{n\neq 0} \chi_{-n}^0 \phi_n^- + \sum_n \chi_{-n}^+ \tilde{\beta}_n . \quad (26)$$

where $\phi_n^\pm = \frac{1}{\sqrt{2}}(\phi_n \pm i\tilde{\phi}_n)$. $Q^{(0)}$ is nilpotent on the entire Fock space.

We proceed now to compute $H^*(Q^{(0)})$, since, if there is a finite number of degrees for each ghost number, according to the Lemmas of ref. 22, $H^*(Q^{(0)})$ is isomorphic to the cohomology of $\hat{Q}$. Recall that states in $\mathcal{F}(J,I)$ are annihilated by

$$\mathcal{L}_0 = \hat{L}_0 + \frac{1}{k+2}[J(J+1) - I(I+1)] ,$$
$$J^{(tot)0}{}_0 = J + I + 1 + \hat{J}^{(tot)0}{}_0 . \quad (27)$$

It is easy to check that $\hat{L}_0$ and $\hat{J}^{(tot)0}{}_0$ are exact under $Q^{(0)}$ and therefore both of them annihilate the states in the cohomology of $Q^{(0)}$ on $\mathcal{F}(J,I)$. Hence, there are no excitations in $H^*(Q^0)$. Moreover, since $\mathcal{L}_0 = 0$, states in the latter must have either

$I = J$ or $I = -(J+1)$. Let us now extract the zero modes contributions to the cohomology. The general structure of these states is

$$|n_\gamma, n_{\tilde\beta}, n_+, n_-\rangle = (\gamma_0)^{n_\gamma}(\tilde\beta_0)^{n_{\tilde\beta}}(\chi_0^+)^{n_+}(\chi_0^-)^{n_-}|I, J\rangle \qquad (28)$$

where obviously $n_+, n_- = 0, 1$ and $n_\gamma, n_{\tilde\beta}$ are non-negative integers. It is now straightforward to deduce the Kernel and the Image of $Q^{(0)}$. It turns out that the only possible state in the relative cohomology of $Q^{(0)}$ is $\chi_0^+|I, J\rangle$, and from the condition $J^{(\text{tot})0}_0 = 0$ we find a state only provided that $I = -J - 1$ and then

$$H^{\text{rel}}(Q^{(0)}) = \{\chi_0^+|-(J+1), J\rangle\} . \qquad (29)$$

The passage from the relative cohomology to the absolute one is then given by

$$H^{\text{abs}}(Q^{(0)}) \simeq H^{\text{rel}}(Q^{(0)}) \oplus \chi_0^0 H^{\text{rel}}(Q^{(0)}) , \qquad (30)$$

in the same way as in ref. 22. We conclude that the cohomology of Q on the full Fock space includes states of arbitrary J with a corresponding $I = -(J+1)$ and with ghost number $G = 0, 1$, where we have shifted the definition of the ghost-number so that the state $\chi_0^+|I, J\rangle$ is at $G = 0$. There is only one state at each ghost number.

For the general case of a twisted $\frac{G}{H}$ the generalization of the bosonization of eqn. (24) is performed in terms of the scalars $\phi_i, i = 1, \ldots, \text{rank } G$ and the commuting $(1, 0)$ systems $(\beta\alpha, \gamma\alpha)$ $(\alpha > 0)$ $(i \leq j)^{11}$. Following a similar derivation as that of the $SL(2, R)$ case one finds[24,22] the corresponding relative cohomology

$$H^{\text{rel}}(Q) = \{ \prod_{\alpha \in H, \alpha > 0} \chi_0^\alpha |\vec{J}, \vec{I}\rangle; \; \vec{J} + \vec{I} + 2\vec{\rho}_H = 0 \} , \qquad (31)$$

where $\vec{\rho}_H$ is half the sum of the positive roots of H. Since the scalars ϕ_i in the J sector have a background charge of $\frac{i}{\sqrt{k+C_G}} \vec{\rho}_G \cdot \partial^2 \vec{\phi}$ while those in the I sector have a background charge of $\frac{i}{\sqrt{k+C_G}}(2\vec{\rho}_H - \vec{\rho}_G) \cdot \partial^2 \vec{\mathcal{H}}$, we find that the weight of this state is $L_0 = \frac{1}{k+c_G}[\vec{J} \cdot (\vec{J} + 2\vec{\rho}_G) - \vec{I} \cdot (\vec{I} + 2(2\vec{\rho}_H - \vec{\rho}_G))] = 0$ for $\vec{J} + \vec{I} + 2\vec{\rho}_H = 0$. This state corresponds to one of the "tachyon" states of the W_N models based on $G = SU(N)^{25}$. The absolute cohomology (without the restriction $\rho_0^i = 0$) is

$$H^{\text{abs}}(Q) \simeq H^{\text{rel}}(Q) \oplus \sum_{\{k_1,\ldots,k_l\}} \chi_0^{k_1} \ldots \chi_0^{k_l} H^{\text{rel}}(Q) \qquad (32)$$

where the sum is over $\{k_1, .., k_l\}$, which are all possible subsets of the set $\{1, .., \text{rank } G\}$. Thus, each state in the relative cohomology gives rise to $2^{\text{rank } G}$ states in the absolute cohomology.

5. IRREDUCIBLE REPRESENTATION AND THE BRST COHOMOLOGY

So far we have analyzed the cohomology on the whole Fock Space. The next step

5. IRREDUCIBLE REPRESENTATION AND THE BRST COHOMOLOGY

So far we have analyzed the cohomology on the whole Fock Space. The next step in the extraction of the physical states is to pass from the cohomology on the Fock space to the irreducible representations of the level k Kac-Moody algebra in the J sector. To simplify the discussion we continue to address the $A_1^{(1)} \frac{G}{G}$ model. In general a representation with highest weight L is reducible iff $2L+1 = r - (s-1)(k+2)$ where r and s are integers with either $r, s \geq 1$ or $r < 0, s \leq 0$[26]. In the $\frac{G}{G}$ model with $G = A_1^{(1)}$ we therefore have reducible representation for $J_{r,s}$ and $I_{r,s}$ where

$$2J_{r,s} + 1 = r - (s-1)(k+2), \qquad 2I_{r,s} + 1 = r + (s-1)(k+2). \qquad (33)$$

Note that $J_{r,s} = -I_{-r,s} - 1$. Completely irreducible representations, which have infinitely many null vectors, appear provided that $k + 2 = \frac{p}{q}$ for p and q positive integers which can be chosen with no common divisor. In this case $I_{r,s} = I_{r+p,s-q}$ and $J_{r,s} = J_{r+p,s+q}$. It is, thus, enough to analyze the domain $1 \leq s \leq q$, and we will choose $1 \leq r \leq p-1$. This choice corresponds to the double line embedding diagram[27]. The states corresponding to $r = p$ have a single line embedding diagram. It was found[27] that for the case of the double line one can construct the irreducible representation which is contained in $\mathcal{F}_{r,s}$, the Fock space built on $|J_{r,s}\rangle$. This is achieved via the cohomology of an operator Q_J which acts on $\mathcal{F}_{r,s}$ the union of the Fock spaces that correspond to $J_{r+2lp,s}$ and $J_{-r+2lp,s}$ for every integer l. It turns out[27] that the relevant information is encoded in $H^0(\mathcal{F}_{r,s}, Q_J)$ and all other levels of the cohomology vanish. The cohomology is only in the J sector and not in the I sector just as there is no use of the cohomology of the Liouville sector in models of $c < 1$ matter coupled to gravity[22]. Thus, the space of physical states of ghost number n is given by

$$H^{(n)}_{rel}[H^{(0)}(\mathcal{F}_{r,s}, Q_J) \times \mathcal{F}_I \times \mathcal{F}^G, Q] \qquad (34)$$

where $\mathcal{F}^G$ is the ghosts' Fock space built on the new vacuum $|0\rangle_G$. Since Q_J acts only in the J sector we can rewrite $H^{(n)}_{rel}$ as

$$H^{(n)}_{rel}[H^{(0)}(\mathcal{F}_{r,s} \times \mathcal{F}_I \times \mathcal{F}^G, Q_J), Q]. \qquad (35)$$

Moreover, since $\{Q, Q_J\} = 0$ one can use theorems[20] about double cohomologies and write this as isomorphic to

$$H^{(n)}[H^{(0)}_{rel}(\mathcal{F}_{r,s} \times \mathcal{F}_I \times \mathcal{F}^G, Q), Q_J]. \qquad (36)$$

The theorems[20] apply only provided that each cohomology separately is different from zero only for one single degree as was shown above. In fact, we have already calculated $H^{(0)}_{rel}(\mathcal{F}_{r,s} \times \mathcal{F}_I \times \mathcal{F}^G, Q)$ since $\mathcal{F}_{r,s}$ is the union of free Fock spaces. Hence the result is that the latter has one state if the Fock space of $J = -I - 1$ is in $\mathcal{F}_{r,s}$, and it is empty otherwise. For each $J_{r,s}$ we get states at $I = -J_{r+2lp,s} - 1$ and $I = -J_{-r+2lp,s} - 1$ where their ghost number is equal to the corresponding degree in the complex of ref. 27. For each such $J_{r,s}$ there is an infinite set of states with $I = I_{-r-2lp,s}$, $G = -2l$ and $I = I_{r-2lp,s}$ $G = 1 - 2l$ for every integer l.

Following the same steps for $G = SL(N,R)$ one finds for each maximal weight J of G_k a rank G dimensional vector of states. This implies also that there is an rank $G - 1$ dimensional lattice of states for each ghost number and J. The latter situation follows from the $N - 1$ dimensional lattice of Fock spaces which are derived by Weyl reflections as well as shifts by linear combinations of roots[23]. The physical states in the $G = SL(2,R)$ model are characterized by $\hat{L}_0$ and $\hat{J}^{(tot)0}_0$ as follows

$$
\begin{array}{ll}
J = J_{r,s}, \quad I = I_{-r-2lp,s}, & J = J_{r,s}, \quad I = I_{r-2lp,s}, \\
G = -2l, & G = 1 - 2l, \\
\hat{L}_0 = l^2 pq + l(qr - sp) + lp, & \hat{L}_0 = l^2 pq - l(qr + sp) + r(s-1) + lp, \\
\hat{J}^{(tot)0}_0 = lp, & \hat{J}^{(tot)0}_0 = lp - r,
\end{array}
\qquad (37)
$$

vanishes and our ghost vacuum has $J_0 = 1$. For integer k we have $J = 0, \ldots, \frac{k}{2}$. Let us now examine the index interpretation of the torus partition function[16]. We want to check now whether it can be rewritten as a trace over the space of the physical states. One has to insert the values of $\hat{L}_0$ and $\hat{J}^0_{(tot)}$ of eqn.(37) into $\text{Tr}[(-)^G q^{\hat{L}_0} e^{i\pi\theta \hat{J}^0_{(tot)}}]$, with $\hat{J}^0_{(tot)}$ and G shifted to the values defined for an $SL(2)$ invariant vacuum. Inserting these values for every l, and adding the values of the level and the eigenvalue of $J^{(tot)0}$ of the $|J\rangle$ vacuum, namely $\hat{L}_0 \to \hat{L}_0 + \frac{J(J+1)}{k+2}$ and $\hat{J}^{(tot)0} \to \hat{J}^{(tot)0} + J$, one finds

$$\text{Tr}[(-)^G q^{\hat{L}_0} e^{i\pi\theta \hat{J}^0_{(tot)}}] = 2iq^{-\frac{1}{4(k+2)}} e^{-i\pi\frac{\theta}{2}} M_{k,J}(\tau,\theta) . \qquad (38)$$

where

$$M_{k,j}(\tau,\theta) = \sum_{l=-\infty}^{\infty} q^{(k+2)(l+\frac{j+\frac{1}{2}}{(k+2)})^2} \sin\{\pi\theta[(k+2)l + j + \tfrac{1}{2}]\} \qquad (39)$$

is the numerator of the character which corresponds to the highest weight state J. We have, thus, rederived using the BRST cohomology and the index of above, the path integral results of ref. 16, for the partition function.

6. COMPARISON WITH GRAVITATIONAL MODELS

The main motivation to analyse the twisted $\frac{G}{H}$ models was the idea that they will be found to be equivalent to certain gravitational models. More specifically, we expect a correspondence between the $A^{(1)}_{N-1}$ twisted $\frac{G}{H}$ models and W_N strings and, in particular, between the case of $G = SL(2)$ and minimal models coupled to gravity. Let us now check whether one can map the topological coset models into string models. In fact, for reasons that will be clarified shortly, it is clear that the comparison to the gravitational models should be done with the topological coset models only after twisting their energy-momentum tensor. For $G = SL(N,R)$ the latter is given by

$$T(z) \to \tilde{T}(z) = T(z) + \sum_{i=1}^{N-1} \partial J^{(tot)i}(z) . \qquad (40)$$

Obviously since $T(z)$ and $\partial J^{(tot)i}(z)$ are BRST exact so is $\tilde{T}(z)$. Thus, the total Virasoro anomaly is unchanged. However, the contribution of each sector to c is modified as follows

$$c_J \to \tilde{c}_J = c_J - d_G C_G k, \quad c_{H(I)} \to \tilde{c}_{H(I)} = c_{H(I)} + d_{H(I)} C_{H(I)}(k + C_G + C_{H(I)}), \quad (41)$$

and the shift in the ghost contribution can be found from a similar expression or from the fact that the sum of the shifts vanishes. We consider here for simplicity the case of $G = SL(N,R)$. The twisted ghost sector includes the ghosts of a W_N gravity, namely, a sequence of ghosts with dimensions $(i, 1-i)$ for $i = 2, \ldots, N$ contributing $\tilde{c}_{Wgh} = -2(N-1)[(N+1)^2 + N^2]$ to $\tilde{c}$. The rest of the ghosts are paired with commuting fields of the same conformal structure coming from the J and I sectors. For $N = 2$ one finds $\tilde{c}_{Wgh} = -26 - 2\#_{\text{pairs}}$ where there are two pairs in the $\frac{G}{G}$ model and one in the $\frac{SL(2)}{U(1)}$ case. The net matter degrees of freedom have the following Virasoro anomaly $c = \tilde{c}_J - \frac{1}{2}[\tilde{c}_{(gh)} - \tilde{c}_{Wgh}] = (N-1)[(2N^2 + 2N + 1) - N(N+1)(t + \frac{1}{t})]$ which is exactly that of a (p,q) minimal W_N matter sector[28] provided $t \equiv k + N = \frac{p}{q}$. This was explicitly verified by analyzing the dimensions and contributions to $\tilde{c}$ of the various free fields in the J sector[11]. The expression for c reduces to that of the p, q minimal model (plus 2) for $N = 2$. Before twisting the contribution of the set of all ϕ^i to c is $c_\phi = (N-1) - \frac{12 \sum_{i,j} g^{ij}}{k+N} = -(N-1)[\frac{(N^2+N)}{t} - 1]$. Due to the twisting the $(\beta(ij), \gamma(ij))^\dagger$ systems acquire dimensions of $(i-j, j-i+1)$ and thus there are $N-1$ systems of dimension $(0, 1)$, $N-2$ pairs of fields of dimension $(-1, 2)$ up to one pair of dimensions $(2-N, N-1)$. The ϕ central charge is modified to

$$\tilde{c}_\phi = (N-1)\left[(2N^2 + 2N + 1) - N(N+1)(t + \frac{1}{t})\right] \quad (42)$$

which is identical to the net matter contribution to c given above and hence the ϕ^i fields are in fact those of the W_N model. A further indication of the latter equivalence is the dimensions of the ϕ fields which correspond to the maximal weights $\lambda^j = \sum_k g^{jk}[(r_k - 1) - t(s_k - 1)]$ which after the twisting are

$$\Delta_{r_1, \ldots, r_{N-1}, s_1, \ldots, s_{N-1}} = \frac{12 \sum_{i,j} g^{ij}(ps_i - qr_i)(ps_j - qr_j) - N(N^2 - 1)(p-q)^2}{24pq} \quad (43)$$

as in the W_N minimal models[28]. If one parametrizes the I sector in the same way as the J sector, then clearly the modified dimensions of the $(\tilde{\beta}(ij), \tilde{\gamma}(ij))$ fields are the same as those without tilde. From the point of view of their contribution to $\tilde{c}$ the ϕ^i fields are then identical to those of W_N gravity. This is achieved by replacing t with $-t$ in $\tilde{c}_\phi$ defined above. With respect to the untwisted T the ghosts (ρ^a, χ^a) are all of dimension $(1, 0)$. The ghost part of $\tilde{T}(z)$ has the form

$$\tilde{T}^{(gh)}(z) = g_{ab} : \rho^a \partial \chi^b :$$
$$+ \sum_{1 \le i \le j \le N-1} (j - i + 1)\partial[: \rho^{-(ij)}(z)\chi^{+(ij)}(z) : - : \rho^{+(ij)}(z)\chi^{-(ij)}(z) :]. \quad (44)$$

It is thus obvious that the members of the Cartan sub-algebra ρ^i, χ^i remain $(1, 0)$ fields. On the other hand the pair $(\chi^{+(ij)}, \rho^{-(ij)})$ carries now dimensions $(i - j - 1, 2 + j - i)$

---

† For the definition of these fields see ref. 11.

and $(\chi^{-(ij)}, \rho^{+(ij)})$ carry dimensions $(j-i+1, i-j)$. Altogether one finds for the (χ, ρ) ghosts $N-1$ pairs of fields of dimension $(0,1)$ coming from the Cartan subalgebra, $N-1$ pairs of dimension $(-1,2)$ $N-2$ pairs of dimension $(-2,3)$ up to a pair of dimension $(1-N, N)$, and similarly $N-1$ pairs of dimension $(1,0)$ up to one pair of $(N-1, 2-N)$. It is now clear that when the dust settles the $\frac{G}{G}$ model of $SL(N,R)$ at level $k = \frac{p}{q} - N$ has the field content of a minimal W_N (p,q) model coupled to W_N gravity plus pairs of "topological sectors" namely pairs of commuting and anti-commuting $(i, 1-i)$ systems for $i = 1, \ldots, N-1$.

Next we want to compare the partition function of the (p,q) model to that of $\frac{G}{G}$ for $G = SL(2)$ and $k = \frac{p}{q} - 2$. Comparing eqn. (38) to the numerator of the character of the minimal model it is clear that correspondence might be achieved only provided one takes $\tau = -\frac{1}{2}\theta$. Recall that in the topological coset models we integrate in the path-integral only over θ (and not over τ) and the result is τ independent[16]. In this case the numerator of the character in the minimal model which is proportional to $\text{Tr}[(-1)^G q^{\hat{L}_0}]$ is mapped into $\text{Tr}[(-1)^G u^{\hat{L}_0 - \hat{J}^{(\text{tot})0}_0}]$ in the $\frac{G}{G}$ model. The integration over the moduli parameter of the torus is therefore replaced by the integration over the moduli of flat gauge connection. We thus need to compare the number of states at a given level and ghost number in the minimal models with the corresponding numbers at the same ghost number and "twisted level". From eqn. (37) we read

$$I = I_{-r-2lp,s} , \; G = -2l , \quad \hat{L}_0 - \hat{J}^{(\text{tot})0}_0 = l^2 pq + l(qr - sp) .$$
$$I = I_{r-2lp,s} , \; G = 1-2l , \quad \hat{L}_0 - \hat{J}^{(\text{tot})0}_0 = l^2 pq - l(qr + sp) + rs . \quad (45)$$

In the minimal models we have states built on vacua labeled by the pair r, s with $1 \leq r \leq p-1$ and $1 \leq s \leq q-1$ with $ps > qr$ which have dimension $h_{r,s} = \frac{(qr-sp)^2-(p-q)^2}{4pq}$. The levels of the excitations are $\hat{L}_0 = \Delta - h_{r,s}$. For $G = 2l+1$ one has $\Delta = A(l) = \frac{[(2pql+qr+sp)^2-(q-p)^2]}{4pq}$ and for $G = 2l$, $\Delta = B(l) = \frac{[(2pql-qr+sp)^2-(q-p)^2]}{4pq}$ [29,30]. Hence, the contribution of the various levels to the partition function are identical to those of $\hat{L}_0 - \hat{J}^{(\text{tot})0}_0$ in eqn. (45) for the same ghost numbers and the respective vacua satisfy $J = \sqrt{\frac{p}{2q}} p_m$ and $I = -\sqrt{\frac{p}{2q}} p_L$ where p_m and p_L are the matter and Liouville momenta respectively. It is thus clear that for a given r, s we get the same number of states with the same ghost number parity in the two models and thus the two partition functions on the torus are in fact identical. To obtain the partition function of the (p,q) models coupled to 2d gravity we have restricted r and s as follows $1 \leq s \leq q-1$ and $1 \leq r \leq p-1$. It is interesting to note that we could include in the sum over r and s which appears in the partition function also the terms with $r = p$. Those terms arise from the states which have a single line as their embedding diagram. The $r = p$ terms cancel between themselves and do not change the result for the partition function.

7. DISCUSSION

The description of gravitational (W_N gravity) models via topological coset models is far from being complete. A comparison of the correlators of the latter with those of other formulations of 2d gravity (W_N gravity) is still missing. Another challenge is to extract information on the behaviour of the models on higher genus Riemann surfaces. The differences between models which correspond to various H for a given

G are only partially understood. Possible flows between these models and in general "renormalization" flows in the space of TCFT's are under current study. The precise relations between the present formulation and the known results on chiral rings of $N = 2^{31}$ theories deserve further investigation.

REFERENCES

1. E. Witten, Topological quantum field theory, *Comm. Math. Phys.* **117** (1988) 353.
2. R. Dijkgraaf, E. Verlinde, and H. Verlinde, Topological strings in $d < 1$, *Nucl. Phys.* **B352** (1991) 59; Notes on topological string theory and $2d$ quantum gravity, Princeton preprint PUPT-1217 (1990), Trieste Spring School 1990, pp. 91–156.
3. J. Sonnenschein and S. Yankielowicz, Novel symmetries of topological conformal field theory, TAUP-1898-91 August 1991.
4. E. Witten, Quantum field theory and the Jones polynomial, *Comm. Math. Phys.* **121** (1989) 351.
5. D. Montano, J. Sonnenschein, The topology of moduli space and quantum field theory, *Nucl. Phys.* **B324** (1989) 348, J. Sonnenschein, Anyonic superconductivity, *Phys. Rev.* **D42** (1990) 2080.
6. F. David, Conformal field theories coupled to 2-d gravity in the conformal gauge, *Mod. Phys. Lett.* **A3** (1988) 1651; J. Distler and H. Kawai, Conformal field theory and 2-d quantum gravity or who's afraid of Joseph Liouville?, *Nucl. Phys.* **B321** (1989) 509.
7. A. M. Polyakov, Quantum gravity in two-dimensions, *Mod. Phys. Lett.* **A2** (1987) 893.
8. D. Gross and A. A. Migdal, Nonperturbative two-dimensional quantum gravity, *Phys. Rev. Lett.* **64** (1990) 127; M. Douglas and S. Shenker, Strings in less than one dimension, *Nucl. Phys.* **B335** (1990) 635; E. Brezin and V. A. Kazakov, Exactly solvable field theories of closed strings, *Phys. Lett.* **236B** (1990) 144.
9. M. Kontsevich, *Comm. Math. Phys.* **147** (1992) 1.
10. O. Aharony, O. Ganor, N. Sochen, J. Sonnenschein and S. Yankielowicz, Physical states in $\frac{G}{G}$ models and two dimensional gravity, Tel Aviv preprint TAUP-1947-92 April 1992.
11. O. Aharony, J. Sonnenschein and S. Yankielowicz, $\frac{G}{G}$ models and W_N strings, *Phys. Lett.* **289B** (1992) 309.
12. O. Aharony, O. Ganor, J. Sonnenschein and S. Yankielowicz, On the twisted $\frac{G}{H}$ topological models, Tel Aviv preprint TAUP-1990-92 August 1992.
13. E. Witten, On Holomorphic Factorization of WZW and Coset Models, *Comm. Math. Phys.* **144** (1992) 189.
14. M. Spiegelglas and S. Yankielowicz, $\frac{G}{G}$ Topological Field Theories by cosetting G_k, *Nucl. Phys.* **B393** (1993) 301; Fusion Rules As Amplitudes in G/G Theories, Technion PH-35-90.
15. K. Bardacki, E. Rabinovici, and B. Saring, String models with $c < 1$ components, *Nucl. Phys.* **B299** (1988) 151.
16. K. Gawedzki and A. Kupiainen, *Phys. Lett.* **215B** (1988) 119, Coset construction from functional integrals, *Nucl. Phys.* **B320** (1989) 649.
17. D. Karabali and H.J. Schnitzer, BRST quantization of the gauged W-Z-W action and coset conformal field theories, *Nucl. Phys.* **B329** (1990) 625.

18. Y. Kazama and H. Suzuki, New $N=2$ superconformal field theories and superstring compactification, *Nucl. Phys.* **B321** (1989) 232.
19. A. Polyakov and P. B. Wiegmann, Theory of nonabelian Goldstone bosons, *Phys. Lett.* **131B** (1983) 121.
20. R. Bott and L. W. Tu, "Differential Forms in Algebraic Topology", Springer-Verlag, New York, 1982.
21. M. Wakimoto, *Comm. Math. Phys.* **104** (1989) 605.
22. P. Bouwknegt, J. McCarthy and K. Pilch, Fock space resolutions of the Virasoro highest weight modules with $c \leq 1$, *Comm. Math. Phys.* **145** (1992) 541.
23. P. Bouwknegt, J. McCarthy and K. Pilch, Free field realizations of WZNW models: BRST complex and its quantum group structure, *Phys. Lett.* **234B** (1990) 297; Quantum group structure in the Fock space resolutions of $sl(n)$ representations, *Comm. Math. Phys.* **131** (1990) 125.
24. B. Feigin and E. Frenkel, Quantization of the Drinfeld-Sokolov reduction, *Phys. Lett.* **246B** (1990) 75.
25. M. Bershadsky, W. Lerche, D. Nemeschansky and N. P. Warner, A BRST operator for non-critical W strings, *Phys. Lett.* **292B** (1992) 35.
26. V. G. Kac and D. A. Kazhdan *Adv. Math* **34** (1979) 79.
27. D. Bernard and G. Felder, Fock representations and BRST cohomology in $sl(2)$ current algebra, *Comm. Math. Phys.* **127** (1991) 145.
28. V.A. Fateev and S.L. Lukyanov, *Int. Jour. of Mod. Phys.***A31** (1988) 507.
29. M. Bershadsky and I. Klebanov, Partition functions and physical states in two-dimensional quantum gravity and supergravity, *Nucl. Phys.* **B360** (1991) 559.
30. B. Lian and G. Zuckerman, New selection rules and physical states in 2-D gravity: conformal gauge, *Phys. Lett.* **254B** (1991) 417.
31. D. Gepner. *Comm. Math. Phys.* **141** (1990) 381.

FINITE W SYMMETRY IN FINITE DIMENSIONAL INTEGRABLE SYSTEMS

T. Tjin

Institute for Theoretical Physics
University of Amsterdam
Valckenierstraat 65
1018 XE Amsterdam
The Netherlands

ABSTRACT

By generalizing the Drinfeld-Sokolov reduction a large class of W algebras can be constructed. We introduce 'finite' versions of these algebras by Poisson reducing Kirillov Poisson structures on simple Lie algebras. A closed and coordinate free formula for the reduced Poisson structure is given. These finitely generated algebras play the same role in the theory of W algebras as the simple Lie algebras in the theory of Kac-Moody algebras and will therefore presumably play an important role in the representation theory of W algebras. We give an example leading to a quadratic sl_2 algebra. The finite dimensional unitary representations of this algebra are discussed and it is shown that they have Fock realizations. It is also shown that finite dimensional generalized Toda theories are reductions of a system describing a free particle on a group manifold. These finite Toda systems have the non-linear finite W symmetry discussed above.

INTRODUCTION

Nonlinear extensions of the Virasoro algebra, generally known as W algebras, have turned up in various areas of mathematical physics (see ref. 1 for a recent review). Unfortunately not too much is known about them, their interpretation, classification and representation theory is still far from complete. In this paper we discuss a very simple class of nonlinear algebras, which are basically 'finite' W algebras and which may serve as instructive playground for the infinite case. However as we shall see they do have some interest of their own.

In ref. 2, W_n algebras were shown to arise as Dirac bracket algebras on submanifolds of Kac-Moody algebras. This gave a clear understanding of how nonlinear W algebras can arise as Poisson reductions of linear current algebras. However, W_n algebras are certainly not the only W algebras known in the literature, which leads one to ask whether others could be constructed in a similar way. The answer to this question turned out to be yes as it was shown in ref. 3 that there are as many different Poisson reductions of an sl_n current algebra leading to W algebras as there are partitions of the number n. The reduction point of view has recently been investigated by a great number of people (for example refs. 2–10, 14, 12, 13). The classical covariant W gravity theories for all the algebras which can be constructed as reductions of Kac-Moody algebras as well as their moduli spaces have been constructed in 15.

Having the picture of W algebras as reductions of Kac-Moody algebras in mind one can ask oneself the question whether this is special for infinitely many dimensions, i.e. can one formulate a similar theory for finite dimensional Lie algebras? The answer to this question is affirmative as was shown in ref. 16. This lead to so called 'finite W algebras'. It turns out that these are a very effective toy model for ordinary W algebras but apart from that they are of some interest themselves. For example they are what one could consider to be the finite algebras underlying the W algebras, just as the simple Lie algebras underly the Kac-Moody algebras. This means that they will play an important role in the representation theory of ordinary W algebras since the subspace of singular vectors of a W algebra will carry a representation of the underlying finite W algebra.

Another area where finite W algebras may play an important role is finite dimensional generalized Toda theory. These were originally introduced as dimensional reductions of self dual Yang-Mills theories in ref. 17. The general solution space of these models was however not constructed (only for some special examples). In this paper we will show that these generalized finite Toda systems are reductions of a system describing a free particle moving on a group manifold and that they have the finite W symmetry discussed above. The general solution of the equations of motion of such a system is easily found and has the form of a transform by the symmetry group of a certain reference solution. This reference solution can be reduced to give a nontrivial solution of the generalized Toda theory and one would expect the orbits of finite W transformations to provide the other solutions.

In this paper we shall give a brief account of the ideas described above deferring details to our forthcoming paper[18].

1 GENERALIZED DRINFELD-SOKOLOV REDUCTIONS

In this section we briefly discuss the results of ref. 3 in which the Drinfeld-Sokolov reductions were generalized.

Let g be simple Lie algebra, $\{t_a\}$ a basis of g and f_{ab}^c the structure constants of g in this basis. Consider the Poisson algebra

$$\{J^a(z), J^b(w)\} = f_c^{ab} J^c(w)\delta(z-w) - kg^{ab}\delta'(z-w), \qquad (1)$$

where g^{ab} is the inverse matrix of $g_{ab} = \text{Tr}(t_a t_b)$. This Poisson structure is actually nothing but the Kirillov Poisson structure on the affine Kac-Moody algebra over g. From now on we fix $g = sl_n$.

Let $i : sl_2 \hookrightarrow sl_n$ be an embedding of sl_2 into sl_n. Under the adjoint action of the embedded sl_2 algebra the algebra sl_n branches into a direct sum of p irreducible

sl_2 multiplets. Let $\{t_{k,m}\}_{m=-j_k}^{j_k}$ be a basis of the kth multiplet where j_k is the highest weight of this multiplet. The numbering is chosen such that $t_{1,\pm 1} = t_\pm$ and $t_{1,0} = t_3$ where $\{t_3, t_\pm\}$ are the Cartan, step up and step down elements of $i(sl_2)$. An arbitrary map $J : S^1 \to sl_n$ can then be written as

$$J(z) = \sum_{k=1}^{p} \sum_{m=-j_k}^{j_k} J^{k,m} t_{k,m}. \qquad (2)$$

Impose now the constraints $\phi^{1,1}(z) \equiv J^{1,1}(z) - 1 = 0$ and $\phi^{k,m}(z) = J^{k,m} = 0$ for $m > 0$, $k \neq 1$. The constraints $\{\phi^{k,m}(z)\}_{m\leq 1}$ are first class which means they generate gauge invariance. This gauge invariance can be completely fixed by gauging away the fields $\{J^{k,m}(z)\}_{m > -j_k}$. After constraining and gauge fixing the currents look like

$$J_{\text{fix}}(z) = \sum_{k=1}^{p} J^{k,-j_k} t_{k,-j_k} + t_+. \qquad (3)$$

The Poisson bracket (1) on the set of 'currents' of the form (2) induces a Poisson bracket (which is in fact a Dirac bracket as first realized in ref. 2) on the set of gauge fixed currents (3). The algebra generated by the fields $\{J^{k,-j_k}(z)\}$ and equipped with the Dirac bracket is then a W algebra with conformal weights $\{\Delta_k = j_k + 1\}$.

The ordinary Drinfeld-Sokolov reductions which yield the Zamolodchikov W_n algebras correspond to the case where one takes the principal embedding of sl_2 into sl_n. The algebra $W_3^{(2)}$ introduced first by Polyakov and Bershadsky corresponds to the only non-principal sl_2 embedding into sl_3. In general however there are as many inequivalent sl_2 embeddings into sl_n as there are partitions of the number n. This gives a large number of possibilities.

2 FINITE W ALGEBRAS

In this section we introduce finite W algebras[16] by reducing finite dimensional simple Lie algebras instead of KM algebras

The starting point is again the Kirillov Poisson structure on a Lie algebra (actually it is on its dual but since simple Lie algebras carry a nondegenerate bilinear form we identify the Lie algebra with its dual). The coordinate free expression of this Poisson bracket is

$$\{F, G\}(x) = (x, [\text{grad}_x F, \text{grad}_x G]), \qquad (4)$$

where $(.,.)$ is the Cartan-Killing form, F, G are smooth functions on g and $\text{grad}_x F$ is defined by

$$\frac{d}{d\epsilon} F(x + \epsilon x')|_{\epsilon=0} = (x', \text{grad}_x F) \quad \text{for all } x' \in g. \qquad (5)$$

Using again the basis $\{t_a\}$ an arbitrary element of g can be written as $x = J^a(x) t_a$ where J^a is a smooth function on g. In terms of these coordinate functions the Kirillov bracket reads

$$\{J^a, J^b\} = f_c^{ab} J^c, \qquad (6)$$

(compare to eq.(1)).

One can go now through the whole procedure again, i.e. choose an sl_2 embedding, impose the constraints and gauge fix. Define the set of gauge fixed elements (which is a submanifold of g) by

$$g_{\text{fix}} = \{t_+ + \sum_{k=1}^{p} y^{k,-j_k} t_{k,-j_k} \mid y^{k,-j_k} \in \mathbf{C}, \mathbf{R}\}. \qquad (7)$$

Again the Kirillov Poisson bracket on g induces a Poisson bracket on g_{fix}. In order to describe this bracket introduce the map

$$L : g \longrightarrow g \qquad (8)$$

which, on $\text{Im}(\text{ad}_{t_+})$ is the inverse of the map $\text{ad}_{t_+} : \text{Im}(\text{ad}_{t_-}) \to \text{Im}(\text{ad}_{t_+})$ and on the complement of $\text{Im}(\text{ad}_{t_+})$ is the zero map. It is shown in ref. 18 that for Q_1 and Q_2 smooth functions on g_{fix} and $y \equiv t_+ + w \in g_{\text{fix}}$ we have

$$\{Q_1, Q_2\}(y) = \left(y, \left[\text{grad}_y Q_1, \frac{1}{1 + L \circ \text{ad}_w} \text{grad}_y Q_2 \right] \right), \qquad (9)$$

where $\text{grad}_y Q \in \text{Ker}(\text{ad}_{t_+})$ is (uniquely) defined by

$$\frac{d}{d\epsilon} Q(y + \epsilon y')|_{\epsilon=0} = (y', \text{grad}_y Q), \qquad (10)$$

for all $y' \in \text{Ker}(\text{ad}_{t_-})$. (This uniquely defines $\text{grad}_y Q$ because $\text{Ker}(\text{ad}_{t_-})$ and $\text{Ker}(\text{ad}_{t_+})$ are nondegenerately paired by the Cartan-Killing form.)

Let us consider an example. The finite versions of W_n, corresponding to the principal sl_2 embeddings, give Abelian Poisson algebras and are therefore not very interesting. The simplest nontrivial case is associated to the nonprincipal sl_2 embedding of sl_2 into sl_3. Under the adjoint action of this embedding sl_3 decomposes into a direct sum of a triplet, two doublets and a singlet (i.e. $k = 1, \ldots, 4$ and $j_1 = 1$, $j_2 = j_3 = \frac{1}{2}$, $j_4 = 0$). The reduced algebra will therefore have 4 generators $J^{1,-1}, J^{2,-1/2}, J^{3,-1/2}$ and $J^{4,0}$ (or equivalently g_{fix} is 4 dimensional). The Poisson brackets (9) in terms of $c = -\frac{4}{3}(J^{1,-1} + 3(J^{4,0})^2)$, $e = \sqrt{\frac{4}{3}} J^{2,1/2}$, $f = \sqrt{\frac{4}{3}} J^{3,-1/2}$ and $h = -4J^{4,0}$ read in this case

$$\begin{aligned} \{h, e\} &= 2e, \\ \{h, f\} &= -2f, \\ \{e, f\} &= h^2 + c, \end{aligned} \qquad (11)$$

and c Poisson commutes with everything. This algebra is obviously a non-linear and centrally extended version of sl_2 and was first constructed in ref. 19 as a solution of the Jacobi identities. We summarize the (real) representation theory of the commutator version of this algebra in the following theorem[16].

Theorem 1 *Let p be a positive integer and x a real number.*

1. *For every pair (p, x) the algebra (11) has a unique highest weight representation $W(p; x)$ of dimension p with highest weight $j(p; x) = p + x - 1$ and central value $c(p; x) = \frac{1}{3}(1 - p^2) - x^2$.*

2. *Let $k \in \{1, \ldots, p-1\}$ then $W(p; \frac{2}{3}k - \frac{1}{3}p)$ is reducible and its invariant subspace is isomorphic as a representation to $W(p - k; -\frac{1}{3}(k + p))$.*

3. *The representation $W(p; x)$ is unitary iff $x > \frac{1}{3}p - \frac{2}{3}$.*

It is well known that it is possible to realize the finite dimensional irreducible representations of any simple Lie algebra on a Fock space. Consider for example the realization on $\mathbf{C}[z]$ of the algebra sl_2

$$\begin{aligned} \sigma_\Lambda(t_+) &= \frac{d}{dz}, \\ \sigma_\Lambda(t_-) &= (\Lambda, \alpha)z - z^2 \frac{d}{dz}, \\ \sigma_\Lambda(t_0) &= (\Lambda, \alpha) - 2z \frac{d}{dz}, \end{aligned} \qquad (12)$$

where Λ is a weight and α is the positive root of sl_2. For Λ a principal dominant weight the representation σ_Λ is reducible and in fact the subspace

$$V = \left\{P(z) \in \mathbb{C}[z] \Big| \left(\frac{d}{dz}\right)^{(\Lambda,\alpha)+1} P(z) = 0\right\} \tag{13}$$

is isomorphic to the $(\Lambda, \alpha) + 1$ dimensional irreducible representation of sl_2 (what we have presented here is the first term of a Fock resolution of the $(\Lambda, \alpha) + 1$ dimensional irrep of sl_2).

Similar realizations exist for the representations $W(p; x)$ of the algebra (11). Define the representation $\hat{\sigma}_\Lambda$ by

$$\hat{\sigma}_\Lambda(h) = (\Lambda, \alpha_1 - \alpha_2) + \frac{1}{3} - 2z\frac{d}{dz},$$

$$\hat{\sigma}_\Lambda(e) = -3(\Lambda, \alpha_2)\frac{d}{dz} - 2z\frac{d^2}{dz^2},$$

$$\hat{\sigma}_\Lambda(f) = (\Lambda, \alpha_1)z - \frac{2}{3}z^2\frac{d}{dz},$$

$$\hat{\sigma}_\Lambda(c) = -(\Lambda, \alpha_1)^2 - (\Lambda, \alpha_2)^2 - (\Lambda, \alpha_1)(\lambda, \alpha_2) + \frac{2}{3}(\Lambda, \alpha_2 - \alpha_1) - \frac{1}{9},$$

where α_1 and α_2 are the simple roots of sl_3 and Λ is a weight of sl_3. It is easy to check that these operators satisfy the algebra (11). For $(\Lambda, \alpha_1) = \frac{2}{3}(p-1)$ and $(\Lambda, \alpha_2) = \frac{2}{3} - \frac{1}{3}p - x$ the Fock representation $\hat{\sigma}_\Lambda$ has a p dimensional invariant subspace

$$V = \left\{P(z) \in \mathbb{C}[z] \Big| \left(\frac{d}{dz}\right)^p P(z) = 0\right\} \tag{14}$$

isomorphic to $W(p; x)$. This provides a Fock realization of the representation $W(p; x)$.

In general (that is for arbitrary embeddings) it is possible to find Fock realizations of finite W algebras by a generalized Miura transformation[18]. Here we shall not go into this however.

3   FINITE W SYMMETRIES IN GENERALIZED TODA THEORIES

Consider the system of a particle moving on a group manifold $G\ (= SL(n))$. The action of such a particle can be taken to be

$$S[g] = \frac{1}{2}\int dt\ \text{Tr}\left(g^{-1}\frac{dg}{dt}g^{-1}\frac{dg}{dt}\right) \tag{15}$$

where $g: \mathbb{R} \to G$ is the world line of the particle. The equations of motion of this action are

$$\frac{d}{dt}\left(g^{-1}\frac{dg}{dt}\right) = 0, \tag{16}$$

or equivalently,

$$\frac{d}{dt}\left(\frac{dg}{dt}g^{-1}\right) = 0. \tag{17}$$

In local coordinates $\{x^i\}$ the action looks like

$$S = \frac{1}{2}\int dt\ g_{ij}\frac{dx^i}{dt}\frac{dx^j}{dt}, \tag{18}$$

where $g_{ij} = R_i^a R_j^b K_{ab}$ and $g^{-1}\frac{dg}{dt} = R_i^a \frac{dx^i}{dt} t_a$ (remember that $\{t_a\}$ is a basis of g and $K_{ab} = \text{Tr}(t_a t_b)$). From this we conclude that the action (28) describes a free particle moving in a curved background.

The action (28) has a left and right G symmetry $g(t) \to a.g(t).b^{-1}$. From the equations of motion we immediately find that the conserved quantities are

$$\bar{J} = \frac{dg}{dt} g^{-1} \equiv \bar{J}^a t_a \quad \text{and} \quad J = g^{-1}\frac{dg}{dt} \equiv J^a t_a. \tag{19}$$

The conserved quantities $\{J^a\}$ form a Poisson algebra

$$\{J^a, J^b\} = f_c^{ab} J^c \tag{20}$$

(and similar equations for $\bar{J}$) which, as we have seen, is nothing but the Kirillov Poisson bracket written out in coordinates. Let's now consider what happens to the theory (15) when we reduce it.

Define $g^{(+)} = \text{span}\{t_{k,m}\}_{m>0}$, $g^{(0)} = \text{span}\{t_{k,0}\}$, $g^{(-)} = \text{span}\{t_{k,m}\}_{m<0}$. Obviously $g = g^{(-)} \oplus g^{(0)} \oplus g^{(+)}$. Let G^-, G^0, G^+ be the corresponding subgroups of G and let $\pi_\pm$ be the projections of g onto $g^{(\pm)}$. The constraints we impose are (as before)

$$\pi_-(J) = t_- \quad \text{and} \quad \pi_+(\bar{J}) = t_+. \tag{21}$$

Inserting the generalized (local) Gauss decomposition $g = g_- g_0 g_+$ where $g \in G$, $g_\pm \in G^\pm$ and $g_0 \in G^0$ into eqns.(21) we find

$$g_0 t_+ g_0^{-1} = g_+^{-1}\frac{dg_+}{dt}, \tag{22}$$

$$g_0^{-1} t_- g_0 = \frac{dg_-}{dt} g_-^{-1}. \tag{23}$$

This means that the constrained currents look like

$$J = g_-^{-1}\left(g_0^{-1}\frac{dg_0}{dt} + t_+\right) g_- + g_-^{-1}\frac{dg_-}{dt}, \tag{24}$$

$$\bar{J} = g_+\left(\frac{dg_0}{dt}g_0^{-1} + t_-\right) g_+^{-1} + \frac{dg_+}{dt}g_+^{-1}. \tag{25}$$

Note that now the equations of motion (16) can be written as

$$\left[\frac{d}{dt}, \frac{d}{dt} + J\right] = 0. \tag{26}$$

Conjugating this equation by g_-, using eqns. (22), (23), (24) and writing out the commutator we find

$$\frac{d}{dt}\left(g_0^{-1}\frac{dg_0}{dt}\right) = [g_0^{-1} t_- g_0, t_+]. \tag{27}$$

This evolution equation describes the gauge invariant part of the constrained theory. The action corresponding to this equation is

$$S[g_0] = \frac{1}{2}\int dt \, \text{Tr}\left(g_0^{-1}\frac{dg_0}{dt} g_0^{-1}\frac{dg_0}{dt}\right) + \int dt \, \text{Tr}\left(g_0^{-1} t_- g_0 t_+\right) \tag{28}$$

which describes a particle moving on the subgroup G_0 of G with some self-interaction. It can be shown that the theory (28) has the nonlinear 'finite' W symmetry corresponding to the sl_2 subalgebra $\{t_3, t_\pm\}$ of g.

Strictly speaking the above arguments only work when the sl_2 subalgebra which one considers provides an 'integral grading' of the Lie algebra g because then there are only first class constraints. However it can be shown that it is always possible to find a set of first class constraints that give the same constrained and gauge fixed manifold g_{fix}. This is done by imposing only 'half' (there are always an even number of second class constraints) of the constraints that turned out to be second class such that they become first class. The other half can then be imposed as gauge fixing conditions (see for a treatment of this in the present context[6]).

In the case where the sl_2 embedding is the 'principal' embedding[20] of sl_2 into $g = sl_n$, equation (27) reduces to the ordinary finite Toda equations

$$\frac{d^2 q_i}{dt^2} + \exp\left(\sum_{j=1}^{n-1} K_{ij} q_j\right) = 0, \qquad (29)$$

where $i = 1, \ldots, rank(g) = n - 1$, K_{ij} is the Cartan matrix of sl_n and $g_0 = \exp(q_i H_i)$.

The generalized finite Toda theories (27) were already derived in ref. 17 as dimensional reductions of the self-dual Yang-Mills equations. For some examples the solutions were constructed, however the general solution space is to our knowledge still not known. The finite W algebras may provide a new tool in this research since they are expected to transform solutions of (27) to (different) solutions of (27) just like the general solution of (16) can be found by letting the symmetry group $G \times G$ act on the simplest solution $g = e^{tX}$.

ACKNOWLEDGEMENTS

I would like to thank Sander Bais, Jan de Boer, Jacob Goeree and Klaas Landsman for many conversations on the subjects presented in this paper.

REFERENCES

[1] P. Bouwknegt, K. J. Schoutens, *W symmetry in CFT*, *Phys. Rep.* **223** (1993) 183.
[2] J. Balog, L. Fehér, P. Forgac, L. O'Raifeartaigh and A. Wipf, *Ann. Phys.* **203** (1990) 76.
[3] F.A. Bais, T. Tjin, P. van Driel, *Nucl. Phys.* **B357** (1991) 632.
[4] L. Fehér, L. O'Raifeartaigh, P. Ruelle and I. Tsutsui, Ann. Phys. **213** (1992); *Phys. Lett.* **B283** (1992) 243.
[5] J. Balog, L. Fehér, P. Forgac, L. O'Raifeartaigh and A. Wipf, *Phys. Lett.* **B227** (1989) 214; *Phys. Lett.* **B244** (1990) 435.
[6] L. Fehér, L. O'Raifeartaigh, P. Ruelle, I. Tsutsui and A. Wipf, On the general structure of Hamiltonian reductions of the WZW theory, Preprint DIAS-STP-91-29, UdeM-LPN-TH-71/91.
[7] L. O'Raifeartaigh, P. Ruelle and I. Tsutsui, *Phys. Lett.* **B258** (1991) 359.
[8] L. O'Raifeartaigh, A. Wipf, *Phys. Lett.* **B251** (1990) 361.
[9] L. O'Raifeartaigh, P. Ruelle, I. Tsutsui, A. Wipf, *Comm. Math. Phys.* **143** (1992) 333.
[10] M. Bershadsky, H. Ooguri, *Comm. Math. Phys.* **126** (1989) 49.
[11] M. Bershadsky, *Comm. Math. Phys.* **139** (1991) 71.
[12] L. Frappat, E. Ragoucy, P Sorba, *W algebras and superalgebras from constrained WZW models: a group theoretical classification*, Preprint Enslapp-AL-391/92.
[13] F. Delduc, E. Ragoucy, P. Sorba, *Comm. Math. Phys.* **146** (1992) 403; *Phys. Lett.* **B279** (1992) 319.
[14] B. L. Feigin, E. Frenkel, *Phys. Lett.* **B246** (1990) 75;
E. Frenkel, *W algebras and Langlands-Drinfeld correspondence*, Harvard Preprint 1991;
E. Frenkel, V. Kac, M. Wakimoto, *Comm. Math. Phys.* **147** (1992) 295.

[15] J. de Boer, J. Goeree, Covariant W gravity and its moduli space, Preprint THU-92/14
F. A. Bais, T. Tjin, P. van Driel, J. de Boer, J. Goeree, W algebras, W gravities and their moduli spaces, Preprints THU-92/26, ITFA-92/24.

[16] T. Tjin, *Phys. Lett.* **B292** (1992) 60.

[17] F.A. Bais, W.P.G. van Veldhoven, *Physica* **139A** (1986) 326.

[18] J. de Boer, J. Goeree, T. Tjin, to be published.

[19] M. Roček, *Phys. Lett.* **B255** (1991) 554.

[20] E. B. Dynkin, *Amer. Math. Soc. Transl.* **6[2]** (1967) 111.

ON THE "DRINFELD-SOKOLOV" REDUCTION OF THE KNIZHNIK-ZAMOLODCHIKOV EQUATION

P. Furlan[1], A.Ch. Ganchev[2], R. Paunov[3] and V.B. Petkova[4]

1) Istituto Nazionale di Fisica Nucleare, Sezione di Trieste, Italy
2) Dipartimento di Fisica Teorica, Universitá di Trieste, Trieste, Italy
3) Institute for Nuclear Research and Nuclear Energy, Sofia 1784, Bulgaria
4) International School for Advanced Studies, via Beirut 2, 34014 Trieste, Italy

INTRODUCTION

Even though the Hamiltonian reduction of two dimensional integrable and conformal theories has been a focus of much attention lately [1] the quantum aspect of the problem is in a more undeveloped stage. It has been shown that the space of states of a WZW model based on a KM algebra $\hat{g}$ reduces to the space of states of a minimal W_g-model [2]. A complete quantum theory is specified by giving not only the space of physical states but also the field operators acting on it, or equivalently all correlators. The study of the reduction of the full quantum WZW model (for the case of $g = sl(2)$) has been initiated by us – [3] and its detailed version [4]. In [5] two of us have pursued further the question – how do the differential equations and relations governing the correlators relate. This report is a survey of the results of [3-5].

After recalling in **1** and **2** a few well known facts, in **3** we give an idea of our approach on the example of two and three point functions – setting $x = z$ reduces the WZW correlators to W correlators where z is the "space" variable while x is an auxiliary "isospin" variable realising the representations of $sl(2)$.

In order to recover the complete Kac table one is forced to consider representations of $\hat{g}$ of fractional level and Kac-Kazdan weights [6] which in general are not integral. In **4** and **5** we consider the simpler case of integral weights. Using the recursion relations for the generalised hypergeometric integrals under consideration one can expand the n-point functions as power series in $(x - z)$. From the form of the co-

---

a – at 1) and 2); b – postdoctoral fellow of 1), present address - 4), permanent address - 3); c – postdoctoral fellow at 4), permanent address - 3); d – visitor at 1) and 4); permanent address - 3)

Low-Dimensional Topology and Quantum Field Theory,
Edited by H. Osborn, Plenum Press, New York, 1993

efficient functions it is immediate that setting $x = z$ we recover the correlators of the reduced theory. Written for the $x - z$ expansion the Knizhnik-Zamolodchikov equation becomes a triangular matrix system. In the case of integral weights $(2j \in \mathbb{Z}_+)$ this matrix system is finite and can be reduced to a scalar differential equation of order $2j+1$ describing the decoupling of the Benoit-S.Aubin [7] Virasoro null vector.

The case of nonintegral weights is far more involved. The first challenge is to define the correlators themselves. The direct generalisation of the Wakimoto bosonization does not work. Instead we generalise the $x - z$ expansion. We propose an ansatz for the coefficient functions which, as in the case of integral weights, satisfy the recursion relations and the triangular KZ equation. This in turn implies that the so defined n-point functions are solutions of the KZ equation. The $x - z$ expansion in the case of nonintegral weights is infinite dimensional since the representations of g are infinite dimensional making the reduction of the KZ system nontrivial. What effectively truncates the KZ system and makes the reduction possible are the algebraic relations on the correlators describing the decoupling of the Malikov-Feigin-Fuks [8] singular vectors. One can say that the triangular KZ equation serves as a bridge between the KM and Virasoro singular vectors.

At the end we discuss the possible generalisations to bigger algebras. In the case of $g = sl(n)$ we show that one can recover the W_g 2-point functions from the g-WZW 2-point functions by setting $x = z$ where x corresponds to the principle $sl(2)$ subalgebra of $sl(n)$.

1. We begin with a quick reminder of the models corresponding to reducible representations of Virasoro [9]. For every central charge of the form

$$c = 13 - 6(\nu + \nu^{-1}) , \qquad (1.1)$$

with $\nu^{-1} \equiv k + 2$, the primary fields have conformal dimensions h_J:

$$h_J = h_J^{\text{sug}} - J \quad \text{where} \quad h_J^{\text{sug}} = J(J+1)\nu, \quad J = j - j'/\nu, \quad 2j, 2j' \in \mathbb{Z}_+ . \qquad (1.2)$$

(If $k+2 = p/p'$, p, p' relatively prime, one has maximally degenerate representations. The corresponding models are called "minimal" – the weights in the Kac table being h_J with $1 \leq 2j^{(\prime)} + 1 \leq p^{(\prime)} - 1$.) The primary fields are Virasoro tensor operators, i.e., they are characterised by the intertwining property:

$$[L_{-n}, \psi^h(z)] = (-z)^{-n}\left((n-1)h + z\frac{\partial}{\partial z}\right)\psi^h(z) . \qquad (1.3)$$

The primary fields also serve as highest weight vectors of Verma modules of descendants, e.g., $(L_{-n_1} \ldots L_{-n_k} \psi^h)(z)$. Using the invariance of the vacuum and the intertwining property the correlator of a descendant and other primary fields can be immediately expressed as:

$$\langle \ldots (L_{-n}\psi^h)(z_a)\ldots\rangle = \mathcal{L}^{(h)}_{-n,a}\langle \ldots \psi^h(z_a)\ldots\rangle \qquad (1.4)$$

where

$$\mathcal{L}^{(h)}_{-n,a} = \sum_{b(\neq a)} \left(\frac{n-1}{z_{ba}^n}h - \frac{1}{z_{ba}^{n-1}}\frac{\partial}{\partial z_b}\right) . \qquad (1.5)$$

The Verma modules of h.w. h_J (1.2) are maximaly degenerate. A null descendant we will denote by $N(J)\psi^{h_J}$ where $N(J)$ is a polynomial in L_{-n}. For example $N(1/2) = (L_{-1})^2 - \nu L_{-2}$. The decoupling of the null vectors leads to BPZ equations [9]

$$\mathcal{N}(J_a)\langle\ldots\psi^h(z_a)\ldots\rangle = 0 , \qquad (1.6)$$

where the differential operator $\mathcal{N}$ is obtained from N by simply substituting in it $\mathcal{L}$ for L. The correlators of Vir minimal models can are solutions of the BPZ equations. But this is not a practical way of describing the correlators and their properties.

Substantial progress was made when the Coulomb gas (free field, bosonization) method was developed [10]. In this approach the primary fields are described by vertex operators

$$\psi^{h_J}(z) = :\exp(i\alpha_J \phi(z)): \qquad (1.7)$$

appropriately screened by screening currents

$$\psi_\pm(u) = :\exp(i\alpha_\pm \phi(u)): \qquad (1.8)$$

integrated along "closed" contours. Here ϕ is a bosonic field with logarithmic 2-point function and the charges are

$$\alpha_\pm = \pm \nu^{\mp 1/2}, \qquad \alpha_J = -j\alpha_- - j'\alpha_+ . \qquad (1.9)$$

An n-point function with the field $\psi^{h_{J_n}}$ at $z_n = \infty$ is described as a multiple contour integral

$$\int_\Gamma \prod_{i=1}^s du_i \prod_{i'=1}^{s'} dv_{i'} \langle h_{J_n} | \psi^{h_{J_{n-1}}}(z_{n-1}) \ldots \psi^{h_{J_1}}(z_1) \psi_-(u_1) \ldots \psi_-(u_s) \psi_+(v_1) \ldots \psi_+(v_{s'}) \rangle \qquad (1.10)$$

where charge balance requires

$$s = j_1 + j_2 + \ldots + j_{n-1} - j_n, \qquad s' = j'_1 + j'_2 + \ldots + j'_{n-1} - j'_n . \qquad (1.11)$$

Using the Wick pairing $:\exp(\alpha\phi(z_1)): :\exp(\beta\phi(z_2)): = z_{12}^{2\alpha\beta} :\exp(\alpha\phi(z_1) + \beta\phi(z_2)):$ of vertex operators we can rewrite the correlator as (a Dotsenko-Fateev integral)

$$\int_\Gamma \Phi_J^{(s,\nu)}(u,z) \Phi_{-\nu J}^{(s',1/\nu)}(v,z) \prod_{i,i'} \frac{1}{(u_i - v_{i'})^2} du_i dv_{i'} , \qquad (1.12a)$$

$$\Phi_J^{(s,\nu)}(u,z) = \prod_{i<j} u_{ij}^{2\nu} \prod_{i=1}^s \prod_{a=1}^{n-1} (u_i - z_a)^{-2 J_a \nu} , \qquad (1.12b)$$

where Γ is a cycle in the twisted homology [11] with local coefficients system defined by the multivalued integrand.

2. Next we will briefly recall some facts about WZW models. The chiral algebra of WZW is the semidirect sum of an affine algebra $\hat{g}$ and the Virasoro algebra with Sugawara central charge

$$[X_n^\alpha, X_m^\beta] = f_\gamma^{\alpha\beta} X_{n+m}^\gamma + k q^{\alpha\beta} n \delta_{n+m,0}, \qquad [X_n^\alpha, L_m] = n X_{n+m}^\alpha , \qquad (2.1a)$$

$$[L_n, L_m] = (n-m)L_{n+m} + \frac{n^3 - n}{12} c \delta_{n,-m} , \qquad (2.1b)$$

where α, β, γ run over a set labeling a basis of the Lie algebra g, $f_\gamma^{\alpha\beta}$ are the structure constants of g, $q^{\alpha\beta}$ is the Killing form, k is the level and $c = c^{\text{sug}}$ is the central charge. Actually we will confine our attention to the case $g = sl(2)$ so that $c^{\text{sug}} = \frac{3k}{2+k}$ and will choose $f_\pm^{0\pm} = \pm 2$, $f_0^{+-} = 1$, $q^{00} = 2 = 2q^{+-} = 2q^{-+}$. It is convenient to use a functional realisation (known also under many other names) for a representation of isospin J in a space of functions C_J of a complex variable x. The generators of $sl(2)$ are given by differential operators

$$S^- = -\frac{\partial}{\partial x}, \quad S^0 = 2x\frac{\partial}{\partial x} - 2J, \quad S^+ = x^2\frac{\partial}{\partial x} - 2xJ. \tag{2.2}$$

In the case of Virasoro minimal models (and more generally W-models) the free field techniques were developed first. In the case of WZW theories it has been the other way around. The simplicity of the relevant differential equation – the Knizhnik-Zamolodchikov equation [12], allowed for its direct solution and investigation of the properties of the correlators [13]. It was only later that the bosonization of KM algebras was developed. The KZ equation reflects the fact that

$$\mathcal{K}_{-1} \equiv L_{-1} - L_{-1}^{\text{sug}} \tag{2.3}$$

produces a null vector. As usual $L_n^{\text{sug}} = \frac{\nu}{2} \sum_m :X_{n-m}^\alpha X_{m\,\alpha}:$. The decoupling of this null vector leads to the KZ equation for the WZW correlators

$$\mathcal{K}_{-1} U^{\text{wzw}} = 0. \tag{2.4}$$

To describe the solutions of this equation, for the case of integral weights, it is nevertheless useful to employ the free field approach [14]. The Wakimoto bosonization amounts to an affinisation of the functional realisation: $x, \partial_x \to \gamma(z), \beta(z)$. The primary fields and screening current are given by

$$\psi_m^j(z) = \gamma^{j-m}(z) :\exp(i\alpha_j\phi(z)):, \qquad \psi_-(u) = \beta(u) :\exp(i\alpha_-\phi(u)):, \tag{2.5}$$

here m labels a state in the representation j. Going to a functional realisation of $sl(2)$ we will trade the label m for the variable x:

$$\psi^j(x,z) = \sum_m \binom{2j}{j-m} x^{j+m} \psi_m^j(z). \tag{2.6}$$

The n-point conformal block with $z_n = \infty$ and $sl(2)$ h.w. state at the nth point can be written as

$$U^{\text{wzw}}(x,z;j) = \sum_{|\mu|=s} \prod_{a=1}^{n-1} \frac{x_a^{\mu_a}}{\mu_a!} \mathbb{I}_\mu(z), \tag{2.7}$$

where $|\mu| = \sum_1^{n-1} \mu_a$, $\mu_a = j_a + m_a$, x and z are $n-1$ vectors, j is an n vector of half-integer isospins and $\mathbb{I}$ is a multiple contour integral [15,16]

$$\mathbb{I}_\mu(z) = \nu^{|\mu|} \frac{s!}{(s-|\mu|)!} \prod_{a=1}^{n-1} \frac{(2j_a)!}{(2j_a-\mu_a)!} \prod z_{ab}^{2\nu j_a j_b} \int_\Gamma \prod_1^s du_i\, \varphi_\mu^{[1,s]}(u,z) \Phi_j^{(s,\nu)}(u,z). \tag{2.8}$$

The multivalued function Φ is the same as (1.12b) while the meromorphic factor φ comes from the expectation value of the β-γ system (the Wick pairing of $\gamma(z_a)$ and $\beta(u_i)$ gives $(z_a - u_i)^{-1}$):

$$\varphi_\mu^{[1,s]}(u,z) = \frac{1}{(z_1-u_1)\dots(z_1-u_{\mu_1})} \cdot \frac{1}{(z_1-u_{\mu_1+1})\dots(z_1-u_{\mu_1+\mu_2})} \cdot \dots + \text{perm.}. \tag{2.9}$$

3. Let us start discussing the question of reduction with the simplest cases – the two and three point functions of $sl(2)$ WZW. In the functional realisation they acquire a very compact form

$$U^{\text{wzw}}(x_1, x_2, z_1, z_2; J) = (x_{12})^{2J} (z_{12})^{-2h_J^{\text{sug}}} \tag{3.1}$$

and

$$U^{\text{wzw}}(x_1, x_2, x_3, z_1, z_2, z_3; J_1, J_2, J_3)$$
$$= \frac{c_{J_1 J_2 J_3} (x_{12})^{J_1+J_2-J_3}(x_{23})^{J_2+J_3-J_1}(x_{31})^{J_3+J_1-J_2}}{(z_{12})^{h_{J_1}^{\text{sug}}+h_{J_2}^{\text{sug}}-h_{J_3}^{\text{sug}}}(z_{12})^{h_{J_1}^{\text{sug}}+h_{J_2}^{\text{sug}}-h_{J_3}^{\text{sug}}}(z_{12})^{h_{J_1}^{\text{sug}}+h_{J_2}^{\text{sug}}-h_{J_3}^{\text{sug}}}}. \tag{3.2}$$

Note that this holds for general spins J_a. Recalling the form of the conformal dimensions in the Kac table, $h_J = h_J^{\text{sug}} - J$, we immediately obtain

$$U^{\text{wzw}}(x, z)|_{x=z} = U^{\text{vir}}(z) . \tag{3.3}$$

In what follows we will try to answer the following questions: how to generalise to higher n-point function; how to generalise to other groups; how to reduce the differential equations?

4. First we will consider the much simpler case of integral weights. In this case the WZW correlators are known (2.7). All we have to do is show that setting $x = z$ we recover the DF integrals (1.12) with $s' = 0$. The tools that we will exploit are certain relations in twisted cohomology [17]. The first one is the Aomoto linear relation. It expresses the fact that the correlator is invariant with respect to translations in x space:

$$\Delta(S^-)U(x, z) = 0 , \tag{4.1}$$

or equivalently that the state at infinity is annihilated by S_n^+ since by duality this operator turns into $\Delta(S^-) \equiv \sum_{a=1}^{n-1} S_a^-$ (S_a acts on the field at z_a). The proof of this relation proceeds by showing

$$\sum_{a=1}^{n-1}(2j_a - \mu_a)\varphi_{\mu+\varepsilon_a}^{[1,s]} = \sum_{i=1}^{s}(\partial_i + (\partial_i \Phi))\varphi_\mu^{[1,s],i} \tag{4.2}$$

where $(\varepsilon_a)_b = \delta_{ab}$. The r.h.s. is a total derivative in the twisted cohomology with local coefficients determined by the multivalued function Φ. Thus integrating over a "closed" contour Γ (a cycle in twisted homology) cancels the r.h.s. and produces [15a]

$$\sum_a \mathit{I}_{\mu+\varepsilon_a} = 0 , \tag{4.3}$$

which is obviously equivalent to (4.1).

A key role in our discussion plays the recursion relations. Denote

$$U_t(x, z) = \sum_{|\alpha|=t} \left(\prod_a \frac{x_a^{\alpha_a}}{\alpha_a!}\right) \mathit{I}_\alpha(z) , \tag{4.4}$$

with $t = 0, 1, 2, \ldots, s$. Obviously $U_s(x, z) = U^{\text{wzw}}(x, z)$, while $U_0(x, z) = \mathit{I}_0$ – the DF integrals (1.13) (since at the moment we are considering the case of integral weights we have a DF integral with only one type of screening charge). The recursion relation states that

$$\Delta(X_1^-) U_{t+1} = \nu(s-t)(c_0-t) U_t, \qquad \Delta(X_1^-) U_0 = 0, \tag{4.5}$$

where $\Delta(X_1^-) \equiv \sum_{a=1}^{n-1} z_a S_a^-$ and $c_0 \equiv \sum_{a=1}^{n} j_a + 1 - \nu^{-1}$.

With the help of these relations we can write U^{wzw} in an $x-z$ expanded form [†]

$$U^{\text{wzw}}(x,z) = \text{const} \sum_{t=0}^{s} \left(\frac{-1}{\nu}\right)^t \frac{(c_0-t)!}{(c_0-s)!} U_t(x-z,z). \tag{4.6}$$

Indeed, expanding the powers of $(x-z)$ in the r.h.s. and using the relations (4.1) and (4.5) we recover the l.h.s. In this $x-z$ expanded form it is obvious that the reduction (3.3) holds. We can view the $x-z$ expansion as a change of basis in the space of correlators. In the case (2.7) the basis is formed by $\mathbb{I}_\alpha$, $|\alpha| = s$, subject to the linear relation (4.3). In the case (4.6) the basis is formed by $\mathbb{I}_\alpha$, $|\alpha| = t$, $t = 0, 1, \ldots, s$, subject to both the linear and recursion relations. The importance of this new basis is that it contains the DF integral $\mathbb{I}_0$ and thus makes the reduction possible.

One can prove that the triangular KZ equation

$$\mathcal{K}_{-1,a} U_t(x,z) = \mathcal{X}_{0,a}^- U_{t+1}(x,z), \tag{4.7}$$

holds. Here $\mathcal{X}_{0,a}^- = -S_a^- = \partial/\partial x_a$. Because of normalization the r.h.s. of (4.7) vanishes for $t = s$ giving in particular the usual KZ equation (2.4).

Having (4.7) we could proceed in a different way to show that (4.6) is true. We could define the correlator by its $x-z$ expansion (the r.h.s. of (4.6)) then the KZ equation (2.4) follows from the fact that (4.7) holds for the separate terms of the expansion. In this manner in **6** we will define the correlators for nonintegral weights.

The following figure summarises the discussion so far:

$$\Delta(X_1^-) \longrightarrow$$

$$U^{\text{wzw}}(x,z) = U_s, \ldots, U_{t+1}, U_t, U_{t-1} \ldots U_0(x,z) = U^{\text{Vir}}(z)$$

$$\longleftarrow \mathcal{K}_{-1}.$$

We could think of $\{U_t\}_{t=0}^s$ as a "multiplet" with $\Delta(X_1^-)$ and $\mathcal{K}_{-1}$ being the lowering and raising operators and the WZW and Vir correlators being the highest and lowest weight vectors. Having cast the KZ equation into a triangular form (4.7) it is natural to consider how does it reduce. Thus we turn now to this question.

5. The equation (4.7) can be generalised to a form [5] that is of key importance in the discussion here and later in **7**:

$$\mathcal{K}_{-n,a} U_t = \mathcal{X}_{-n+1,a}^- U_{t+1}, \tag{5.1}$$

where $\mathcal{X}_{-n,a}^- = \sum_{b(\neq a)} z_{ba}^{-n} S_b^-$ and

$$\mathcal{K}_{-n,a} = \mathcal{L}_{-n,a}^{(h^{\text{sug}})} - \mathcal{L}_{-n,a}^{\text{sug}}\big|_{\text{vac}}, \quad n > 0, \tag{5.2}$$

[†] An $x-z$ expansion for the 4-point functions of $su(2)$ WZW was considered in [18] though there the coefficient functions were not determined.

with $L^{sug}_{-n,a}|_{vac} = \frac{\nu}{2}\sum_{k=0}^{n} :X^{\alpha}_{-n+k}X_{-k\,\alpha}:$.

Applying $(\mathcal{X}^{-}_{0,a})^t$ to both sides of (5.1), commuting it through $\mathcal{K}_{-n}$, eliminating recursively all terms $(\mathcal{X}^{-}_{0,a})^t \mathcal{X}^{-}_{-n+1,a} W_{t+1}$ and using $\mathbb{I}_t \equiv \mathbb{I}_{t\varepsilon_a} = (\mathcal{X}^{-}_{0,a})^t W_t$ we obtain

$$\mathbb{I}_{t+1} = (\mathcal{K}_{-1} + \nu t \mathcal{X}^0_{-1})\mathbb{I}_t + \sum_{p=1}^{t}(-\nu)^p \prod_{i=0}^{p-1}\gamma_{t-i}(J)(\mathcal{K}_{-p-1} + \nu(t-p)\mathcal{X}^0_{-p-1})\mathbb{I}_{t-p} \quad (5.3)$$

where $\gamma_t(J) = t(2J - t + 1)$.

For the case $j' = 0$ (or $j = 0$) there is an explicit formula for the Vir singular vectors [7] which was cast into a triangular matrix form in [19] exploiting an auxiliary $sl(2)$ multiplet. Let us introduce generators $\mathbb{J}^{0,\pm}$ with the same commutation relations as $S^{0,\pm}$ acting on a basis $\{v_t, t = 0, 1, ...\}$ as follows: $\mathbb{J}^- v_t = v_{t+1}$, $\mathbb{J}^+ v_t = \gamma_t(j)v_{t-1}$, $\mathbb{J}^0 v_t = 2(j - t)v_t$. If F_0 is the h.w. vector of a Vir Verma module of h.w. h_j then the singular vector F_{2j+1} can be readily expressed from the matrix system [19]

$$\mathcal{L}F \equiv \left(-\mathbb{J}^- + \sum_{p=0}^{2j}(-\nu\mathbb{J}^+)^p L_{-p-1}\right)F = v_0 F_{2j+1}, \quad F = \sum_{t=0}^{2j} v_{2j-t} F_t, \quad (5.4)$$

recovering the formula of [7].

It is immediate to rewrite the system (5.3) in a form similar to (5.4)

$$\mathcal{K}\mathbb{I} \equiv \left(-\mathbb{J}^- + \sum_{p=0}^{2j}(-\nu\mathbb{J}^+)^p\left(\mathcal{K}_{-p-1} + \nu\frac{\mathcal{X}^0_{-p-1}}{2}(\mathbb{J}^0 + 2j)\right)\right)\mathbb{I} = v_0 \mathbb{I}_{2j+1}. \quad (5.5)$$

where $\mathbb{I}(v, z) = \sum_{t=0}^{2j} v_{2j-t}\mathbb{I}_t(z)$ and we keep the r.h.s. although it is identically zero (the same formula with summation not restricted from above and with zero r.h.s. holds for arbitrary J).

Reducing this triangular matrix system we obtain the BPZ equation reflecting the decoupling of the Benoit-S. Aubin singular vector. In fact the functions $\{F_t\}$ can be realised as z-dependent linear combinations of the integrals $\{\mathbb{I}_t\}$:

$$F = g\mathbb{I} \quad \text{with} \quad g = \left[\exp(-\nu\mathbb{J}^+(L_{-1,a} - \frac{1}{2}\mathcal{X}^0_{-1,a}))\right] \cdot 1 \quad (5.6)$$

and the "gauge transformation" g intertwines the matrix systems (5.4) and (5.5)

$$g\mathcal{K}\mathbb{I} = \mathcal{L}g\mathbb{I}. \quad (5.7)$$

6. Finally we come to the more interesting and challenging problems arising in the case of nonintegral weights: $J = j - j'/\nu$, $j' \neq 0$ The first one is how to define the n-point functions. All the results on solutions of the KZ equation (WZW correlators) have been concerned with the case of integral weights. In the language of bosonisation the difficulty here is that the second screening charge

$$\psi_+(v) = \beta^{-(k+2)}(v) : \exp(i\alpha_+\phi(v)) :, \quad (6.1)$$

which will also be necessary to restore the charge balance, is not well defined. The expectation value $\varphi_\mu(v, z)$ of the β-γ system will be nonmeromorphic. This will change the local system and will eventually lead to different braiding and fusion properties. (for some examples see [20] and [4]).

The approach we have taken is to generalise the $x-z$ expansion. Namely we have proposed an ansatz [3] and have shown [4] that it is a solution of the KZ equation. Set

$$\varphi_\sigma^{(s,s')}(u,v,z) = \frac{(-s'/\nu)!s!}{S!} \sum_\tau \binom{S-|\sigma|}{s-|\tau|} \prod_{a=1}^{n-1} \binom{\sigma_a}{\tau_a} \varphi_\tau^{[1,s]}(u,z)\varphi_{\sigma-\tau}^{[-1/\nu,s']}(v,z) ,$$
(6.2a)

$$\varphi_\mu^{[L,m]}(u,z) = \frac{(L!)^m(mL-|\mu|)!}{(mL)!}$$

$$\times \sum_{N:\sum_i N_{ia}=\mu_a} \frac{\prod_a \mu_a!}{\prod_{i=1}^m (L-\sum_{a=1}^{n-1} N_{ia})! \prod_{a=1}^{n-1} N_{ia}!(u_i-z_a)^{N_{ia}}} ,$$
(6.2b)

where $S = s - s'/\nu$, the factorials are in general Γ-functions, N_{ia} are nonnegative integers and the sum is finite. Define

$$\mathbb{I}_\sigma^{(s,s')} = \nu^{|\sigma|} \frac{S!}{(S-|\sigma|)!} \prod_{a=1}^{n-1} \frac{(2J_a)!}{(2J_a-\sigma_a)!} \prod z_{ab}^{2\nu J_a J_b}$$
(6.3)
$$\times \int_\Gamma \varphi_\sigma^{(s,s')}(u,v,z) \Phi_J^{(s,s')}(u,v,z) \prod_1^s du_i \prod_1^{s'} dv_{i'}$$

where $\Phi_J^{(s,s')}$ is the integrand of (1.12a). The main technical result of [4] is that the linear relation (4.1), the recursion relation (4.5) (with $c_0 = \sum_{a=1}^n J_a + 1 - \nu^{-1}$ and s substituted by S) and the triangular KZ equation (4.7) hold also for the integrals (6.3).

Thus we conclude that

$$U^{wzw}(x,z) = \text{const} \sum_{t=0}^\infty \left(\frac{-1}{\nu}\right)^t \frac{(c_0-t)!}{(c_0-S)!} \left(\sum_{|\alpha|=t} \left(\prod_a \frac{(x_a-z_a)^{\alpha_a}}{\alpha_a!} \right) \mathbb{I}_\alpha^{(s,s')}(z) \right)$$
(6.4)

is a solution of the KZ equation (2.4). The braiding and fusion properties of this correlators reproduce the ones of their minimal counterparts. Setting $x = z$ we recover the general DF integrals describing the minimal correlators. The infinite sum over t reflects the fact that for nonintegral weights the representations of $sl(2)$ are infinite dimensional.

7. Let us discuss now the reduction of the KZ equation for nonintegral weights. There is marked difference from the case of integral weights when the KZ system was a finite triangular system. In the case of general weights it is again triangular but not finite any more. Nevertheless it is still possible to "reduce" the KZ equation to a BPZ equation*. The point is that due to the KM singular vectors there are relations among the integrals (6.3) which allow to truncate the infinite KZ system and subsequently reduce it. This idea was put forward in [3] and [4] while in [5] it was developed into

---

* In the case of h_J with $j \neq 0 \neq j'$ the formulas for the Virasoro singular vectors [19,21] are much less explicit than in the case when either j or j' vanishes.

an algorithm for transforming KM singular vectors into Virasoro ones via the KZ equation.

For spins of the form $J = j - j'/\nu$ with $r = 2j + 1$, $r' = 2j' + 1 \in \mathbb{Z}_+$, $\nu \in \mathbb{C}$, $\nu \neq 0$, the KM Verma module of highest weight $2J$ contains a singular vector of weight $2(J - r)$ (this is part of the Kac-Kazhdan theorem [6]). There is an explicit expression for the singular vector $P_{(r,r')}|J\rangle$ given by the Malikov-Feigin-Fuchs formula [8]:

$$P_{(r,r')} = (X_0^-)^{r+(r'-1)/\nu}(X_{-1}^+)^{r+(r'-2)/\nu} \\ \cdots (X_0^-)^{r-(r'-3)/\nu}(X_{-1}^+)^{r-(r'-2)/\nu}(X_0^-)^{r-(r'-1)/\nu} . \quad (7.1)$$

This expression is a monomial but X_0^-, X_{-1}^+ are raised to, in general, complex powers. It can be rewritten as a polynomial in X_0^-, X_{-n}^α, $n = 1, 2 \ldots$, with coefficients that are polynomials in $1/\nu$. Recently a different algorithm to obtain these singular vectors was proposed [22].

Inserting such a KM descendant field into a correlator one can "pull outside" P and obtain (in analogy with the way the BPZ equations (1.6) were produced)

$$\mathcal{P}_{(r,r')}U^{\mathrm{wzw}}(x,z) = 0 . \quad (7.2)$$

In the basis $\{U_t\}$ this relation becomes [5]

$$\frac{(\mathcal{X}_0^-)^{rr'}}{(r(r'-1))!}U_{rr'} + \sum_{k=0}^{r(r'-1)-1}\frac{\nu^{r(r'-1)-k}}{k!}\mathcal{P}_{(r,r')}^{(k)}U_{r+k} = 0, \qquad P^{(k)} \equiv \left(\frac{\partial}{\partial X_{-1}^+}\right)^k P . \quad (7.3)$$

Note that the "derivatives" of P are well defined. Taking all the derivatives in x (coming from $\mathcal{X}_{-n}^{0\pm}$) one is left with a linear relation (with rational in z coefficients) between the integrals $\mathbb{I}_0, \mathbb{I}_1, \ldots, \mathbb{I}_{rr'}$. Thus in particular the infinite KZ system is effectively truncated.

Using REDUCE it has been checked on a number of examples that the triangular KZ equation (4.7) transforms the KM null decoupling relation (7.3) into a Virasoro null decoupling equation (BPZ equation). The core of the algorithm consists of transforming every $X_{-n+1}^-U_{t+1}$ from (7.3) into $(L_{-n}^{(h_{rr'})} - \frac{(n-1)}{2}X_{-n}^0 - L_{-n}^{\mathrm{sug}}|_{\mathrm{vac}})U_t$ — we are just applying equation (5.1). Finally one is left with $L^{(h_{rr'})}$'s acting on U_0 - a BPZ equation of order rr'.

8. At the end we point to a possible generalisation of our approach to bigger algebras. It is known that for a g-WZW model there are different reductions corresponding to the different embeddings of $sl(2)$ into g [23]. This fact suggests that the constraint $x = z$ considered above can be generalised equating the "space" coordinate z with the "isospin" coordinate x of the special $sl(2)$ which defines the reduction. The standard W_n-algebras [24] correspond to a reduction of $sl(n)$ along the principle embedding defined by the Weyl vector ρ. For these W-algebras the "Kac table" can be written in a way generalising (1.2)

$$h_\Lambda = h_\Lambda^{\mathrm{sug}} - \rho \cdot \Lambda , \qquad h_\Lambda^{\mathrm{sug}} = \frac{\Lambda \cdot (\Lambda + 2\rho)}{2(k+n)} , \quad (8.1a)$$

$$\Lambda = \sum_{i=1}^{n-1} n_i \Lambda_i , \qquad n_i = (r_i - 1) - (r_i' - 1)(k + n) , \quad (8.1b)$$

where Λ_i are the fundamental weights and $\rho = \sum \Lambda_i$.

The two (and three) point functions in the functional realisation of $sl(n)$ – parametrized by an upper triangular matrix with ones on the diagonal, i.e., a complex coordinate for every negative root, have been written down in [25]

$$U(x_1, x_2, z_1, z_2, \Lambda) = \text{const } z_{12}^{-2h_\Lambda^{\text{sug}}} \prod_{i=1}^{n-1} \left(\Delta^{(i)}(x_1, x_2)\right)^{n_i} \tag{8.2}$$

where

$$\Delta^{(k)}(x_1, x_2) = \sum_{i,j} \varepsilon^{i_1 \ldots i_k \, j_1 \ldots j_{n-k}} \prod_{r=1}^{k} (x_1)_{i_r, n-k+r} \prod_{s=1}^{n-k} (x_2)_{j_s, k+s} \,. \tag{8.3}$$

For simplicity consider the case of $sl(3)$. Then

$$\Delta^{(1)}(x, y) = (x_{13} - y_{13}) - y_{12}(x_{23} - y_{23}), \qquad \Delta^{(2)}(x, y) = -(x_{13} - y_{13}) + x_{12}(x_{23} - y_{23}). \tag{8.4}$$

The principle embedding leads to the constraint

$$(x_a)_{12} = (x_a)_{23} = z_a \,, \qquad (x_a)_{13} = z_a^2/2 \,, \tag{8.5}$$

thus

$$\Delta^{(i)}(x_1, x_2) \to (z_1 - z_2)^2 \,, \tag{8.6}$$

and taking into account (8.1a) the two point functions of WZW reduce to the ones of the W-model. For general n the constraint "$x = z$" leads to

$$x_{ij} = \frac{z^{j-i}}{(j-i)!} \tag{8.7}$$

and

$$\left.\Delta^{(i)}(x_1, x_2)\right|_{x=z} = \text{const } (z_1 - z_2)^{i(n-i)} \,. \tag{8.8}$$

Since $\Lambda \cdot \rho = \frac{1}{2}\sum_{i=1}^{n-1} i(n-i) n_i$ the reduction for 2-point functions indeed holds.

We remark that for the partition functions of WZW models a constraint similar to our $x = z$ produces the partition functions of W-models [26].

REFERENCES

[1] V. Drinfeld and V. Sokolov, *J. Sov. Math.* **30** (1984) 1975; J. Balog et al., *Ann. of Phys.* **203** (1990) 76; L. Fehér et al., Dublin preprint, DIAS-STP-91-29.
[2] M. Bershadsky and H. Ooguri, *Comm. Math. Phys.* **126** (1989) 49; B. Feigin and E. Frenkel, *Phys. Lett.* **246B** (1990) 75; also P. Bouwknegt, J. McCarthy, K. Pilch, CERN-TH.6267/91, Proceedinggs of 'Strings and Symmetries 1991', Stony Brook 1991.
[3] P. Furlan, R. Paunov, A.Ch. Ganchev and V.B. Petkova, *Phys. Lett.* **267B** (1991) 63.
[4] P. Furlan, R. Paunov, A.Ch. Ganchev and V.B. Petkova, CERN-TH.6289/91.
[5] A.Ch. Ganchev and V.B. Petkova, *Phys. Lett.* **293B** (1992) 56.
[6] V. G. Kac and D. A. Kazhdan, *Adv. Math.* **34** (1979) 97.
[7] L. Benoit and Y. Saint-Aubin, *Phys. Lett.* **215B** (1988) 517.
[8] F.G. Malikov, B.L. Feigin and D.B. Fuks, *Funct. Anal. Prilozhen.* **20**, no. 2 (1987) 25.
[9] A. Belavin, A. Polyakov and A. Zamolodchikov, *Nucl. Phys.* **B241** (1984) 333.

[10] B.L. Feigin and D.B. Fuchs, *Funk. Anal. Prilpzh.* **16**, No 2 (1982) 47; V.S. Dotsenko and V.A. Fateev, *Nucl. Phys.* **B240** (1984) 312; **B251** (1985) 691; A. Tsuchyia and Y. Kanie, *Publ. RIMS* **22** (1986) 259; G. Felder, *Nucl. Phys.* **B317** (1989) 215.
[11] G. Felder and R. Silvotti, *Phys. Lett.* **231B** (1989) 411.
[12] V.G. Knizhnik and A.B. Zamolodchikov, *Nucl. Phys.* **B247** (1984) 83.
[13] P. Christe and R. Flume, *Nucl. Phys.* **B282** (1987) 466;
P. Christe, PhD Thesis, Bonn (1986).
[14] A.Gerasimov *et al.*, *Int. J. Mod. Phys.* **A5** (1990) 2495; B.L. Feigin and E.V. Frenkel, in: V.G. Knizhnik Memorial volume, eds. L. Brink *et al.*, World Scientific, Singapore, 1990; D. Bernard and G. Felder, *Comm. Math. Phys.* **127** (1990) 145; A. Tsuchyia and Y. Kanie, *Adv. Stud. Pure Math.* **16** (1988) 297; P. Bouwknegt, J. McCarthy, K. Pilch, *Prog. Theor. Phys. (Suppl.)* **102** (1990) 67.
[15] E. Date, M. Jimbo, A. Matsuo and T. Miwa, in: Yang-Baxter Equations, Conformal Invariance and Integrability in Statistical Mechanics and Field Theory, World Scientific (1989); A. Matsuo, *Comm. Math. Phys.* **134** (1990) 65; G. Kuroki, *Comm. Math. Phys.* **142** (1991) 511; F. Falceto, K. Gawedzki and A. Kupiainen, *Phys. Lett.* **260B** (1991) 101; H. Awata, A. Tsuchiya and Y. Yamada, Tsukuba preprint, *Nucl. Phys.* **B365** (1991) 680.
[16] V.V. Schechtman and A.N. Varchenko, *Inven. Math.* **106** (1991) 139.
[17] K. Aomoto, *J. Math. Soc. Japan* **39** (1987) 191.
[18] A.B. Zamolodchikov and V.A. Fateev, *Sov. J. Nucl. Phys.* **43** (1986) 657.
[19] M. Bauer, Ph. Di Francesco, C. Itzykson and J.-B. Zuber, *Nucl. Phys.* **B362** (1991) 515.
[20] Vl.S. Dotsenko, *Nucl. Phys.* **B358** (1991) 67.
[21] A. Kent, *Phys. Lett.* **273B** (1992) 56.
[22] M. Bauer, N. Sochen, *Phys. Lett.* **275B** (1992) 82.
[23] F.S. Bais, T. Tjin, P. van Driel, *Nucl. Phys.* **B357** (1991) 632; L. Fehér *et al.*, *Ann. of Phys.* **213** (1991) 1.
[24] A. Zamolodchikov, *Teor. Mat. Fiz.* **59** (1985) 108; A. Zamolodchikov, V. Fateev, *Nucl. Phys.* **B280** (1987) 644; V.A. Fateev, S.L. Lykyanov, *Int. J. Mod. Phys.* **A3** (1988) 507.
[25] W. Rühl, *Ann. of Phys.* **206** (1991) 368.
[26] S. Mukhi and S. Panda, *Nucl. Phys.* **B338** (1990) 263; E.V. Frenkel, V.G. Kac and M. Wakimoto, *Comm. Math. Phys.* **147** (1992) 295.

NONCRITICAL DIMENSIONS FOR CRITICAL STRING THEORY: LIFE BEYOND THE CALABI–YAU FRONTIER

Rolf Schimmrigk

Institut für Theoretische Physik
Universität Heidelberg
Philosophenweg 16
6900 Heidelberg, FRG

ABSTRACT

A recently introduced framework for the compactification of supersymmetric string theory involving noncritical manifolds of complex dimension $2k + D_{\mathrm{crit}}$, $k \geq 1$, is reviewed. These higher dimensional manifolds are spaces with quantized positive Ricci curvature and therefore do not, a priori, describe consistent string vacua. It is nevertheless possible to derive from these manifolds the massless spectra of critical string ground states. For a subclass of these noncritical theories it is also possible to explicitly construct Calabi–Yau manifolds from the higher dimensional spaces. Thus the new class of theories makes contact with the standard framework of string compactification. This class of manifolds is more general than that of Calabi–Yau manifolds because it contains spaces which correspond to critical string vacua with no Kähler deformations, i.e. no antigenerations, thus providing mirrors of rigid Calabi–Yau manifolds. The constructions reviewed here lead to new insights into the relation between exactly solvable models and their mean field theories on the one hand and Calabi–Yau manifolds on the other, leading, for instance, to a modification of Gepner's conjecture. They also raise fundamental questions about the Kaluza–Klein concept of string compactification, in particular regarding the rôle played by the dimension of the internal theories.

1. INTRODUCTION

String theory remains the only viable candidate for a unified theory of quantum gravity. One of the attractions of this theory is the fact that it describes a rather tight framework. A consequence is that there are severe restrictions on the internal part of the theory which to a large extent determines the observable low energy physics in four dimensions. Based on the conventional framework formulated in ref. 1 it is believed that in left–right symmetric compactifications without torsion the internal space of the heterotic string is described by a space which has to be a compact manifold which is

- complex,

- Kähler, and admits a

- covariantly constant spinor,

i.e. has vanishing first Chern class, so–called Calabi–Yau manifolds.

Such manifolds are particularly simple, a fact that is encoded concisely in the spectrum of the theory described, in part, by the cohomology of the space. Because the space is complex the real cohomology can be decomposed, via the Hodge decomposition, into complex cohomology groups. Thus the Betti numbers $b_i = \dim_{\mathbb{R}} H^i(\mathcal{M}, \mathbb{R})$ can be expressed in terms of the Hodge numbers $h^{(p,q)} = \dim_{\mathbb{C}} H^{p,q}(\mathcal{M}, \mathbb{C})$:

$$b_i = \sum_{p+q=i} h^{(p,q)}. \tag{1}$$

Because the manifold is Kähler the Hodge numbers are symmetric, $h^{(p,q)} = h^{(q,p)}$, and because the first Chern class vanishes it follows that $h^{(p,0)} = 0 = h^{(0,p)}$ for $p = 1, 2$ and $h^{(3,0)} = 1 = h^{(0,3)}$. Hence the cohomology of the internal space, summarised in the Hodge diamond

$$\begin{array}{ccccccc}
 & & & 1 & & & \\
 & & 0 & & 0 & & \\
 & 0 & & h^{(1,1)} & & 0 & \\
1 & & h^{(2,1)} & & h^{(2,1)} & & 1 \\
 & 0 & & h^{(1,1)} & & 0 & \\
 & & 0 & & 0 & & \\
 & & & 1 & & & \\
\end{array}$$

contains only two independent elements $h^{(1,1)} = h^{(2,2)}$ and $h^{(2,1)} = h^{(1,2)}$ which parametrise the number of antigenerations and generations, respectively, that are observed in low energy physics.

It is also believed that this class of string vacua features an unexpected symmetry, mirror symmetry, which has been discovered in the context of Landau–Ginzburg vacua in ref. 2. Independent evidence for this symmetry has been found in the context of orbifolds of exactly solvable tensor models by Greene and Plesser[3]. The effect of this symmetry is that for each string vacuum with some number $h^{(1,1)}$ of antigenerations and some number $h^{(2,1)}$ of generations there exists a mirror vacuum for which these number are exchanged: the spectrum of the mirror vacuum consists of $h^{(2,1)}$ antigenerations and $h^{(1,1)}$ generations. Mirror symmetry thus flips the Hodge diamond along the offdiagonal.

Mirror symmetry is by now well established: beyond the class of exactly solvable models discussed in ref. 3, in which mirror symmetry is understood best, lies the much larger class of Landau–Ginzburg theories constructed explicitly in ref. 2[1]which clearly indicates that mirror symmetry is a property of string theory. That this symmetry is not accidental in this wider context has been proven in ref. 4 where it was shown that by

---

[1]The construction of all quasihomogeneous $N = 2$ Landau–Ginzburg with an isolated singularity was recently completed in ref. 5, 6.

a combination of orbifolding and fractional transformations a mirror construction can be established between a priori independent pairs of Landau–Ginzburg theories with opposite spectrum. Mirror symmetry is at present being used as a hypothesis to obtain results in algebraic geometry and has been shown to be correct in all computations that have been performed so far[10–12].

Mirror symmetry creates a puzzle. There exist well-known Calabi–Yau vacua which are rigid, i.e. they do not have string modes corresponding to complex deformations of the manifold. Since mirror symmetry exchanges complex deformations and Kähler deformations of a manifold it would seem that the mirror of a rigid Calabi–Yau manifold cannot be Kähler and hence does not describe a consistent string vacuum. In fact, it appears, using Zumino's result[13] that $N = 2$ supersymmetry of a σ-model requires that the target manifold is Kähler, that the mirror vacuum cannot even be $N = 1$ spacetime supersymmetric. It follows that the class of Calabi–Yau manifolds is not the appropriate setting by a long shot in which to discuss mirror symmetry and the question arises what the proper framework might be.

In this review I discuss recent work[14] which shows the existence of a new class of manifolds which generalises the class of Calabi–Yau spaces of complex dimension D_{crit} in a natural way. The manifolds involved are of complex dimension $(2k + D_{crit})$ and have a positive first Chern class which is quantized in multiples of the degree of the manifold. Thus they do not describe, a priori, consistent string ground states. Surprisingly however, it is possible to derive from these higher dimensional manifolds the spectrum of critical string vacua. This can be done not only for the generations but also for the antigenerations. For particular types of these new manifolds it is also possible to construct the corresponding D_{crit}-dimensional Calabi–Yau manifold directly from the $(2k + D_{crit})$-dimensional space.

This new class of manifolds is, however, not in one to one correspondence with the class of Calabi–Yau manifolds as it also contains manifolds which describe string vacua that do not contain massless modes corresponding to antigenerations. It is precisely this new type of manifold that is needed in order to construct mirrors of rigid Calabi–Yau manifolds without generations.

2. HIGHER DIMENSIONAL MANIFOLDS WITH QUANTIZED POSITIVE FIRST CHERN CLASS

Consider the class of manifolds of complex dimension N embedded in a weighted projective space $\mathbb{P}_{(k_1,...,k_{N+2})}$ as hypersurfaces

$$M_{N,d} \equiv \{p(z_1,\ldots,z_{N+2}) = 0\} \cap \mathbb{P}_{(k_1,...,k_{N+2})} \quad (2)$$

defined as the zero set of some transverse polynomial p of degree d. Here the k_i are the weights of the ambient weighted projective space. The set of hypersurfaces determined by such polynomials will be denoted by

$$\mathbb{P}_{(k_1,k_2,......,k_{N+2})}[d] \quad (3)$$

and called a configuration. Assume that for the hypersurfaces (3) the weights k_i and the degree d are related via the constraint

$$\sum_{i=1}^{N+2} k_i = Qd, \quad (4)$$

where Q is a positive integer. Relation (4) is the defining property of the class of spaces to be considered below. It is a rather restrictive condition in that it excludes many types of varieties which are transverse and even smooth but are not of physical relevance[2]. A simple example is the Fermat hypersurface

$$\mathbb{P}_{(420,280,210,168,140,120,105)}[840] \ni \{p = \sum_{i=1}^{7} z_i^{i+1} = 0\} \tag{5}$$

which is a rather nice, transverse, i.e. quasismooth manifold. It is also interesting from a different point of view. A curious aspect of Calabi–Yau hypersurfaces is that they are automatically what is called *well formed*, i.e. they do not contain orbifold singularities that are surfaces (in the case of threefolds). More generally this fact translates into the statement that the only resolutions that have to be performed are so–called small resolutions, i.e. the singular set are of codimension larger than one. The same is true for the higher dimensional manifolds defined above whereas the manifold (5) contains the singular 4–fold $S = \mathbb{P}_{(210,140,105,84,70,60)}[420]$.

Alternatively, manifolds of the type above may be characterised via a curvature constraint. Because of (4) the first Chern class is given by

$$c_1(M_{N,d}) = (Q-1)\, c_1(\mathcal{N}) \tag{6}$$

where $c_1(\mathcal{N}) = dh$ is the first Chern class of the normal bundle $\mathcal{N}$ of the hypersurface $M_{N,d}$ and h is the pullback of the Kähler form $H \in H^{(1,1)}\left(\mathbb{P}_{(k_1,...,k_{N+2})}\right)$ of the ambient space. Hence the first Chern class is quantized in multiples of the degree of the hypersurface $M_{N,d}$. For $Q=1$ the first Chern class vanishes and the manifolds for which (4) holds are Calabi–Yau manifolds, defining consistent ground states of the supersymmetric closed string. For $Q > 1$ the first Chern class is nonvanishing and therefore these manifolds cannot possibly describe vacua of the critical string, or so it seems.

It turns out that these spaces are closely related to string vacua of complex critical dimension

$$D_{\text{crit}} = N - 2(Q-1) \tag{7}$$

i.e. the critical dimension is offset by twice the coefficient of the first Chern class of the normal bundle of the hypersurface. The evidence for this is twofold. First it is possible to derive from these higher dimensional manifolds the massless spectrum of critical vacua. Furthermore it is shown that for certain subclasses of hypersurfaces of type (4) it is possible to construct Calabi–Yau manifolds M_{CY} of dimension D_{crit} and complex codimension

$$\text{codim}_{\mathbb{C}}(M_{CY}) = Q \tag{8}$$

directly from these manifolds. In terms of the critical dimension and the codimension the class of manifolds to be investigated below can be described as the projective configurations

$$\mathbb{P}_{(k_1,...,k_{(D_{\text{crit}}+2Q)})}\left[\frac{1}{Q}\sum_{i=1}^{D_{\text{crit}}+2Q} k_i\right]. \tag{9}$$

As mentioned already in the introduction the class of spaces defined by (9) contains manifolds with no antigenerations and hence it is necessary to have some way other than Calabi–Yau manifolds to represent string ground states if one wants to compare them

---

[2]It will become clear below that this definition is rather natural in the context of the theory of Landau–Ginzburg string vacua with an arbitrary number of scaling fields. A particular simple manifold in this class, the cubic sevenfold $\mathbb{P}_8[3]$, has been the subject of recent investigations[7–9].

with the higher dimensional manifolds. One possible way to do this is to relate them to Landau–Ginzburg theories: any manifold of type (3) can be viewed as a projectivisation via a weighted equivalence defined on an affine noncompact hypersurface defined by the same polynomial

$$\mathbb{C}_{(k_1,...,k_{N+2})}[d] \ni \{p(z_1, ..., z_{N+2}) = 0\}. \tag{10}$$

Because the polynomial p is assumed to be transverse in the projective ambient space the affine variety has a very mild singularity: it has an isolated singularity at the origin defining what is called a catastrophe in the mathematics literature.

The complex variables z_i parametrising the ambient space are to be viewed as the field theoretic limit $\varphi_i(z,\bar{z}) = z_i$ of the lowest components of the order parameters $\Phi_i(z_i, \bar{z}_i, \theta_i^\pm, \bar{\theta}_i^\pm)$, described by chiral $N = 2$ superfields of a 2–dimensional Landau–Ginzburg theory defined by the action

$$\int d^2z d^2\theta d^2\bar{\theta}\ K(\Phi_i, \bar{\Phi}_i) + \int d^2z d^2\theta\ W(\Phi_i) + c.c.\,, \tag{11}$$

where K is the Kähler potential and W is the superpotential. It was the important insight of Martinec[15] and Vafa and Warner[16] that such Landau–Ginzburg theories are useful for the understanding of string vacua and also that much information about such ground states is already encoded in the associated catastrophe (10). A crucial piece of information about a vacuum, e.g., is its central charge. Using a result from singularity theory, it is easy to derive that the central charge of the conformal fixed point of the LG theory is

$$c = 3\sum_{i=1}^{N+2}(1 - 2q_i), \tag{12}$$

where $q_i = k_i/d$ are the U(1) charges of the superfields. It is furthermore possible to derive the massless spectrum of the GSO projected fixed of the LG theory, defining the string vacuum, directly from the catastrophe (10) via a procedure described by Vafa[17].

The manifolds (9) therefore correspond to LG theories of central charge

$$c = 3(N - 2(Q - 1)) = 3D_{\text{crit}} \tag{13}$$

where the relation (7) has been used.

In certain benign situations the subring of monomials of charge 1 in the chiral ring describes the generations of the vacuum[18]. For this to hold at all it is important that the GSO projection is the canonical one with respect to the cyclic group $\mathbb{Z}_d$, the order of which is the degree d of the superpotential[3] Thus the generations are easily derived for this subclass of theories in (9) because the polynomial ring is identical to the chiral ring of the corresponding Landau–Ginzburg theory. In general a more sophisticated analysis, involving the resolution of higher dimensional singularities, will have to be done[20].

It remains to extract the second cohomology. In a Calabi–Yau manifold there are no holomorphic 2–forms and hence all of the second cohomology is in $H^{(1,1)}$. Because of Kodaira's vanishing theorem the same is true for manifolds with positive first Chern class and therefore for the manifolds under discussion. At first sight it might appear hopeless to find a construction corresponding to the analysis of (2,1)–forms because of the following example which involves the orbifold of a 3–torus.

---

[3]It does not hold for projections that involve orbifolds with respect to different groups such as those discussed in ref. 19. This is to be expected as these modified projections can be understood as orbifolds of canonically constructed vacua. The additional moddings generate singularities the resolution of which introduces, in general, additional modes in both sectors, generations and antigenerations.

Consider the orbifold $T_1^3/\mathbb{Z}_3^2$ where the two actions are defined as $(z_1, z_4) \longrightarrow (\alpha z_1, \alpha^2 z_4)$, all other coordinates invariant and $(z_1, z_7) \longrightarrow (\alpha z_1, \alpha^2 z_7)$, all other invariant. Here α is the third root of unity. The resolution of the singular orbifold leads to a Calabi–Yau manifold with 84 antigenerations and no generations[21]. This is precisely the mirror flipped spectrum of the exactly solvable tensor model 1^9 of 9 copies of $N=2$ superconformal minimal models at level $k=1$ [22] which can be described in terms of the Landau–Ginzburg potential $W = \sum z_i^3$ which belongs to the configuration $\mathbb{C}_{(1,1,1,1,1,1,1,1,1)}[3]$. After imposing the GSO projection by modding out a $\mathbb{Z}_3$ symmetry this Landau–Ginzburg theory leads to the same spectrum as the 1^9 theory.

This Landau–Ginzburg theory clearly is a mirror candidate for the resolved torus orbifold just mentioned[7–9] and the question arises whether a manifold corresponding to this LG potential can be found. Since the theory does not contain modes corresponding to (1,1)–forms it appears that the manifold cannot be Kähler and hence not projective. Thus it appears that the 7-dimensional manifold $\mathbb{P}_8[3]$ whose polynomial ring is identical to the chiral ring of the LG theory is merely useful as an auxiliary device in order to describe one sector of the critical LG string vacuum. Even though there exists a precise identity between the Hodge numbers in the middle cohomology group of the higher dimensional manifold and the middle dimension of the cohomology of the Calabi–Yau manifold this is not the case for the second cohomology group.

3. NONCRITICAL MANIFOLDS AND CRITICAL VACUA

It turns out however, that by looking at the manifolds (9) in a slightly different way it is nevertheless possible to extract the second cohomology in a canonical manner (even if there is *none*). The way this works is as follows: the manifolds of type (9) will, in general, not be described by smooth spaces but will have singularities which arise from the projective identification. The basic idea now is to associate the existence of antigenerations in a *critical* string vacuum with the existence of singularities in these higher dimensional *noncritical* spaces.

Because the structure of these geometrical singularities depends on the precise form of the polynomial constraint it is difficult to prove the correctness of this idea in full generality. Instead I will, in the following, make this idea more precise and illustrate how it works with a few particularly simple classes of theories, leaving a more detailed investigation of other types of manifolds for a more extensive discussion[20]. As an unexpected bonus this derivation will provide new insight into the Landau–Ginzburg/Calabi–Yau connection.

Consider again the simple example related to the tensor model 1^9. Its LG theory is described by $\mathbb{C}_9^*[3]$ the naive compactification of which leads to

$$\mathbb{P}_8[3] \ni \{\sum_{i=1}^9 z_i^3 = 0\}. \tag{14}$$

Counting monomials leads to the spectrum of 84 generations found previously for the corresponding string vacuum and because this manifold is smooth *no* antigenerations are expected in this model! Hence there does not exist a Calabi–Yau manifold that describes the ground state[4]. A second theory in the space of all LG vacua with no

---

[4]It would seem that a generalisation of this 7-dimensional smooth manifold is the infinite class of models $\mathbb{C}_{(1,1,1,1,1,1,1,1,1+3q)}[3+q]$, but since the manifolds (9) are required to be transverse the only possibility is $q=0$.

antigenerations is

$$(2^6)^{(0,90)}_{A^6} \equiv \mathbb{C}^*_{(1,1,1,1,1,1,2)}[4] \ni \{\sum_{i=1}^{6} z_i^4 + z_7^2 = 0\} \tag{15}$$

with an obviously smooth manifold $\mathbb{P}_{(1,1,1,1,1,1,2)}[4]$.

Vacua without antigenerations are rather exceptional however; the generic ground state will have both sectors, generations and antigenerations. The idea described above to derive the antigenerations works for higher dimensional manifolds corresponding to different types of critical vacua but in the following we will illustrate it with two types of such manifolds. A more detailed analysis can be found in ref. 20.

To be concrete consider the exactly solvable tensor theory $(1 \cdot 16^3)_{A_2 \otimes E_7^3}$ with 35 generations and 8 antigenerations which corresponds to a Landau–Ginzburg theory belonging to the configuration

$$\mathbb{C}^*_{(2,3,2,3,2,3,3)}[9]^{(8,35)} \tag{16}$$

and which induces, via projectivisation, a 5–dimensional weighted hypersurface

$$\mathbb{P}_{(2,2,2,3,3,3,3)}[9] \ni \{p = \sum_{i=1}^{3}(y_i^3 x_i + x_i^3) + x_4^3 = 0\}. \tag{17}$$

with orbifold singularities

$$\mathbb{Z}_3 \;:\; \mathbb{P}_3[3] \ni \{p_1 = \sum_{i=1}^{4} x_i^3 = 0\},$$

$$\mathbb{Z}_2 \;:\; \mathbb{P}_2. \tag{18}$$

The $\mathbb{Z}_3$–singular set is a smooth cubic surface which supports seven (1,1)–form as can be easily shown. The $\mathbb{Z}_2$ singular set is just the projective plane and therefore adds one further (1,1)–form. Hence the singularities induced on the hypersurface by the singularities of the ambient weighted projective space give rise to a total of eight (1,1)–forms. A simple count leads to the result that the subring of monomials of charge 1 is of dimension 35. Thus we have derived the spectrum of the critical theory from the noncritical manifold (17).

It is presumably possible to derive this result via a surgery process on the singular space (17) but more important is, at this point, that the idea introduced above of relating the spectrum of the string vacuum to the singularity structure of the noncritical manifold also makes it possible to derive from these higher dimensional manifolds the Calabi–Yau manifold of critical dimension! Thus a canonical prescription is obtained which also allows to pass from the Landau–Ginzburg theory to its geometrical counterpart.

This works as follows: Recall that the structure of the singularities of the weighted hypersurface just involved part of the superpotential, namely the cubic polynomial p_1 which determined the $\mathbb{Z}_3$ singular set described by a surface. The superpotential thus splits naturally into the two parts

$$p = p_1 + p_2 \tag{19}$$

where p_2 is the remaining part of the polynomial. The idea now is to consider the product $\mathbb{P}_3[3] \times \mathbb{P}_2$ where the factors are determined by the singular sets of the higher dimensional space and to impose on this 4–dimensional space a constraint described by the remaining part of the polynomial which did not take part in constraining the singularities of this ambient space. In the case at hand this leaves a polynomial of bidegree $(3,1)$ and hence we are lead to a manifold embedded in

$$\begin{array}{c} \mathbb{P}_2 \\ \mathbb{P}_3 \end{array} \begin{bmatrix} 3 & 0 \\ 1 & 3 \end{bmatrix} \tag{20}$$

defined by polynomials

$$\begin{aligned} p_1 &= y_1^3 x_1 + y_2^3 x_2 + y_3^3 x_3, \\ p_2 &= \sum_{i=1}^{4} x_i^3, \end{aligned} \tag{21}$$

which is precisely the manifold constructed in ref. 23, the exactly solvable model of which was later found in ref. 24. Thus we have found how to construct from the noncritical manifold (17) the critical Calabi–Yau manifold.

A subclass of manifolds of a different type which can be discussed in this framework rather naturally is defined by

$$\mathbb{P}_{(2k,K-k,2k,K-k,2k_3,2k_4,2k_5)}[2K] \tag{22}$$

where $K = k + k_3 + k_4 + k_5$ and it is assumed, for simplicity, that K/k and K/k_i are integers. The potentials are

$$W = \sum_{i=1}^{2}(x_i^{K/k} + x_i y_i^2) + x_3^{K/k_3} + x_4^{K/k_4} + x_5^{K/k_5}. \tag{23}$$

The singularities in these manifolds are of two types,

$$\begin{array}{ll} \mathbb{Z}_2 : & \mathbb{P}_{(k,k,k_3,k_4,k_5)}[K], \\ \mathbb{Z}_{K-k} : & \mathbb{P}_1. \end{array} \tag{24}$$

where the constraint of degree K is given by

$$p_1 = \sum_{i=1}^{5} x^{K/k_i}. \tag{25}$$

The $\mathbb{Z}_2$-singular set is 3–fold with positive first Chern class embedded in weighted $\mathbb{P}_4$ whereas the $\mathbb{Z}_{K-k}$ singular set is just the sphere $S^2 \sim \mathbb{P}_1$.

To construct the corresponding critical manifolds note that the structure of the singularities of the weighted hypersurface just involved part of the superpotential, namely the quartic polynomial p_1 which determined the $\mathbb{Z}_2$ singularity set described by a 3–fold. The superpotential thus splits naturally into the two parts $p = p_1 + p_2$ where p_2 is the remaining part of the polynomial. The idea now is to consider again the product $\mathbb{P}_{(k,k,k_3,k_4,k_5)}[K] \times \mathbb{P}_1$ of singular sets of the higher dimensional space and to impose on this 4–dimensional space a constraint described, as before, by the remaining part p_2 of the polynomial which did not take part in constraining the singularities of this ambient space. In the case at hand this leaves a polynomial of bidegree $(k,2)$ and hence we are lead to a manifold embedded in

$$\begin{array}{c} \mathbb{P}_1 \\ \mathbb{P}_{(k,k,k_3,k_4,k_5)} \end{array} \begin{bmatrix} 2 & 0 \\ k & K \end{bmatrix} \tag{26}$$

defined by polynomials

$$\begin{aligned} p_1 &= y_1^2 x_1 + y_2^2 x_2 \\ p_2 &= x_1^{K/k} + x_2^{K/k} + x_3^{K/k_3} + x_4^{K/k_4} + x_5^{K/k_5}. \end{aligned} \quad (27)$$

That this correspondence is in fact correct can be inferred from the work of ref. 25 where it was shown that these codimension–2 weighted CICYs correspond to $N = 2$ minimal exactly solvable tensor models of the type

$$\left[2\left(\frac{K}{k}-1\right)\right]_D^2 \cdot \prod_{i=3}^{5}\left(\frac{K}{k_i}-2\right)_A. \quad (28)$$

where the subscripts indicate the affine invariants chosen for the individual levels.

The general picture that emerges from these constructions then is the following: embedded in the higher dimensional manifold is a submanifold which is fibered where the base and the fibres are determined by the singular sets of the ambient manifold. The Calabi–Yau manifold itself is a hypersurface embedded in this fibered submanifold. A heuristic sketch of the geometry is shown in the Figure 1.

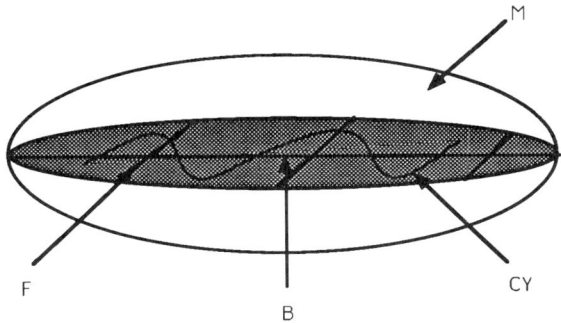

Figure 1. The Calabi–Yau manifold CY is embedded in the fibered submanifold $B \times F$ of the higher dimensional manifold M. Both the base B and the fibres F of the bundle (indicated by the grey surface) are determined by the singular sets of the manifold M.

The examples above illustrate the simplest situation that can appear. In more complicated manifolds the singularity structure will consist of hypersurfaces whose fibers and/or base themselves are fibered, leading to an iterative procedure. The submanifold to be considered will, in those cases, be of codimension larger than one and the Calabi–Yau manifold will be described by a submanifold with codimension larger than one as well. To illustrate this point consider the 7-fold

$$\mathbb{P}_{(1,1,6,6,2,2,2,2,2)}[8] \ni \left\{ \sum_{i=1}^{2}\left(x_i^2 y_i + y_i z_i + z_i^4\right) + z_3^4 + z_4^4 + z_5^4 = 0 \right\} \quad (29)$$

which leads to the $\mathbb{Z}_2$ fibering $\mathbb{P}_1 \times \mathbb{P}_{(3,3,1,1,1,1,1)}[4]$ which in turn leads to the $\mathbb{Z}_3$ fibering $\mathbb{P}_1 \times \mathbb{P}_1 \times \mathbb{P}_4[4]$. Following the splits of the potential thus leads to the Calabi–Yau configuration

$$\begin{matrix} \mathbb{P}_1 \\ \mathbb{P}_1 \\ \mathbb{P}_4 \end{matrix} \begin{bmatrix} 2 & 0 & 0 \\ 1 & 1 & 0 \\ 0 & 1 & 4 \end{bmatrix} \ni \left\{ \begin{matrix} p_1 = \sum_{i=1}^{2} x_i^2 y_i = 0 \\ p_2 = \sum_{i=1}^{2} y_i z_i = 0 \\ p_3 = \sum_{j=1}^{5} z_i^4 = 0 \end{matrix} \right\} \quad (30)$$

which is of codimension 3. This example also shows that there are nontrivial relations between these higher dimensional manifolds. The way to see this is via the process of splitting and contraction of Calabi–Yau manifolds introduced in ref. 26. It can be shown in fact that the Calabi–Yau manifold (30) is an ineffective split of a Calabi–Yau manifold in the class (22). Thus there also exists a corresponding relation between the higher dimensional manifolds.

4. GENERALISATION TO ARBITRARY CRITICAL DIMENSIONS

Even though the examples discussed in the previous section were all concerned with 6-dimensional Calabi–Yau manifolds and the way they are embedded in the new class of spaces, it should be clear that the ideas presented are not specific to this dimension. Instead of considering compactifications down to the physical dimension, 4, one might contemplate compactifying down to 2, 6 or 8 dimensions, or else, discuss the class of manifolds above not in the context of compactification at all.

To illustrate this point consider the infinite class of manifolds

$$\mathbb{P}_{(2,n-1,2,n-1,2,\ldots,2)}[2n] \qquad (31)$$

of complex dimension $n+1$, defined by polynomials

$$p = \sum_{i=1}^{2}(x_i^n + x_i y_i^2) + x_3^n + \cdots + x_{n+1}^n. \qquad (32)$$

According to the considerations above these spaces are related to Calabi–Yau manifolds embedded in

$$\begin{matrix} \mathbb{P}_1 \\ \mathbb{P}_n \end{matrix} \begin{bmatrix} 2 & 0 \\ 1 & n \end{bmatrix} \qquad (33)$$

via the equations

$$\begin{aligned} p_1 &= y_1^2 x_1 + y_2^2 x_2, \\ p_2 &= \sum_{i=1}^{n+1} x_i^n. \end{aligned} \qquad (34)$$

The simplest example is, of course, the case $n = 2$ where the higher dimensional manifold is a 3–fold described by

$$\mathbb{P}_{(2,1,2,1,2)}[4] \ni \{\sum_{i=1}^{2}(z_i^2 + z_i y_i^2) + z_3^2 = 0\} \qquad (35)$$

with a $\mathbb{Z}_2$ singular set isomorphic to the sphere $\mathbb{P}_2[2] \sim \mathbb{P}_1$ which contributes one (1,1)–form, the remaining one being provided by the $\mathbb{P}_1$ defined by the remaining coordinates. The singularity structure of the 3–fold then relates this space to the complex torus described by the algebraic curve

$$\begin{matrix} \mathbb{P}_1 \\ \mathbb{P}_2 \end{matrix} \begin{bmatrix} 2 & 0 \\ 1 & 2 \end{bmatrix}. \qquad (36)$$

It should be remarked that the Landau–Ginzburg theory corresponding to this theory derives from an exactly solvable tensor model $(2^2)_{D^2}$ described by two $N = 2$ superconformal minimal theories at level $k = 2$ equipped with the affine D–invariant.

It is of interest to consider the cohomology groups of the 3–fold itself. With the third Chern class $c_3 = 2h^3$ the Euler number of the singular space is

$$\chi_s = \int c_3 = 1 \qquad (37)$$

and hence the Euler number of the resolved manifold is

$$\tilde{\chi} = 1 - (2/2) + 2 \cdot 2 = 4. \tag{38}$$

Since the singular set is a sphere its resolution contributes just one (1,1)–form and hence the second Betti number becomes $b_2 = 2$. With $\tilde{\chi} = 2(1 + h^{(1,1)}) - 2h^{(2,1)}$ it follows that

$$h^{(2,1)} = 1. \tag{39}$$

The case $n = 3$ is particularly illuminating because it involves a higher dimensional manifold that is smooth

$$\mathbb{P}_5[3] \tag{40}$$

and hence it is easy to determine the cohomology groups of this space and to compare it with the spectrum of $K3$

$$K3 = \begin{matrix} \mathbb{P}_1 \\ \mathbb{P}_3 \end{matrix} \begin{bmatrix} 2 & 0 \\ 1 & 3 \end{bmatrix}, \tag{41}$$

which consists of $h^{(0,0)} = h^{(2,2)} = h^{(2,0)} = h^{(0,2)} = 1$ and $h^{(1,1)} = 20$, all other Hodge numbers are zero. Hence the Euler number becomes $\chi(K3) = 24$.

The Euler number for the noncritical manifold is easily computed to be $\chi = 27$. Since the manifold is Kähler $h^{(p,q)} = h^{(q,p)}$ and because of Poincaré duality $b_p = b_{8-p}$. Because the manifold has positive first Chern class it follows from Kodaira's vanishing theorem that $h^{(p,0)} = 0$ for $p \neq 0$ and via Lefshetz' hyperplane theorem it is known that below the middle dimension all the cohomology is inherited from the ambient space and therefore the only nonvanishing cohomology groups lead to $h^{(0,0)} = h^{(1,1)} = 1$. It can be shown that $h^{(3,1)} = h^{(1,3)} = 1$[5] and therefore the only remaining cohomology is in $H^{(2,2)}$. Since

$$\chi = 2(b_0 + b_2) + b_4 = 6 + h^{(2,2)} \tag{42}$$

it follows that $h^{(2,2)} = 21 = 20 + 1$. Thus we have obtained the spectrum of $K3$ plus one additional mode which always appears in this construction.

This example is also useful because it indicates a generalisation of the considerations of the previous section. The surprising new feature of this manifold is that even though the higher dimensional manifold did not have any orbifold singularities it was nevertheless possible to split it in such a way as to construct a Calabi–Yau manifold from it. This was possible because the defining equation was not of Fermat type but involved couplings between the fields. Because of this the manifold featured a new $\mathbb{Z}_2$ symmetry not present in the Fermat hypersurface and it is this new symmetry that dictated how to perform the split. This indicates that even for smooth higher dimensional manifolds it is possible to relate them to Calabi–Yau manifolds once one moves away from the symmetric point. Table 1 containes a few other examples of how to relate different singular 4–folds to $K3$ representations.

An example involving a 4–dimensional critical manifold of a different type is defined by the polynomial

$$p = \sum_{i=1}^{3} \left(x_i^3 + x_i y_i^3 \right) + \sum_{j=4}^{5} x_j^6 \tag{43}$$

which corresponds to the tensor model $(16^3 \cdot 4)_{E^3 \otimes D^2}$ with central charge $c = 12$ and belongs to the configuration

$$\mathbb{P}_{(6,4,6,4,6,4,3,3)}[18]. \tag{44}$$

---

[5]I'm grateful to P.Candelas and T.Hübsch for explanations regarding this computation.

Table 1: Examples of dimension $D_{\mathbb{C}} = 2$.

ECFT	Projective Manifold	Singularities	CY
$(1^6)_{D^2 \otimes A^4}$	$\mathbb{P}_5[3]$ $\sum(z_i^3 + z_i y_i^2) + z_3^3 + z_4^3$	$\mathbb{Z}_2 : \mathbb{P}_1$	$\begin{matrix} \mathbb{P}_1 \\ \mathbb{P}_3 \end{matrix} \begin{bmatrix} 2 & 0 \\ 1 & 3 \end{bmatrix}$
$(6^2 \cdot 2)_{D^2 \otimes A}$	$\mathbb{P}_{(2,3,2,3,2,4)}[8]$ $\sum(z_i^4 + z_i y_i^2) + z_3^4 + z_4^2$	$\mathbb{Z}_2 : \mathbb{P}_{(1,1,1,2)}[4]$ $\mathbb{Z}_3 : \mathbb{P}_1$	$\begin{matrix} \mathbb{P}_1 \\ \mathbb{P}_{(1,1,1,2)} \end{matrix} \begin{bmatrix} 2 & 0 \\ 1 & 4 \end{bmatrix}$
$(10^2 \cdot 1)_{D^2 \otimes A}$	$\mathbb{P}_{(2,5,2,5,4,6)}[12]$ $\sum(z_i^6 + z_i y_i^2) + z_3^3 + z_4^2$	$\mathbb{Z}_2 : \mathbb{P}_{(1,1,2,3)}[6]$ $\mathbb{Z}_5 : \mathbb{P}_1$	$\begin{matrix} \mathbb{P}_1 \\ \mathbb{P}_{(1,1,2,3)} \end{matrix} \begin{bmatrix} 2 & 0 \\ 1 & 6 \end{bmatrix}$

The critical manifold derived from this 6–fold belongs to the configuration class

$$\begin{matrix} \mathbb{P}_2 \\ \mathbb{P}_{(2,2,2,1,1)} \end{matrix} \begin{bmatrix} 3 & 0 \\ 2 & 6 \end{bmatrix} \tag{45}$$

which is indeed a Calabi–Yau deformation class.

Again it should be emphasized that the construction is not restricted to the infinite series defined in (32) as the final example illustrates. A five–dimensional critical vacuum is obtained by considering the Landau–Ginzburg potential

$$W = \sum_{j=1}^{2} \left(u_i^3 + u_i v_i^2 \right) + \sum_{i=3}^{5} \left(u_i^3 + u_i w_i^3 \right) \tag{46}$$

which corresponds to the exactly solvable model $(16^3 \cdot 4^2)_{E_7^3 \otimes D^2}$. The nine–dimensional noncritical manifold

$$\mathbb{P}_{(3,2,3,2,3,2,3,3,3)}[9] \tag{47}$$

leads, via its singularity structure, to the five–dimensional critical manifold

$$\begin{matrix} \mathbb{P}_1 \\ \mathbb{P}_2 \\ \mathbb{P}_4 \end{matrix} \begin{bmatrix} 2 & 0 & 0 \\ 0 & 3 & 0 \\ 1 & 1 & 3 \end{bmatrix}. \tag{48}$$

It is again crucial that a non–Fermat polynomial was chosen for the last four coordinates in the noncritical manifold.

5. CONCLUSION

Mirror symmetry cannot be understood in the framework of Calabi–Yau manifolds. Thus, beyond the class of such spaces, there must exist a space of a new type of noncritical manifolds which contain information about critical vacua, such as the mirrors of rigid Calabi–Yau manifolds. Mirrors of spaces with both sectors, antigeneration and generations, are again of Calabi–Yau type and hence the noncritical manifolds which correspond to such ground states should make contact with Calabi–Yau manifolds in some manner.

What has been shown in ref. 14 is that the class (9) of higher dimensional Kähler manifolds with positive first Chern class, quantized in a particular way, generalises the framework of Calabi–Yau vacua in the desired way: For particular types of such noncritical manifolds Calabi–Yau manifolds of critical dimension are embedded algebraically in a fibered submanifold. For string vacua which cannot be described by Kähler manifolds and which are mirror candidates of rigid Calabi–Yau manifolds the higher dimensional manifolds still lead to the spectrum of the critical vacuum and a rationale emerges that explains why a Calabi–Yau representation is not possible in such theories. Thus these manifolds of dimension $c/3 + 2k$ define an appropriate framework in which to discuss mirror symmetry.

There are a number of important consequences that follow from the results of the previous sections. First it should be realised that the relevance of noncritical manifolds suggests the generalisation of a conjecture regarding the relation between (2,2) superconformal field theories of central charge $c = 3D$, $D \in \mathbb{N}$, with N=1 spacetime supersymmetry on the one hand and Kähler manifolds of complex dimension D with vanishing first Chern class on the other. It was suggested by Gepner that this relation is 1-1. It follows from the results above that instead superconformal theories of the above type are in correspondence with Kähler manifolds of dimension $c/3 + 2k$ with a first Chern class quantized in multiples of the degree.

A second consequence is that the ideas of section 3 lead, for a large class of Landau–Ginzburg theories, to a new canonical prescription for the construction of the critical manifold, if it exists, directly from the 2D field theory.

Recently Batyrev[27] introduced a new construction of mirrors of Calabi–Yau manifolds based on dual polyhedra. His method appears to apply only to manifolds defined by one polynomial in a weighted projective space or products thereof. The method of toric geometry that is used in ref/ 27 is however not restricted to Calabi–Yau manifolds[28] and therefore the constructions described in sections 3 and 4 lead to the exciting possibility of extending Batyrev's results to Calabi–Yau manifolds of codimension larger than one by proceeding via noncritical manifolds.

A final remark is that in this framework the role played by the dimension of the manifolds becomes of secondary importance. This is as it should be, at least for an effective theory, which tests only matter content and couplings. It is, however, somewhat mysterious that via ineffective splittings manifolds of different dimension describe one and the same critical vacuum.

It is clear that the emergence in string theory of manifolds with quantized first Chern class should be understood better. The results described here are a first step in this direction. They indicate that these manifolds are not just auxiliary devices but may be as physical as Calabi–Yau manifolds of critical dimension. In order to probe the structure of these models in more depth it is important to get further insight into the complete spectrum of these theories and to compute the Yukawa couplings of the fields. It is clear from the results presented here that the spectra of the higher dimensional manifolds contain additional modes beyond those that are related to the generations and antigenerations of the critical vacuum and the question arises what physical interpretation these fields have.

A better grasp on the complete spectrum of these spaces should also give insight into a different, if not completely independent, approach toward a deeper understanding of these higher dimensional manifold, which is to attempt the construction of consistent σ–models defined via these spaces. Control of the complete spectrum will shed light on the precise relation between the σ–models associated to the noncritical manifolds and critical σ–models.

6. ACKNOWLEDGEMENTS

I'm grateful to CERN for hospitality and the theoretical theorists there for lively discussions, in particular Per Berglund, Philip Candelas, Wolfgang Lerche and Jan Louis. It is a pleasure to thank Herbert Clemens, Tristan Hübsch, Cumrun Vafa and Nick Warner for conversations and R. Hartshorne for asking the right question. I'm also grateful to the Aspen Center of Physics for hospitality.

REFERENCES

[1] P. Candelas, G. Horowitz, A. Strominger and E. Witten, Vacuum configurations for superstrings, *Nucl. Phys.* **B258** (1985) 49.

[2] P. Candelas, M. Lynker and R. Schimmrigk, Calabi-Yau manifolds in weighted P(4), *Nucl. Phys.* **B341** (1990) 383.

[3] B.R. Greene and R. Plesser, Duality in Calabi-Yau moduli space, *Nucl. Phys.* **B338** (1990) 15.

[4] M. Lynker and R. Schimmrigk, Landau-Ginzburg theories as orbifolds, *Phys. Lett.* **249B** (1990) 237.

[5] A. Klemm and R. Schimmrigk, Landau–Ginzburg String Vacua, CERN preprint CERN–TH–6459/92 and Heidelberg preprint HD–THEP–92–13.

[6] M. Kreuzer and H. Skarke, No Mirror Symmetry among Landau–Ginzburg Vacua, CERN preprint CERN–TH–6461/92.

[7] P. Candelas, Talk at the Workshop on Geometry and Quantum Field Theory, Baltimore, March 1992.

[8] P. Candelas, E. Derrick and L. Parkes, in preparation.

[9] C. Vafa, Topological Mirrors and Quantum Rings, Harvard preprint HUTP–91/A059.

[10] P. Candelas, X. de la Ossa, P. Green and L. Parkes, An exactly soluble superconformal theory from a mirror pair of Calabi-Yau manifolds, *Phys. Lett.* **258B** (1991) 118; A pair of Calabi-Yau manifolds as an exactly soluble superconformal theory, *Nucl. Phys.* **B359** (1991) 21.

[11] G. Ellingsrud and S.A. Stromme, The Number of twisted cubic curves on the general quintic 3–fold.

[12] S. Katz, private communication.

[13] B. Zumino, *Phys. Lett.* **87B** (1979) 203.

[14] R. Schimmrigk, Critical String Vacua from Noncritical Manifolds: A Novel Framework for String Compactification, Heidelberg preprint HD–THEP–92–29.

[15] E. Martinec, Algebraic geometry and effective lagrangians, *Phys. Lett.* **217** (1989) 431

[16] C. Vafa and N. Warner, Catastrophes and the classification of conformal theories, *Phys. Lett.* **218** (1989) 51.

[17] C. Vafa, String vacua and orbifoldized L-G models, *Mod. Phys. Lett.* **A4** (1989) 1169.

[18] P. Candelas, Yukawa couplings between (2,1) forms, *Nucl. Phys.* **B298** (1988) 488.

[19] P. Berglund, B.R. Greene and T. Hübsch, Classical vs. Landau–Ginzburg Geometry of Compactification, *Mod. Phys. Lett.* **A7** (1992) 1855.

[20] R. Schimmrigk, work in progress.

[21] B.R. Greene, C. Vafa and N. Warner, Calabi-Yau manifolds and renormalization group flows, *Nucl. Phys.* **B324** (1989) 371.

[22] D. Gepner, Space-time supersymmetry in compactified string theory and superconformal models, *Nucl. Phys.* **B296** (1988) 757.

[23] R. Schimmrigk, A new construction of a three generation Calabi–Yau manifold, *Phys. Lett.* **193B** (1987) 175.

[24] D. Gepner, String Theory on Calabi–Yau Manifolds: The Three Generation Case, Princeton University preprint, December 1987.

[25] R. Schimmrigk, Heterotic RG flow fixed points with nondiagonal affine invariants, *Phys. Lett.* **229B** (1989) 227.

[26] P. Candelas, A. Dale, C.A. Lütken and R. Schimmrigk, Complete intersection Calabi-Yau manifolds, *Nucl. Phys.* **B298** (1988) 493.

[27] V.V. Batyrev, Dual Polyhedra and Mirror Symmetry for Calabi–Yau Hypersurfaces in Toric Varieties, University of Essen preprint.

[28] V.V. Batyrev, Variations of the Mixed Hodge Structure of Affine Hypersurfaces in Algebraic Tori, University of Essen preprint.

W_∞ ALGEBRA IN TWO-DIMENSIONAL BLACK HOLES

Tohru Eguchi[1], Hiroaki Kanno[2] and Sung-Kil Yang[3]

[1]Department of Physics, Faculty of Science
University of Tokyo, Tokyo 113, Japan
[2]DAMTP, University of Cambridge
Cambridge CB3 9EW, England
[3]Institute of Physics, University of Tsukuba
Tsukuba, Ibaraki 305, Japan

ABSTRACT

We study the $SL(2;R)/U(1)$ coset model of two-dimensional black hole and its relation to the Liouville theory coupled to the $c = 1$ matter. We uncover a basic isomorphism in the algebraic structures of these theories and show that the black hole model has the same physical spectrum as the $c = 1$ model, i.e. tachyons, W_∞ currents and the ground ring elements. We also identify the operator responsible for the creation of the mass of the black hole.

Recently the study of the two-dimensional black hole has been receiving much attentions[1-7]. It is an attractive idea to look into the mysteries of black holes and try to provide new insights into their physics making use of some of the tools developed in string theory.

In this article we discuss the $SL(2;R)/U(1)$ coset model of two-dimensional black hole[1] and its relation to the Liouville theory coupled to $c = 1$ matter. Making use of free field realization of $SL(2;R)$ current algebra we first show that the stress tensor of the coset theory is identical to that of the $c = 1$ Liouville theory up to BRST exact terms and thus there exists a basic isomorphism in the algebraic structure between these two theories. We next determine the physical spectrum (BRST invariants) of the black hole theory which turns out to be identical to that of the $c = 1$ two-dimensional gravity. Besides the standard tachyon states, there also exist discrete states in two-dimensional black hole which possess the same algebraic structure, the W_∞-algebra and ground ring, as in the $c = 1$ theory. Thus the black hole and $c = 1$ theory in fact seem to originate from the identical conformal field theory. We then try to identify the operator which is responsible for the creation of the mass of the black hole. We shall show that the mass perturbation operator is given by $\beta\exp(-2\sqrt{2}\phi)$ (β,γ,ϕ are the free fields used to realize $SL(2;R)$ current algebra), the screening operator of $SL(2;R)$ algebra. In the Liouville theory the perturbing operator is given by the cosmological constant $\exp(-\sqrt{2}\phi)$ and thus the black hole and $c=1$ gravity theories are the same conformal theories but perturbed by different marginal operators.

Let us start from the free field realization of $SL(2;R)$ current algebra[8] given by

$$J_+ = \beta, \tag{1}$$

$$J_- = \beta\gamma^2 + \sqrt{2k'}\gamma\partial\phi + k\partial\gamma, \quad k' = k-2, \tag{2}$$

$$J_3 = -\beta\gamma - \sqrt{\frac{k'}{2}}\partial\phi. \tag{3}$$

Here β,γ are the commuting ghost fields with the dimensions $h=1,0$ and we have the operator product expansions (OPE's) $\beta(z)\gamma(w) \sim 1/(z-w)$, $\phi(z)\phi(w) \sim -\log(z-w)$. k is the level of the $SL(2;R)$ current algebra,

$$J_+(z)J_-(w) \sim \frac{k}{(z-w)^2} - \frac{2J_3(w)}{z-w} + \cdots, \tag{4}$$

$$J_3(z)J_\pm(w) \sim \pm\frac{J_\pm(w)}{z-w} + \cdots, \tag{5}$$

$$J_3(z)J_3(w) \sim \frac{-k/2}{(z-w)^2} + \cdots. \tag{6}$$

Then the Sugawara construction of the stress tensor reads

$$T_{SL(2;R)}(z) = \frac{1}{k-2}\left\{\frac{1}{2}J_+(z)J_-(z) + \frac{1}{2}J_-(z)J_+(z) - J_3(z)J_3(z)\right\} \tag{7}$$

$$= \beta\partial\gamma - \frac{1}{2}(\partial\phi)^2 - \frac{1}{\sqrt{2k'}}\partial^2\phi. \tag{8}$$

Let us next gauge away the $U(1)$ degrees of freedom and construct a coset theory $SL(2;R)/U(1)$ making use of the BRST method. We introduce a gauge boson X with $X(z)X(w) \sim -\log(z-w)$ and form a BRST operator for the $U(1)$ symmetry,

$$Q_{U(1)} = \oint dz\, \xi(z)\left(J_3(z) - i\sqrt{\frac{k}{2}}\partial X(z)\right). \tag{9}$$

Here ξ,η are anti-commuting ghosts with $h=0,1$, respectively, and obey $\xi(z)\eta(w) \sim 1/(z-w)$. Since the total J_3 current, $J_3^{tot} = J_3 - i\sqrt{\frac{k}{2}}\partial X$, has no central term,

$$J_3^{tot}(z)J_3^{tot}(w) \sim \text{regular}, \tag{10}$$

$Q_{U(1)}$ is nilpotent

$$Q_{U(1)}^2 = 0. \tag{11}$$

The process of gauging has introduced the new fields X,ξ,η and then the total stress tensor is given by

$$T_{SL(2;R)/U(1)} = \beta\partial\gamma - \eta\partial\xi - \frac{1}{2}(\partial\phi)^2 - \frac{1}{\sqrt{2k'}}\partial^2\phi - \frac{1}{2}(\partial X)^2. \tag{12}$$

When we choose $k=9/4$ or $k'=1/4$, $T_{SL(2;R)/U(1)}$ has the central charge $c=26$ and describes a critical string theory.

Now let us eliminate the β, γ, η, ξ fields from (12) so that the stress tensor has the form of the Liouville field (ϕ) coupled with the matter (X). It is easy to derive the following identity which is of basic importance in our analysis.

$$T_{SL(2;R)/U(1)} = -\frac{1}{2}(\partial\phi')^2 - \frac{1}{\sqrt{2k'}}(\partial^2\phi') - \frac{1}{2}(\partial X')^2 + \{Q_{U(1)}, -\partial(\log\gamma)\cdot\eta\}, \quad (13)$$

$$\phi' = \phi + \sqrt{\frac{k'}{2}}\log\gamma, \quad (14)$$

$$X' = X + i\sqrt{\frac{k}{2}}\log\gamma, \quad (15)$$

In (13) β, γ, η, ξ have in fact disappeared from the stress tensor except for the shift of the fields ϕ, X and the $Q_{U(1)}$-exact term. We should note that the commuting ghosts β, γ may be bosonized as

$$\beta = -i\partial v \exp(iv - u), \quad (16)$$
$$\gamma = \exp(u - iv), \quad (17)$$

with $u(z)u(w) \sim -\log(z-w)$, $v(z)v(w) \sim -\log(z-w)$. Thus $\log\gamma = u - iv$. Note that it is possible to take fractional or negative powers of γ while the power of β is restricted to positive integers.

Now we consider the critical case $k = 9/4$ and further introduce the diffeomorphism ghosts b, c (with $h = 2, -1$),

$$T^{tot}_{SL(2;R)/U(1)} = -\frac{1}{2}(\partial\phi')^2 - \sqrt{2}\partial^2\phi' - \frac{1}{2}(\partial X')^2 - 2b\cdot\partial c - \partial b\cdot c + \{Q_{U(1)}, -(\partial\log\gamma)\cdot\eta\} \quad (18)$$

BRST operator for diffeomorphism invariance is defined as usual by

$$Q_{\text{diff}} = \oint dz\, c(z)\left(T_{SL(2;R)/U(1)}(z) + \frac{1}{2}T_{\text{gh}}\right) \quad (19)$$

where $T_{\text{gh}} = -2b\cdot\partial c - \partial b\cdot c$. Note that $Q^2_{\text{diff}} = 0$, $\{Q_{\text{diff}}, Q_{U(1)}\} = 0$. Let us define our BRST operator as the sum

$$Q = Q_{\text{diff}} + Q_{U(1)}. \quad (20)$$

Then $T^{tot}_{SL(2;R)/U(1)}$ is further rewritten as

$$T^{tot}_{SL(2;R)/U(1)} = -\frac{1}{2}(\partial\phi')^2 - \sqrt{2}\partial^2\phi' - \frac{1}{2}(\partial X')^2 - 2b'\cdot\partial c - \partial b'\cdot c + \{Q, -(\partial\log\gamma)\cdot\eta\}, \quad (21)$$

where

$$b' = b + \eta\,\partial\log\gamma. \quad (22)$$

Thus up to a BRST exact term, the stress tensor has exactly the same form as in the $c = 1$ Liouville theory

$$T^{tot}_{c=1} = -\frac{1}{2}(\partial\phi)^2 - \sqrt{2}\partial^2\phi - \frac{1}{2}(\partial X)^2 - 2b\cdot\partial c - \partial b\cdot c, \quad (23)$$

at the vanishing value of the cosmological constant ($\mu = 0$). The shift of the fields eq. (14),(15) have a simple physical interpretation; ϕ' and X' are the charge-neutral

161

versions of ϕ and X. In fact it is easy to check that ϕ' and X' are neutral with respect to the $U(1)$ charge $\oint J_3^{\text{tot}}(z)\,dz$ and thus describe the degree of freedom which live in the coset space $SL(2;R)/U(1)$. We notice that since γ does not generate a singularity when contracted with itself, ϕ', X', b', c have exactly the same OPE as ϕ, X, b, c. (21), (23) exhibit the basic identity in the algebraic structure of the gauged WZW and $c=1$ theories.

Let us next turn to the discussion of the physical spectrum of two-dimensional black hole. Physical states are defined as cohomology classes of the BRST operator,

$$Q|\text{phys}\rangle = 0, \quad |\text{phys}\rangle \simeq |\text{phys}\rangle + Q|\text{any}\rangle, \tag{24}$$

where Q is given by (20). It turns out that the BRST invariant states in $SL(2;R)/U(1)$ theory can be obtained from the known BRST invariants in the $c=1$ theory simply by making use of the transformations (14),(15),(22) at $k=9/4$,

$$\phi' = \phi + \frac{1}{2\sqrt{2}}\log\gamma, \quad X' = X + i\frac{3}{2\sqrt{2}}\log\gamma, \quad b' = b + \eta\partial\log\gamma. \tag{25}$$

Tachyons

In the $c=1$ theory tachyon wave functions are simply given by the vertex operators

$$\exp(ip_X X)\exp(p_L \phi), \tag{26}$$

with p_X, p_L satisfying the on-shell condition,

$$\pm p_X = p_L + \sqrt{2}. \tag{27}$$

In the black hole theory tachyons are given by

$$\gamma^{\frac{1}{2\sqrt{2}}(-3p_X + p_L)} \exp(ip_X X)\exp(p_L \phi). \tag{28}$$

(28) is simply the charge-neutral version of (26).

W_∞ currents

In the $c=1$ Liouville theory certain special states appear at discrete values of the momentum[9,10,11,12,13,14]; current operators which form W_∞ (wedge) algebra are given by

$$W_{j,j}^{(+)} = \exp(i\sqrt{2}jX)\exp(\sqrt{2}(j-1)\phi), \quad j = 0, \tfrac{1}{2}, 1, \cdots, \tag{29}$$

$$W_{j,m}^{(+)} = \left(\oint dw \exp(-i\sqrt{2}X(w))\right)^{(j-m)} W_{j,j}^{(+)}, \quad -j \leq m \leq j. \tag{30}$$

There also exist the "negative" current operators

$$W_{j,j}^{(-)} = \exp(i\sqrt{2}jX)\exp(-\sqrt{2}(j+1)\phi), \quad j = 0, \tfrac{1}{2}, 1, \cdots \tag{31}$$

$$W_{j,m}^{(-)} = \left(\oint dw \exp(-i\sqrt{2}X(w))\right)^{(j-m)} W_{j,j}^{(-)}, \quad -j \leq m \leq j. \tag{32}$$

Corresponding operators in the black hole theory are again obtained by the simple substitution (25). Explicitly some examples are given by

$$W_{0,0}^{(+)} = W_{0,0}^{(-)} = \gamma^{-1/2} \exp(-\sqrt{2}\phi) , \qquad (33)$$

$$W_{1/2,1/2}^{(+)} = \gamma^{-1} \exp\left(\frac{i}{\sqrt{2}}X\right) \exp\left(-\frac{1}{\sqrt{2}}\phi\right) , \quad W_{1/2,-1/2}^{(+)} = \gamma^{1/2} \exp\left(\frac{-i}{\sqrt{2}}X\right) \exp\left(-\frac{1}{\sqrt{2}}\phi\right) , \qquad (34)$$

$$W_{1,1}^{(+)} = \gamma^{-3/2} \exp(i\sqrt{2}X) , \quad W_{1,0}^{(+)} = \frac{i}{\sqrt{2}}\left(\partial X + \frac{3i}{2\sqrt{2}}\partial\gamma \cdot \gamma^{-1}\right) ,$$

$$W_{1,-1}^{(+)} = \gamma^{3/2} \exp(-i\sqrt{2}X) , \qquad (35)$$

$$W_{1/2,1/2}^{(-)} = \gamma^{-3/2} \exp\left(\frac{i}{\sqrt{2}}X\right) \exp\left(-\frac{3}{\sqrt{2}}\phi\right) , \quad W_{1/2,-1/2}^{(-)} = \exp\left(\frac{-i}{\sqrt{2}}X\right) \exp\left(-\frac{3}{\sqrt{2}}\phi\right) , \qquad (36)$$

$$W_{1,1}^{(-)} = \gamma^{-5/2} \exp(i\sqrt{2}X) \exp(-2\sqrt{2}\phi) ,$$

$$W_{1,0}^{(-)} = \beta \exp(-2\sqrt{2}\phi) \simeq \gamma^{-1}\frac{i}{\sqrt{2}}\left(\partial X + \frac{3i}{2\sqrt{2}}\partial\gamma \cdot \gamma^{-1}\right) \exp(-2\sqrt{2}\phi) ,$$

$$W_{1,-1}^{(-)} = \gamma^{1/2} \exp(-i\sqrt{2}X) \exp(-2\sqrt{2}\phi) , \qquad (37)$$

It is easy to check that these are in fact invariant under our BRST operator (20) (to be precise, $cW_{j,m}^{(\pm)}$ are the invariant operators. $W_{j,m}^{(\pm)}$ is BRST invariant up to a total derivative). Since ϕ', X' have the same OPE as ϕ, X, our $W^{(+)}$ currents above also form the same W_∞ algebra as known in the $c=1$ gravity theory.

Ground Ring

In the $c = 1$ Liouville theory the ground ring elements appear as partners of W_∞ currents at a neighbouring ghost number[12]. The black hole theory also possesses the ground ring elements whose generators as given by

$$x = \left\{(bc - \frac{i}{\sqrt{2}}\partial X + \frac{1}{\sqrt{2}}\partial\phi)\gamma^{-1/2} + (\partial\gamma + \eta c \cdot \partial\gamma)\gamma^{-3/2}\right\} \exp\left(\frac{i}{\sqrt{2}}X + \frac{1}{\sqrt{2}}\phi\right) , \qquad (38)$$

$$y = \left\{(bc + \frac{i}{\sqrt{2}}\partial X + \frac{1}{\sqrt{2}}\partial\phi)\gamma + (-\frac{1}{2}\partial\gamma + \eta c \cdot \partial\gamma)\right\} \exp\left(-\frac{i}{\sqrt{2}}X + \frac{1}{\sqrt{2}}\phi\right) , \qquad (39)$$

It is also easy to check that x, y are invariant under the BRST operator (20) and their explicit expressions (38) (39) agree exactly with those obtained from the known expressions of the $c = 1$ theory after substitution (25). Other elements of the ring are obtained by multiplying the generators (38) (39) and we obtain a ring isomorphic to $\mathbb{C}[x, y]$ as in the $c = 1$ theory.

What do these results mean on the relation between the gauged WZW and the $c = 1$ gravity theories? They seem to imply that these theories in fact originate from the common algebraic structure or conformal field theory but perhaps perturbed into different directions. In the Liouville theory the perturbing operator will be the worldsheet cosmological constant $\exp(-\sqrt{2}\phi)$ and the effects of the perturbation on the W_∞ algebra and the ground ring have been studied in detail[15,16,17]. In the case of the black hole model, however, the perturbing operator responsible for the non-zero mass of the black hole has not been clearly identified (for previous attempts, see refs.

18,19). In our search for the mass operator we take the following heuristic approach based on the analysis of a non-linear σ-model: we first write down an action which reproduces the stress-tensor of the coset theory (13) as

$$S_0 = \frac{1}{8\pi} \int d^2x (\partial \phi' \bar{\partial} \phi' + \partial X' \bar{\partial} X') - \frac{1}{4\pi\sqrt{2k'}} \int d^2x \sqrt{g} R^{(2)} \phi' , \qquad (40)$$

Here $R^{(2)}$ is the scalar curvature of the Riemann surface and we have dropped the $Q_{U(1)}$-exact term. We then look for an $(1,1)$ operator which, when added to (40), transforms the flat target-space metric into that of a black hole. Such an operator must be of the form

$$F_{\phi\phi}(\phi', X')\partial\phi'\bar{\partial}\phi' + F_{\phi X}(\phi', X')(\partial\phi'\bar{\partial}\phi' + \bar{\partial}\phi'\partial X') + F_{XX}(\phi', X')\partial X'\bar{\partial}X' .$$

The new target-space metric becomes $G_{\phi\phi} = 1 + F_{\phi\phi}, G_{\phi X} = F_{\phi X}, G_{XX} = 1 + F_{XX}$ which would describe a curved space-time. Thus the perturbing operator V must be linear in the derivative of the fields ϕ', X' (for both left and right components) and this requirement uniquely determines the operator V. In fact V, being a marginal perturbation, is given by the $(1,1)$ form $W_{j,m}^{(+)}(z)W_{j,m}^{(+)}(\bar{z})$ or $W_{j,m}^{(-)}(z)W_{j,m}^{(-)}(\bar{z})$. $W_{j,m}^{(\pm)}$ currents, however, are differential polynomials in ϕ', X' of degree $j^2 - m^2$ (multiplied by a vertex operator) and hence only $W_{1,0}^{(+)}, W_{1,0}^{(-)}$ are linear in the derivative of the fields. Now $W_{1,0}^{(+)}(z)W_{1,0}^{(+)}(\bar{z}) \simeq \partial X'\bar{\partial}X'$ and hence is a trivial change of the radius of X'. Therefore $V = W_{1,0}^{(-)}(z)W_{1,0}^{(-)}(\bar{z})$ is the unique candidate for the black hole mass operator. As shown in (37), $W_{1,0}^{(-)}$ is just the screening operator of the $SL(2;R)$ current algebra

$$W_{1,0}^{(-)} = \beta \exp\left(-\sqrt{\frac{2}{k'}}\phi\right) \qquad (41)$$

and expected to appear naturally in the theory. It has in fact been suggested to be the black hole mass operator by Bershadsky and Kutasov in ref. 18. Up to a $Q_{U(1)}$-exact term $W_{1,0}^{(-)}$ may be written as

$$W_{1,0}^{(-)} = -\left(\sqrt{\frac{k'}{2}}\partial\phi' + i\sqrt{\frac{k}{2}}\partial X'\right) \exp\left(-\sqrt{\frac{2}{k'}}\phi'\right) \qquad (42)$$

which we employ for the discussion of the mass perturbation. Then the perturbed action is given by

$$S = \frac{1}{8\pi} \int d^2x \left(\partial\phi\bar{\partial}\phi + \partial X\bar{\partial}X\right) - \frac{1}{4\pi\sqrt{2k'}} \int d^2x \sqrt{g}R^{(2)}\phi$$
$$+ \frac{\mu}{8\pi} \int d^2x \left(\sqrt{\frac{k'}{2}}\partial\phi + i\sqrt{\frac{k}{2}}\partial X\right)\left(\sqrt{\frac{k'}{2}}\bar{\partial}\phi + i\sqrt{\frac{k}{2}}\bar{\partial}X\right) \exp\left(-\sqrt{\frac{2}{k'}}\phi\right) . \qquad (43)$$

(We drop the prime on ϕ, X hereafter.) (43) gives a non-linear σ-model with the target-space metric

$$G_{\phi\phi} = 1 + \frac{k'}{2}\mu \exp\left(-\sqrt{\frac{2}{k'}}\phi\right), \quad G_{XX} = 1 - \frac{k}{2}\mu \exp\left(-\sqrt{\frac{2}{k'}}\phi\right),$$
$$G_{X\phi} = i\sqrt{\frac{k'}{2}}\sqrt{\frac{k}{2}}\mu \exp\left(-\sqrt{\frac{2}{k'}}\phi\right), \qquad (44)$$

together with the linear dilaton field

$$\Phi = \sqrt{\frac{2}{k'}}\phi . \qquad (45)$$

It is easy to check that (44),(45) satisfy the equation of the vanishing one-loop β-function,

$$R_{ab} = D_a D_b \Phi , \quad a,b = \phi, X \qquad (46)$$

in the leading order in $1/k$. Thus the mass perturbation (42) is in fact a marginal perturbation of the black hole theory.

In the leading order in $1/k$ the metric (44) is given by

$$G_{\phi\phi} = 1 + \frac{k}{2}\mu\exp\left(-\sqrt{\frac{2}{k}}\phi\right), \quad G_{XX} = 1 - \frac{k}{2}\mu\exp\left(-\sqrt{\frac{2}{k}}\phi\right),$$

$$G_{X\phi} = i\frac{k}{2}\mu\exp\left(-\sqrt{\frac{2}{k}}\phi\right) . \qquad (47)$$

After diagonalization $(d\psi = dX + ik\mu/2 \exp(-\sqrt{2/k}\phi)(1-k\mu/2\exp(-\sqrt{2/k}\phi))^{-1}d\phi)$ the metric (47) becomes the "two-dimensional Schwarzschild metric" discovered in ref. 20,

$$ds^2 = \left(1 - \frac{k\mu}{2}\exp\left(-\sqrt{\frac{2}{k}}\phi\right)\right)^{-1} d\phi^2 + \left(1 - \frac{k\mu}{2}\exp\left(-\sqrt{\frac{2}{k}}\phi\right)\right)d\psi^2 . \qquad (48)$$

Its scalar curvature is given by

$$R = \mu\exp\left(-\sqrt{\frac{2}{k}}\phi\right) , \qquad (49)$$

and thus the curvature singularity exists at $\phi = -\infty$. We also note that the event horizon occurs at $\phi = \phi^* = \sqrt{k/2}\log\mu k/2$. Thus the region $\phi^* < \phi < \infty$ describes the asymptotically flat region of the black hole, while $-\infty < \phi < \phi^*$ describes the region between the event horizon and singularity. We note that if one replaces k by $k' = k - 2$ in (44) (as in ref. 1), it becomes equivalent to (47) and becomes a black hole in all orders in $1/k$. The coordinate transformation which brings the action (43) into that of the gauged WZW model is also easy to work out. By a simple change of variables[18]

$$\phi = \frac{1}{\sqrt{2}}\log\cosh r + \phi^* , \quad \theta = X + i\log\tanh r . \qquad (50)$$

(43) is transformed into

$$S_{WZW} = \frac{k}{4\pi}\int d^2x \, gl(\partial r \bar{\partial} r + \tanh^2 r \partial\theta\bar{\partial}\theta) - \frac{1}{4\pi}\int d^2x \, \sqrt{g}R^{(2)}\log\cosh r , \qquad (51)$$

in the leading order in $1/k$. (51) is the effective action of the gauged WZW model obtained after eliminating the gauge field[1]. The presence of i (imaginary unit) in formulas (44) (50) is somewhat disturbing. However, it disappears when we go to the Minkowski convention ($X \to iX$) and there the change of variables seems well-defined.

We should remark that our results on the physical spectrum of the coset model $SL(2;R)/U(1)$ do not agree with those obtained by Distler and Nelson[21] based on a different method for quotienting the $U(1)$ symmetry. Extra physical states found in ref. 21 in the coset model become BRST trivial states in our formulation.

In this paper we have considered the cohomology H_Q of the total BRST operator $Q = Q_{\text{diff}} + Q_{U(1)}$. We believe that the non-zero cohomology of $Q_{U(1)}$ is concentrated at the ghost number zero sector and hence H_Q agrees with that of $H_{Q_{\text{diff}}}(H_{Q_{U(1)}})$ where Q_{diff} cohomology is taken after computing the $Q_{U(1)}$ cohomology.

T.E. would like to thank Profs. M. Atiyah and P. Goddard for their hospitality at Newton Institute for Mathematical Sciences, University of Cambridge. The research of T.E. and S.K.Y. is partly supported by Grant-in-Aid for Scientific Research on Priority Area "Infinite Analysis".

REFERENCES

1. E. Witten, On string theory and black holes, *Phys. Rev.* **D44** (1991) 314.
2. R. Dijkgraaf, H. Verlinde and E. Verlinde, String propagation in a black-hole geometry, *Nucl. Phys.* **B371** (1992) 269.
3. T. Eguchi, Topological field theories and the space-time singularity, *Mod. Phys. Lett.* **A7** (1992) 85.
4. C. Callan, S. Giddings, J. Harvey and A. Strominger, Evanescent black-holes, *Phys. Rev.* **D45** (1992) R1005.
5. T. Banks, A. Dabholkar, M. Douglas and M. O'Loughlin, Are horned particles the end-point of Hawking evaporation?, *Phys. Rev.* **D45** (1992) 3607.
6. A. Bilal and C. Callan, Liouville models of black hole evaporation, *Nucl. Phys.* **B394** (1993) 73.
7. S. de Alwis, Quantization of a theory of $2d$ dilaton gravity, *Phys. Lett.* **B289** (1992) 278.
8. M. Wakimoto, Fock representations of the affine lie-algebra $sl(1)$, *Comm. Math. Phys.* **104** (1986) 605.
9. A.M. Polyakov, Self-tuning fields and resonant correlations in 2d-gravity, *Mod. Phys. Lett.* **A6** (1991) 635.
10. D. Gross, I.R. Klebanov and M. Newman, The 2-point correlation-function of the one-dimensional matrix model, *Nucl. Phys.* **B350** (1991) 621.
11. I.R. Klebanov and A.M. Polyakov, Interaction of discrete states in 2-dimensional string theory, *Mod. Phys. Lett.* **A6** (1991) 3273.
12. E. Witten, Ground ring of 2-dimensional string theory, *Nucl. Phys.* **B373** (1992) 187.
13. B. Lian and G. Zuckerman, 2d gravity with $c = 1$ matter, *Phys. Lett.* **B266** (1991) 21.
14. P. Bouwknegt, J. McCarthy and K. Pilch, BRST analysis of physical states for 2d gravity coupled to $C \leq 1$ matter, *Comm. Math. Phys.* **145** (1992) 541.
15. S. Kachru, Quantum rings and recursion-relations in 2d quantum-gravity, *Mod. Phys. Lett.* **A7** (1992) 1419.
16. J.L.F. Barbón, Perturbing the ground ring of 2-D string theory, CERN preprint CERN-TH 6379/92, January, 1992.
17. Vl.S. Dotsenko, The operator algebra of the discrete state operators in 2d gravity with nonvanishing cosmological constant, *Mod. Phys. Lett.* **A7** (1992) 2505.

18. M. Bershadsky and D. Kutasov, Comment on gauged WZW theory, *Phys. Lett.* **B266** (1991) 345.
19. E. Martinec and S.L. Shatashvili, Black-hole physics and Liouville theory, *Nucl. Phys.* **B368** (1992) 338.
20. G. Mandal, A.M. Sengupta and S. Wadia, Classical-solutions of 2-dimensional string theory, *Mod. Phys. Lett.* **A6** (1991) 1685.
21. J. Distler and P. Nelson, New discrete states of strings near a black-hole, *Nucl. Phys.* **B374** (1992) 123.

GRADED LIE DERIVATIVES AND SHORT DISTANCE EXPANSIONS IN TWO DIMENSIONS

Rainer Dick

Sektion Physik
Universität München
D-W-8000 München 2
Germany

1 INTRODUCTION

The exploration of algebraic structures in two–dimensional field theories underwent a rapid development in recent years, particularly through the investigation of W-algebras. It was demonstrated by Zamolodchikov that there exist algebras which close nonlinearly into finitely many (quasi–)primary fields[1]. For recent work on these W_N-algebras see ref. 2–5 and references therein. A certain $N \to \infty$ limit of these algebras, denoted meanwhile commonly as $w_{1+\infty}$, has been studied by Bakas[6] (see also ref. 7 for studies of homology complexes of a subalgebra of $w_{1+\infty}$), and Pope, Romans and Shen succeeded[8] in the construction of a unique W_∞-algebra with the property to admit central terms for all integer weights ≥ 2. For a discussion of field theoretic realisations of these algebras and their relations see ref. 9 and references therein. Both $w_{1+\infty}$ and the related algebra w_∞ are subalgebras of the algebra of area–preserving diffeomorphisms in two dimensions, and it has been pointed out by Witten that these algebras are realized as symmetries of the ground ring of two–dimensional string theory[10], implying in particular the existence of operator representations of these algebras in the framework of Liouville theory coupled to $c = 1$ matter fields. A very nice account of this has also been given by Eguchi, who studied the equivalence of Liouville theory coupled to matter fields and $SU(2)/U(1)$ coset theory[11].

Motivated by the complete characterization of central extensions of primary field modules of the Virasoro algebra[12] (see also chapter 5 in ref. 13), we will find that the construction of an algebra which provides a setting for all these central extensions yields an algebra v_∞, which is very similar to w_∞, but has a richer sector of central charges[14]: While w_∞ admits only one charge, v_∞ turns out to admit six central charges. Furthermore, its non–extended version can also be considered as a subalgebra of the algebra of area–preserving diffeomorphisms, and therefore it might be of some relevance in further investigations of the ground ring. However, when my collaborator Martin Weigt and

I started to think about an algebra that could accommodate all the central charges of modules constructed in ref. 12, we didn't think about area–preserving diffeomorphisms, and therefore we encountered the algebra of area–preserving diffeomorphisms as a generalized Lie derivative of primary fields in two dimensions.

In this paper we will follow the original approach to v_∞ and start with an investigation of modules of the Virasoro algebra spanned by the vector spaces of primary fields. Therefore, an excursion on the covariantization of the notion of primary fields on two–manifolds will be given in section 2. Section 3 contains an exhaustive list of central charges which may appear in extended primary field modules of the Virasoro algebra, while the generalized Lie derivatives will be introduced in section 4.

2 COVARIANT PRIMARY FIELDS ON TWO–MANIFOLDS

Primary fields play a prominent rôle in low–dimensional field theory. Initially primary fields Φ of conformal weight $(\lambda, \bar\lambda)$ are defined by their transformation behaviour under a holomorphic change of charts $z \to u(z)$:

$$\Phi(u, \bar u) = \Phi(z, \bar z) \cdot \left(\frac{\partial u}{\partial z}\right)^{-\lambda} \cdot \left(\frac{\partial \bar u}{\partial \bar z}\right)^{-\bar\lambda} . \tag{1}$$

Note that I employ the usual convention to denote the weight for the complex conjugate sector of coordinates by $\bar\lambda$, which generically is not the complex conjugate of λ. The complex conjugate of λ will be denoted by λ^*.

Obviously this definition of primary fields works only in a conformal gauge, i.e. in an atlas with holomorphic transition functions. This causes no problem for integer values of λ and $\bar\lambda$, because the corresponding primary fields might be considered as remnants of tensor fields in the conformal gauge. However, such an interpretation is not possible for fractional conformal weights, and we will therefore generalize equation (1) for diffeomorphisms $z \to u(z, \bar z)$, i.e. we will define primary fields for arbitrary atlases on smooth two–manifolds, thereby introducing a covariant definition of half–differentials. A more detailed exposition is contained in chapters 1 and 2 of ref. 13.

To covariantize the notion of primary field we will introduce particular non–holonomic bases of vector fields and differentials over two–manifolds:

$$\mathcal{D}_z = \partial_z - \mu_z{}^{\bar z} \cdot \partial_{\bar z} , \tag{2}$$

$$\mathcal{D}z = \frac{1}{1 - \mu_{\bar z}{}^z \cdot \mu_z{}^{\bar z}} (dz + \mu_{\bar z}{}^z \cdot d\bar z) , \tag{3}$$

$$\partial_z = \frac{1}{1 - \mu_{\bar z}{}^z \cdot \mu_z{}^{\bar z}} (\mathcal{D}_z + \mu_z{}^{\bar z} \cdot \mathcal{D}_{\bar z}) , \tag{4}$$

$$dz = \mathcal{D}z - \mu_{\bar z}{}^z \cdot \mathcal{D}\bar z . \tag{5}$$

In the construction of the non–holonomic bases (2,3) we employed the Beltrami–differential $\mu_z{}^{\bar z}$. This is defined in terms of the metric on the two–manifold via

$$(ds)^2 = \frac{2 g_{z\bar z}}{1 + \mu_{\bar z}{}^z \cdot \mu_z{}^{\bar z}} \cdot |dz + \mu_{\bar z}{}^z \cdot d\bar z|^2 . \tag{6}$$

This implies a subtle transformation behaviour under orientation preserving diffeomorphisms $z, \bar z \to u, \bar u$, $|\partial_z u| > |\partial_{\bar z} u|$:

$$\mu_{\bar u}{}^u = \frac{\mu_z{}^{\bar z} \cdot \partial_{\bar u} z - \partial_{\bar u} u}{\partial_{\bar z} \bar u - \mu_z{}^{\bar z} \cdot \partial_z \bar u} = \frac{\partial_{\bar u} z + \mu_{\bar z}{}^z \cdot \partial_{\bar u} \bar z}{\partial_u z + \mu_{\bar z}{}^z \cdot \partial_u \bar z} .$$

As a consequence of the transformation properties of the Beltrami–differential the bases introduced in eqs. (2,3) provide factorized representations of the diffeomorphism group of the two–manifold:

$$\mathcal{D}_u = (\mathcal{D}_u z)\,\mathcal{D}_z\,, \qquad \mathcal{D}u = \mathcal{D}z\,\mathcal{D}_z u\,, \qquad \mathcal{D}_z u = (\mathcal{D}_u z)^{-1}\,. \qquad (7)$$

This permits for a consistent covariant definition of primary fields:

A field Φ over a two-manifold is denoted as *primary* of weight $(\lambda, \bar\lambda)$*if its local representations $\Phi(z, \bar z)$ transform under a change of coordinates $z, \bar z \to u, \bar u$ according to

$$\Phi(u, \bar u) = \Phi(z, \bar z) \cdot (\mathcal{D}_z u)^{-\lambda} \cdot (\mathcal{D}_{\bar z} \bar u)^{-\bar\lambda}\,.$$

In particular any tensor representation of the diffeomorphism group factorizes into appropriate primary fields with integer weights upon expansion with respect to the non–holonomic bases (2,3), but the crucial point is that also fractional weights can be defined without conformal gauge fixing. However, there is one restriction on the admissible values of the weight $(\lambda, \bar\lambda)$: In a region of three intersecting patches z_I, z_J, z_K the product of transition functions for $z_I \to z_J \to z_K \to z_I$ must yield the identity. While this yields no restriction on the scaling dimension $\Delta = \lambda + \bar\lambda^*$, it implies that both the real and the imaginary part of the spin $s = \lambda - \bar\lambda^*$ are restricted to integer or half-integer values. This is a consequence of the fact that the Z_2–cohomology of a two-manifold is trivial, while the Z_q–cohomology is generically non-trivial for $q > 2$.

For an introduction to two-dimensional field theory in this framework, see chapter 2 in ref. 13. We will conclude this section by explaining what we mean by a chiral λ–differential. We introduce a covariant primary derivative ∇_z which maps primary fields of weight $(\lambda, \bar\lambda)$ into primary fields of weight $(\lambda + 1, \bar\lambda)$ by employing the anholonomy coefficients of the primary bases (2,3):

$$[\mathcal{D}_z, \mathcal{D}_{\bar z}] = C^{\bar z}{}_{\bar z z}\,\mathcal{D}_{\bar z} - C^z{}_{z\bar z}\,\mathcal{D}_z\,, \qquad d\mathcal{D}z = C^z{}_{z\bar z}\,\mathcal{D}z \wedge \mathcal{D}\bar z\,.$$

A unique covariant primary derivative which satisfies a torsion constraint and leaves the metric invariant under parallel translation is then given by

$$\nabla_z \Phi(z, \bar z) = \mathcal{D}_z \Phi(z, \bar z) - \lambda \Phi(z, \bar z)\,\mathcal{D}_z \ln(\varrho_{z\bar z}) + (\lambda - \bar\lambda)\,C^{\bar z}{}_{\bar z z}\,\Phi(z, \bar z) \qquad (8)$$

where we introduced a primary field $\varrho_{z\bar z}$ of weight $(1,1)$ related to the metric via

$$\varrho_{z\bar z} = g_{z\bar z}\,\frac{(1 - \mu_z{}^{\bar z} \cdot \mu_{\bar z}{}^z)^2}{1 + \mu_z{}^{\bar z} \cdot \mu_{\bar z}{}^z}\,. \qquad (9)$$

A chiral λ–differential is a primary field of weight $(\lambda, 0)$ which satisfies the constraint

$$\nabla_{\bar z} \Phi(z, \bar z) = 0 \qquad (10)$$

and has poles of at most finite order. In the conformal gauge these are the holomorphic or meromorphic λ–differentials.

---

*Note again: the bar over $\bar\lambda$ denotes no complex conjugation. We do not require the weights to be real.

3 CENTRAL EXTENSIONS OF THE MODULES $\mathcal{M}_\lambda$ OF CHIRAL FIELDS

Henceforth we will switch to a conformal gauge for convenience, although it is clear from section 2 that all statements about primary fields can also be inferred in a covariant setting. Conformal gauge means that we use an atlas with vanishing Beltrami–differentials, this implying holomorphic transition functions (see ref. 13 and references therein for a more detailed explanation). Therefore, z and $\bar{z}$ always denote coordinates in a patch of a conformal atlas in the sequel.

Primary fields are the natural objects to study in two–dimensional field theory because they exhibit a factorized transformation behaviour under diffeomorphisms, as we have explained before. In the conformal gauge the remnant of this transformation behaviour is a simple transformation behaviour under holomorphic transformations, implying that primary fields provide modules of the Virasoro algebra. This yields non–perturbative information on correlation functions of primary fields. If the action is conformally invariant then this is expressed in a particularly simple and well–known form through the conformal Ward–identity:

$$\left\langle \frac{1}{2\pi i} \oint [dz\, \varepsilon^z\, T_{zz} - d\bar{z}\, \varepsilon^{\bar{z}}\, T_{\bar{z}\bar{z}}] \cdot \prod_{j=1}^{N} \Phi(z_j, \bar{z}_j) \right\rangle \tag{11}$$

$$= \left\langle \sum_{j=1}^{N} \Phi(z_1, \bar{z}_1) \cdot \ldots \cdot \Phi(z_{j-1}, \bar{z}_{j-1}) \cdot \delta_\varepsilon \Phi(z_j, \bar{z}_j) \cdot \Phi(z_{j+1}, \bar{z}_{j+1}) \cdot \ldots \cdot \Phi(z_N, \bar{z}_N) \right\rangle .$$

Here T_{zz} is the holomorphic component of the energy–momentum tensor, ε^z denotes a deformation of the conformal atlas: $z \to z + \varepsilon^z(z, \bar{z})$, and the Lie derivative of primary fields is given by

$$\delta_\varepsilon \Phi = \varepsilon^z\, \partial_z \Phi + \varepsilon^{\bar{z}}\, \partial_{\bar{z}} \Phi + \lambda \Phi\, \partial_z \varepsilon^z + \bar{\lambda} \Phi\, \partial_{\bar{z}} \varepsilon^{\bar{z}} . \tag{12}$$

The conformal Ward–identity yields information on the singular sector of short distance expansions of (quasi–)primary fields, implying e.g. for a chiral λ–differential:

$$T_{zz}(z)\, \Phi(w) = \frac{\lambda}{(z-w)^2}\, \Phi\!\left(w + \frac{z-w}{\lambda}\right) + \text{regular terms} . \tag{13}$$

This short distance expansion partially fixes the algebra of moments of the fields involved:

$$L_m = \frac{1}{2\pi i} \oint dz\, z^{m+1}\, T_{zz}(z) , \tag{14}$$

$$a_n = \frac{(-)^\lambda}{2\pi i} \oint dz\, z^{n+\lambda-1}\, \Phi(z) , \tag{15}$$

$$[L_m, a_n] = \big((\lambda - 1)m - n\big) \cdot a_{m+n} , \tag{16}$$

corresponding to the Lie derivative of the $(1-\lambda)$–differential $z^{n+\lambda-1}(-dz)^{1-\lambda}$ along the vector field $-z^{m+1}\partial_z$. This module of the Virasoro algebra, which is spanned by the $(1-\lambda)$–differentials, will be denoted as the module of $(1-\lambda)$–differentials $\mathcal{M}_{1-\lambda}$ henceforth.

Obviously, the information we have on the algebra of moments and the related short distance expansions is incomplete, because we know for sure only the commutators involving moments of the energy–momentum tensor. A few years ago, Zamolodchikov in a seminal paper initiated attempts to gain further knowledge on short distance expansions[1]. This usually involves some guesswork: Either one has to specify which chiral fields are supposed to appear in the short distance expansion of products of given chiral fields, or one has to suppose a certain non–central extension of equation (16). In this section I will adopt a conservative point of view by avoiding any guesswork. Instead, I'd like to emphasize that we know for sure that the action of the moments of the energy–momentum tensor on the moments of (quasi–)primary fields is given *essentially* by equation (16), but we also know for sure that there exists a non–trivial central extension in the module of chiral vector fields, because this module is exactly the Virasoro algebra, and that the corresponding central charge serves to partially classify conformal field theories in two dimensions. Therefore, our conservative, but necessarily restricted attempt to gain further insights into the classification of field theories in two dimensions starts from the following problem: Which modules $\mathcal{M}_\lambda$ admit non–trivial central extensions, and what are these extensions?

In mathematical terms we study the following problem: Introduce a central charge $C_{-1,\lambda}$ into the module of λ–differentials $\mathcal{M}_\lambda$:

$$L_m \circ a_n = -(n + \lambda m) a_{m+n} + C_{-1,\lambda}(m,n) \tag{17}$$

such that the extended object remains a module of the Virasoro algebra. This means to solve the consistency condition

$$(m-n) C_{-1,\lambda}(m+n,k) = (k+\lambda m) C_{-1,\lambda}(n,m+k) - (k+\lambda n) C_{-1,\lambda}(m,n+k) \tag{18}$$

modulo the trivial solutions

$$C_{-1,\lambda}^{\text{trivial}}(m,n) = -(n+\lambda m) f_{-1,\lambda}(m+n) \ .$$

The trivial solutions are the analogue of trivial extensions of a Lie algebra in the terminology of physics or of split extensions in the terminology of mathematics.

The details of the solution of this cohomology problem can be found in ref. 12 or ref. 13. The result is that there exist non–trivial solutions if and only if $\lambda \in \{-1, 0, 1\}$:

$$\lambda = -1: \quad L_m \circ a_n = (m-n) a_{m+n} + c_{-1,-1} m^3 \delta_{m+n,0} \ , \tag{19}$$

$$\lambda = 0: \quad L_m \circ a_n = -n a_{m+n} + c_{-1,0} m^2 \delta_{m+n,0} \ , \tag{20}$$

$$\lambda = 1: \quad L_m \circ a_n = -(m+n) a_{m+n} + (c_{-1,1} m + d_{-1,1}) \delta_{m+n,0} \ . \tag{21}$$

From the short distance expansion one finds that $c_{-1,0}$ appears like a background charge, see e.g. ref. 15, and the above result clarifies the algebraic significance of background charges as central extension of the module $\mathcal{M}_0$, which is realized through moments of conserved currents. In a similar vein the charges $c_{-1,1}$ and $d_{-1,1}$ can be realized in models with a background charge by the action of the modified energy–momentum tensor on a scalar field with logarithmic two–point function.[†] From a more radical point of view, however, one might be tempted to conclude from canonical quantization that quantization means to create central terms in algebras of fields where this is consistently possible. Therefore, in the next section we will try to construct a non–central extension of the Virasoro algebra which contains all the charges we found in equations (19-21).

4 GRADED LIE DERIVATIVES, w_∞-ALGEBRAS, AND CENTRAL CHARGES

In this section I will refer to a recent collaboration with Martin Weigt[14].

[†]I'm grateful to W. Nahm for a comment in this direction.

In order to proceed beyond the results of section 3, we now employ some guesswork: While it is clear that the action of moments of the energy–momentum tensor on moments of (quasi–)primary fields in a conformally invariant field theory is given by equations (17,19-21), usually realized as commutators, we would like to know the action of moments in other cases as well. We are particularly interested in the problem whether there exists an extended chiral algebra which accommodates the Virasoro algebra, some of the modules $\mathcal{M}_\lambda$, all the central charges found in section 3, and eventually more charges. This can hardly be settled in a systematic manner, but there is an Ansatz which naturally comes to mind: The action of the Virasoro algebra on primary fields is given by Lie derivatives, so we consider a generalized Lie derivative of a chiral ν–differential along a chiral μ–differential, yielding a $(\mu + \nu + 1)$–differential:

$$[\phi_\mu, \phi_\nu] = \nu\, \phi_\nu\, \partial\, \phi_\mu - \mu\, \phi_\mu\, \partial\, \phi_\nu \,. \tag{22}$$

Insertion of the bases of chiral μ–differentials on the twice–punctured sphere: $\phi_{\mu,m} = z^{m-\mu}(-dz)^\mu$, yields particularly the algebra of area–preserving diffeomorphisms:

$$[\phi_{\mu,m}, \phi_{\nu,n}] = (\mu n - \nu m)\, \phi_{\mu+\nu+1, m+n} \,. \tag{23}$$

Various subalgebras of this algebra appeared recently in the literature[6-8]: The restriction $\mu \leq -1$ yields the subalgebra w_∞, while the restriction $\mu \leq 0$ yields the subalgebra $w_{1+\infty}$. However, the most interesting case for us will correspond to the restriction $\mu \geq -1$, which yields a subalgebra denoted[14] as v_∞. Both the algebra (23) and the various subalgebras have quite distinct sectors of central charges.

To introduce a central term $C(\mu, m|\nu, n)$ into equation (23) means to solve the consistency condition

$$(\mu n - \nu m)\, C(\lambda, l | \mu + \nu + 1, m + n) + \text{cycl.}\,(\lambda, l; \mu, m; \nu, n) = 0 \,, \tag{24}$$

$$C(\mu, m|\nu, n) = -C(\nu, n|\mu, m) \,,$$

modulo the trivial solutions

$$C^{\text{trivial}}(\mu, m|\nu, n) = (\mu n - \nu m) \cdot f(\mu + \nu + 1, m + n) \,. \tag{25}$$

The details of how to solve this cohomology problem starting from the results of section 3 can be found in ref. 14. The result for w_∞ is known: The Virasoro charge is the only non–trivial central extension of w_∞. In the other cases one finds for $w_{1+\infty}$ that Planck's constant yields the only admissible non–trivial central extension, while for the full algebra of area–preserving diffeomorphisms one finds two central charges, both of which resemble Planck's constant. They are denoted below as $\hbar$ and $\hbar'$. However, for v_∞ one finds *six* central charges, including the quadruplet of central charges of section 3 and $\hbar$:

$\mu \leq -1 : w_\infty$

$$C(\mu, m|\nu, n) = c_{-1,-1}\, m^3\, \delta_{\mu+1,0}\, \delta_{\nu+1,0}\, \delta_{m+n,0} \tag{26}$$

$\mu \leq 0 : w_{1+\infty}$

$$C(\mu, m|\nu, n) = \hbar\, m\, \delta_{\mu,0}\, \delta_{\nu,0}\, \delta_{m+n,0} \tag{27}$$

area-preserving diffeomorphisms:

$$C(\mu, m|\nu, n) = (\hbar\, m + \hbar'\, \mu)\, \delta_{\mu+\nu,0}\, \delta_{m+n,0} \tag{28}$$

$\mu \geq -1 : v_\infty$

$$C(\mu, m | \nu, n) = \Big[\hbar\, m\, \delta_{\mu,0}\, \delta_{\nu,0} + c_{-1,-1}\, m^3\, \delta_{\mu+1,0}\, \delta_{\nu+1,0} \qquad (29)$$
$$+ c_{-1,0}\, m^2 \{\delta_{\mu+1,0}\, \delta_{\nu,0} - \delta_{\mu,0}\, \delta_{\nu+1,0}\}$$
$$+ c_{-1,1}\, m \{\delta_{\mu+1,0}\, \delta_{\nu-1,0} + \delta_{\mu-1,0}\, \delta_{\nu+1,0}\}$$
$$+ d_{-1,1} \{\delta_{\mu+1,0}\, \delta_{\nu-1,0} - \delta_{\mu-1,0}\, \delta_{\nu+1,0}\}$$
$$+ c_{0,1} \{\delta_{\mu,0}\, \delta_{\nu-1,0} - \delta_{\mu-1,0}\, \delta_{\nu,0}\}\Big]\delta_{m+n,0}.$$

The fact that the algebra of area–preserving diffeomorphisms and its various subalgebras admit such remarkably different sectors of central charges is a consequence of the fact that a larger algebra on the one hand has more place to accommodate central charges, but on the other hand also yields more Jacobi identities to kill them.

In conformally invariant field theories physical fields with half–odd integer spin correspond to fermionic degrees of freedom, as follows immediately from the modular invariant two–point functions. In this case the algebra of area–preserving diffeomorphisms (23) can be supersymmetrized by supplying it with the anti–commutator

$$\{\phi_{\mu,m}, \phi_{\nu,n}\} = 2\, \phi_{\mu+\nu, m+n} \qquad (30)$$

if both weights μ and ν are half–odd integer. Restricting the conformal weights again to values ≥ -1 yields the algebra super-v_∞, and it turns out that three charges of the bosonic sector survive the supersymmetric consistency conditions, while another charge appears in the centrally extended anti–commutator:

μ or ν integer:

$$[\phi_\mu, \phi_\nu] = \nu\, \phi_\nu\, \partial\, \phi_\mu - \mu\, \phi_\mu\, \partial\, \phi_\nu \qquad (31)$$
$$+ \frac{1}{2\pi i}\Big[\hbar\, \delta_{\mu,0}\, \delta_{\nu,0} \oint \phi_\nu \partial \phi_\mu + c_{-1,-1}\, \delta_{\mu+1,0}\, \delta_{\nu+1,0} \oint \phi_\nu \partial^3 \phi_\mu$$
$$- c_{-1,0} \{\delta_{\mu+1,0}\, \delta_{\nu,0} - \delta_{\mu,0}\, \delta_{\nu+1,0}\} \oint \phi_\nu \partial^2 \phi_\mu\Big]$$

μ and ν half–odd integer:

$$\{\phi_\mu, \phi_\nu\} = 2\phi_\nu \phi_\mu \qquad (32)$$
$$- \frac{2}{\pi i}\Big[\hbar\, \delta_{\mu-\frac{1}{2},0}\, \delta_{\nu-\frac{1}{2},0} \oint \phi_\nu \phi_\mu + c_{-1,-1}\, \delta_{\mu+\frac{1}{2},0}\, \delta_{\nu+\frac{1}{2},0} \oint \phi_\nu \partial^2 \phi_\mu$$
$$- c_{-1,0} \{\delta_{\mu+\frac{1}{2},0}\, \delta_{\nu-\frac{1}{2},0} - \delta_{\mu-\frac{1}{2},0}\, \delta_{\nu+\frac{1}{2},0}\} \oint \phi_\nu \partial \phi_\mu$$
$$+ c_{-\frac{1}{2},\frac{3}{2}} \{\delta_{\mu+\frac{1}{2},0}\, \delta_{\nu-\frac{3}{2},0} + \delta_{\mu-\frac{1}{2},0}\, \delta_{\nu-\frac{1}{2},0} + \delta_{\mu-\frac{3}{2},0}\, \delta_{\nu+\frac{1}{2},0}\} \oint \phi_\nu \phi_\mu\Big].$$

The centrally extended super-v_∞ yields the following short distance expansions of Boson–Boson or Boson–Fermion products:

$$\Phi_\mu(z)\, \Phi_\nu(w) = \frac{\mu+\nu-2}{(z-w)^2}\, \Phi_{\mu+\nu-2}\left(\frac{z+w}{2}\right) + \frac{\mu-\nu}{2(z-w)}\, \partial\Phi_{\mu+\nu-2}\left(\frac{z+w}{2}\right) \qquad (33)$$
$$+ \frac{\hbar}{(z-w)^2}\, \delta_{\mu-1,0}\, \delta_{\nu-1,0} + \frac{6 c_{-1,-1}}{(z-w)^4}\, \delta_{\mu-2,0}\, \delta_{\nu-2,0}$$
$$+ \frac{2 c_{-1,0}}{(z-w)^3} \{\delta_{\mu-1,0}\, \delta_{\nu-2,0} - \delta_{\mu-2,0}\, \delta_{\nu-1,0}\} + \text{regular terms}.$$

175

while the product of fermionic fields is

$$\Phi_\mu(z)\,\Phi_\nu(w) = \frac{2}{w-z}\Phi_{\mu+\nu-1}\left(\frac{z+w}{2}\right) \qquad (34)$$
$$+ \frac{4\hbar}{w-z}\delta_{\mu-\frac{1}{2},0}\delta_{\nu-\frac{1}{2},0} + \frac{8c_{-1,-1}}{(w-z)^3}\delta_{\mu-\frac{3}{2},0}\delta_{\nu-\frac{3}{2},0}$$
$$- \frac{4c_{-1,0}}{(w-z)^2}\left\{\delta_{\mu-\frac{1}{2},0}\delta_{\nu-\frac{3}{2},0} - \delta_{\mu-\frac{3}{2},0}\delta_{\nu-\frac{1}{2},0}\right\}$$
$$+ \frac{4c_{-\frac{1}{2},\frac{3}{2}}}{w-z}\left\{\delta_{\mu+\frac{1}{2},0}\delta_{\nu-\frac{3}{2},0} + \delta_{\mu-\frac{1}{2},0}\delta_{\nu-\frac{1}{2},0} + \delta_{\mu-\frac{3}{2},0}\delta_{\nu+\frac{1}{2},0}\right\}$$
$$+ \text{regular terms}.$$

In this form the short distance expansions suppose existence of chiral operators of arbitrary negative weight, albeit with vanishing two–point functions of the fields with negative weights. However, it is consistently possible to truncate the short distance expansions and the underlying algebra to conformal weights $\frac{1}{2}, 1, \frac{3}{2}$ and 2, and a thorough analysis reveals that in this case the set of central charges following from equations (31)–(34) is already complete.

REFERENCES

[1] A.B. Zamolodchikov, *Theor. Math. Phys.* **65** (1985) 1205.
[2] R. Blumenhagen, M. Flohr, A. Kliem, W. Nahm, A. Recknagel and R. Varnhagen, *Nucl. Phys.* **B361** (1991) 255.
[3] L. Fehér, L. O'Raifeartaigh, P. Ruelle, I. Tsutsui and A. Wipf, *Ann. Phys.* **213** (1992) 1.
[4] L. Frappat, E. Ragoucy and P. Sorba, ENSLAPP–AL–391/92 (July 1992).
[5] T. Tjin, A geometric construction of W–algebras, contribution to this workshop.
[6] I. Bakas, *Phys. Lett.* **B228** (1989) 57.
[7] D.B. Fuks, *Funct. Anal. Appl.* **19** (1985) 305.
[8] C.N. Pope, L.J. Romans and X. Shen, *Phys. Lett.* **B236** (1990) 173, *Nucl. Phys.* **B339** (1990) 191.
[9] E. Bergshoeff, P.S. Howe, C.N. Pope, E. Sezgin, X. Shen and K.S. Stelle, *Nucl. Phys.* **B363** (1991) 163.
[10] E. Witten, *Nucl. Phys.* **B373** (1992) 187.
[11] T. Eguchi, w_∞–algebra in 2D black holes, contribution to this workshop.
[12] R. Dick, *in:* "Nonperturbative Methods in Low Dimensional Quantum Field Theories", ed. G. Domokos *et al.*, World Scientific, Singapore 1991, p. 455.
[13] R. Dick, *Fortschr. Phys.* **40** (1992) 519.
[14] R. Dick and M. Weigt, LMU–TPW 92–11 (June 1992).
[15] D. Friedan, E. Martinec and S. Shenker, *Nucl. Phys.* **B271** (1986) 93.

2D BLACK HOLES AND 2D GRAVITY

Farhad Ardalan

Institute for Studies in Theoretical Physics and Mathematics
P.O.Box 19395-1795
Tehran, Iran
and
Physics Department
Sharif University of Technology
P.O.Box 11365-9161
Tehran, Iran

ABSTRACT

The $SL(2,R)/U(1)$ coset model, with $U(1)$ an element of the third conjugacy class of $SL(2,R)$ subgroups, is considered. The resulting theory is seen to collapse to a one dimensional Liouville field theory. Then the 2 dimensional black hole $SL(2,R)/U(1)$, with $U(1)$ a non-compact subgroup boosted by a Lorentz transformation, is considered. In the limit of high boost, the resulting black hole is found to tend to the Liouville field coupled to a $C = 1$ matter field. The limit of the vertex operators of the 2 dimensional black hole also tend to those of the $C = 1$ two dimensional gravity.

The two dimensional black hole of the $SL(2,R)/U(1)$ coset model has been obtained by gauging a noncompact $U(1)$ subgroup of the $SL(2,R)$ WZW theory[1]. Gauging a compact subgroup yields a coset model which results in the same black hole solution when it is properly analytically continued. These two subgroups represent two of the three classes of conjugacies of subgroups of $SL(2,R)$.

It is therefore intriguing to study the result of gauging a subgroup in the remaining third conjugacy class. The situation is of course reminiscent of the three classes of isotropy subgroups of the four dimensional Lorentz group, for massive, tachyonic, and massless states. The third conjugacy class corresponds to the massless class. It will be found that the $SL(2,R)$ WZW model gauged by this third class of subgroups results in a one dimensional theory which is a Liouville field theory.

To understand this reduction of the two dimensional space to the one dimensional Liouville field theory, a $U(1)$ subgroup of $SL(2,R)$ is gauged which is boosted by a Lorentz transformation, with boost parameter t. It is found that is $t \to \infty$, the 2d black hole in fact collapses to one dimensional Liouville; the collapse being the consequence of a sudden enlargement of the orbit of the gauge group.

For large values of t, the remaining degree of freedom resembles a $C = 1$ matter conformal field theory coupled to the Liouville. Therefore in the large t limit a concrete connection between the 2d black hole theory and the $C = 1$ matter coupled to 2d gravity is established. The connection is utilised to study the relation between the vertex operators of the two theories; in particular the amplitudes for the scattering of strings in the two backgrounds are related. It is hoped that this connection could shed some light on the question of the exact relation between the spectrum of the two theories[2,3,4].

To begin with, let us take σ_1, $i\sigma_2$ and σ_3 as the generators of $SL(2,R)$. The $i\sigma_2$ generates the compact $U(1)$ subgroup which upon gauging gives the Euclidean black hole; and σ_3 generates the noncompact subgroup whose gauging is responsible for the Lorentzian black hole[1]. An element of the third conjugacy class of $U(1)$ subgroups is generated by $\sigma = \sigma_3 + i\sigma_2$, which upon gauging in the usual manner[1], yields the following action:

$$\begin{aligned}I(g,A) &= I_{WZW} + \frac{k}{2\pi}\int d^2z\,\mathrm{tr}(\bar{A}g^{-1}\partial g + A\bar{\partial}gg^{-1} + g^{-1}Ag\bar{A}) ,\\ I_{WZW} &= \frac{k}{4\pi}\int_\Sigma d^2z\,\mathrm{tr}(g^{-1}\partial gg^{-1}\partial g) - \frac{k}{12\pi}\int_B d^2z\,\mathrm{tr}(g^{-1}dg \wedge g^{-1}dg \wedge g^{-1}dg) ,\end{aligned} \quad (1)$$

where g is an element of $SL(2,R)$ and A is the gauge field lying in the Lie algebra generated by σ. Note that the usual $A\bar{A}$ term does not appear here because the element σ is nilpotent. It is easily verified that this action is invariant under the axial vector gauge transformation,

$$g \longrightarrow hgh, \quad A \longrightarrow h(A+\partial)h^{-1}, \quad \bar{A} \longrightarrow h^{-1}(\bar{A}+\bar{\partial})h . \quad (2)$$

Now in the parametrization

$$g = \begin{pmatrix} a & u \\ -v & b \end{pmatrix}, \quad uv + ab = 1 , \quad (3)$$

of g, the gauge freedom (2) allows for a single gauge fixing $a + b = 0$, after whose imposition, and also upon integration of the action over the gauge fields $A, \bar{A}$, leads to the effective action,

$$I_{\mathrm{eff}} = \frac{k}{8\pi}\int d^2z\,(\partial\varphi\bar{\partial}\varphi + \alpha\varphi) , \quad (4)$$

which is that of a Liouville field. α is a constant, which is left undetermined due to ambiguities in the determinant in the path integral, and $e^\varphi = a - b - u - v$. A striking feature of this action is the absence of an additional field expected on the basis of a simple counting of the degrees of freedom.

The Liouville field theory has been considered from the point of view of WZW gauging in previous works[5,6]. However, the gauging used in these attempts are of the form $g \longrightarrow h_L g h_R$, where h_L and h_R are independent and correspond to positive and negative roots of $SL(2,R)$, respectively. Therefore it is natural that two degrees of freedom are eliminated by gauging the original $SL(2,R)$ theory. This is to be contrasted with the gauging above which is expected to eliminate only one degree of freedom.

To understand this reduction of the degrees of freedom, let us consider the 2d black hole anew; this time with a $U(1)$ subgroup which is a Lorentz transformation of the original noncompact $U(1)$ subgroup, i.e. the group generated by σ_3^t, where

$$\sigma_3^t = e^{-\frac{1}{2}t\sigma_1}\sigma_3 e^{\frac{1}{2}t\sigma_1} = \cosh \tfrac{1}{2}t\,\sigma_3 + \sinh \tfrac{1}{2}t\,i\sigma_2 \,. \tag{5}$$

As $t \to \infty$, σ_3^t becomes proportional to σ defined above. Therefore it is expected that as $t \to \infty$, the 2d black hole obtained from gauging the $U(1)$ generated by σ_3^t, should reduce to the Liouville theory as obtained above; moreover the asymptotic behaviour should explain the reduction of the degrees of freedom. Note that, for finite t, this gauged $SL(2,R)$ theory is equivalent to the original black hole theory with $t = 0$, because subgroups with different t are in the same conjugacy class and the resultant coset theories should be equivalent.

To describe the black hole geometry of the gauged theory, a gauge fixing is required. Recall that for the $U(1)$ gauging, prior to boosting, two regions of the (u, v) plane are distinguished by two distinct gauge conditions. The first region, consisting of the region $I - IV$ of ref. 1 is determined by the gauge condition $a = b$, in the parameterization (3); and the second region, consisting of the region V and VI of the above reference is determined by the gauge condition $a + b = 0$. Similarly for the boosted theory there are two regions, the first one is determined by the gauge condition $a - b = (u + v)\tanh t$; and the second one is determined by the condition $a + b = 0$. Now, using the $U(1)$ generator (5) and integrating the gauge fields leads to a black hole geometry in which the horizon which originally lied on the u and v axes will now tend towards each other as t gets larger and the singularity moves away from the origin of (u, v) plane. The effective action in the large t limit for the gauge condition $a + b = 0$, with $x = u - v$, will then read,

$$I_{\text{eff}} = \frac{k}{8\pi}\int d^2z\,[\partial\varphi\bar{\partial}\varphi + \alpha\varphi - 4e^{-2t}\partial x\bar{\partial}x e^{-2\varphi} - 4e^{-2t}(x^2+4)e^{-2\varphi}(\partial\varphi\bar{\partial}\varphi + \beta)], \tag{6}$$

where again α and β are constants to be determined e.g. à la DDK[8]. Terms decreasing faster than e^{-2t} have been dropped in (6).

As $t \to \infty$, only the first two terms in (6) survive and the action becomes identical to the Liouville action (4) obtained as a result of gauging σ. Now it is possible to trace the cause of the elimination of the extra degree of freedom x when $t \to \infty$: as was pointed out before, when t is finite there are two distinct regions of (u, v) plane determined by gauges $a - b = (u + v)\tanh t$ and $a + b = 0$. When t increases the region described by the second gauge fixing condition, region V of ref. 1, narrows and tends to a one dimensional set, parametrized by φ above. What then happens to the other region determined by the first gauge condition? The answer is that, at exactly $t = \infty$ (σ gauge), the condition $a + b = 0$ suffices to describe all the regions, i.e. at $t = \infty$ gauge orbits of the two regions connect to yield a single gauge orbit, thus making the theory one dimensional.

Eq.(6) also allows an interesting description of the black hole for finite but large t. The point is that if the field x is required to vary rapidly compared with the Liouville field φ, then the last two terms in (6) can be ignored compared with the first three; and one obtains a Liouville field φ coupled to a $C = 1$ matter field x. This novel connection between two dimensional gravity and two dimensional black hole theories i.e. the former theory being the limit of a large boost on the $U(1)$ gauge group of the latter, for quasistatic Liouville background, may provide a means of a detailed and explicit comparison of various aspects of the two theories, in particular their spectrum.

To this end and as a first step, the vertex operators of the two theories will be related in the following.

Vertex operators of the 2d black hole theory can be written in terms of the matrix elements of a group element $g(u,v)$ between states labelled by the Casimir operator $\bar{J}_0^2$ of $SL(2,R)$ and eigenvalues of the third component of the zero modes of the current,

$$J_0^3|\lambda,\omega\rangle = \omega|\lambda,\omega\rangle,$$
$$\bar{J}_0^2|\lambda,\omega\rangle = (-\tfrac{1}{4}-\lambda^2)|\lambda,\omega\rangle, \quad \omega,\lambda \text{ real},\tag{7}$$

where $L_0 \to -\dfrac{1}{k-2}\bar{J}_0^2 - \dfrac{1}{k}(J_0^3)^2$ and $c = \dfrac{3k}{k-2}-1 = 26$ giving $k = \tfrac{9}{4}$.

It can be shown that the only nonzero matrix elements are[7],

$$V(u,v) = \langle\lambda,\omega|g(u,v)|\lambda,-\omega\rangle,\tag{8}$$

which in the region I of ref. 1, and in terms of the parameters r and τ defined by

$$u = \sinh\tfrac{1}{2}r\, e^{\tfrac{1}{2}\tau}, \quad v = -\sinh\tfrac{1}{2}r\, e^{-\tfrac{1}{2}\tau},\tag{9}$$

become

$$V(r,\tau) = e^{-i\omega\tau}\langle\lambda,\omega|g(r)|\lambda,-\omega\rangle,\tag{10}$$

where $g(r)$ is a Lorentz transformation generated by σ_1. The matrix elements in (10) are Legendre functions whose asymptotic behaviour leads to

$$V(r,\tau) \sim e^{-\tfrac{r}{2}}[e^{-i(\lambda r+\omega\tau)} + S(\lambda,\omega)e^{i(\lambda r-\omega\tau)}],\tag{11}$$

with, on mass shell $(L_0-1)|\lambda,\omega\rangle = 0$ requiring $\omega^2 = 9\lambda^2$,

$$S(\lambda,3\lambda) = \frac{\Gamma(1+i\lambda)\Gamma(\tfrac{1}{2}-4i\lambda)^2}{\Gamma(1-2i\lambda)\Gamma(\tfrac{1}{2}-2i\lambda)^2},\tag{12}$$

giving the amplitude for string scattering in the black hole background[7].

Similarly, one can obtain the vertex operators and scattering amplitude for the 2d gravity theory by considering Liouville field to be an $SL(2,R)/U(1)$ theory as discussed in refs. 5,6,7 and take the $C=1$ matter to be some timelike variable. In the case at hand these vertex operators may be obtained as the limit of the corresponding black hole vertex operators in (10) with the boosted $U(1)$ subgroup. The boosted vertex operators will then be matrix elements of a group element g between states which are eigenvectors of the boosted $U(1)$,

$$V_t(u,v) = \langle\lambda,\omega_t|g(u,v)|\lambda,-\omega_t\rangle,\tag{13}$$

$$|\lambda,\omega_t\rangle = e^{-\tfrac{1}{2}t\sigma_1}|\lambda,\omega\rangle.\tag{14}$$

In the region determined by the gauge condition $a+b=0$, as $t \to \infty$ we expect to get 2d gravity vertex operators from (13). In fact as

$$|\lambda,\omega_t\rangle \longrightarrow |\lambda,\chi\rangle,\tag{15}$$

for $|\lambda,\chi\rangle$ an eigenvector of σ with eigenvalue χ. A simple calculation leads to these vertex operators,

$$V_t(u,v) \longrightarrow e^{-2i\chi e^{-\varphi}x}e^{-\varphi}K_{i\lambda}(2xe^{-\varphi}),\tag{16}$$

where $K_{i\lambda}$ is the Bessel's function of the third kind.

A few comments are now in order: first, the result (6) confirms the identification of x with the $C = 1$ matter field; second, χ, the limit of ω_t, replaces the cosmological constant in the conventional treatment of Liouville theory[7]; finally, the energy of the $2d$ gravity state in $\chi e^{-\varphi}$ which confirms the quasi-static nature of the Liouville field in this formulation.

The scattering amplitude can also be read off eq.(16) with the result,

$$S = \chi^{2i\lambda} \frac{\Gamma(1+i\lambda)}{\Gamma(1-i\lambda)}, \qquad (17)$$

in agreement with the expected result for $2d$ gravity[7]. Further study of the relation between $2d$ gravity and $2d$ black hole theory using the map defined in this work, in particular the relation between discrete states of the two theories, is in progress.

ACKNOWLEDGEMENTS

The author wishes to thank H. Arfaei and M. Alimohammadi for collaboration at the early stages of this investigation; also gratefully acknowledges useful discussions with L. O'Raifeartaigh and P. Sorba on the $SL(2,R)$ formulation of Liouville theory.

REFERENCES

1. E. Witten, *Phys. Rev* **D44** (1991) 314.

2. J. Distler and P. Nelson, *Nucl. Phys.* **B374** (1992) 123.

3. S. Chaudhuri and J. Lykken, *Nucl. Phys.* **B396** (1993) 270.

4. T. Eguchi, H. Kanno, and S. Yang, contribution to this workshop.

5. A. Alekseev and S. Shatashvili, *Nucl. Phys.* **B323** (1989) 719.

6. L. O'Raifeartaigh, P. Ruelle, and I. Tsutsui, *Phys. Lett.* **B258** (1991) 359.

7. R. Dijkgraaf, E. Verlinde, and H. Verlinde, *Nucl. Phys.* **B371** (1992) 269.

8. F. David, *Mod. Phys. Lett.* **A3** (1988) 1651; J. Distler and H. Kawai, *Nucl. Phys.* **B321** (1989) 509.

THE STRUCTURE OF FINITE DIMENSIONAL AFFINE HECKE ALGEBRA QUOTIENTS AND THEIR REALIZATION IN 2D LATTICE MODELS

D. Levy

School of Physics and Astronomy
Tel-Aviv University
Ramat-Aviv, Tel-Aviv 69978
Israel

ABSTRACT

The affine Hecke algebra of type A_{N-1} is introduced and its finite dimensional representations are discussed. It is demonstrated, in a particular finite dimensional quotient, how generic and non-generic irreducible representations are obtained by diagonalizing the maximal commutative subalgebra. Examples of 2D models of statistical mechanics, in which the affine Hecke algebra is realized, are given. The twisted xxz quantum chain serves as an example of how a translation invariant model, which also gives a representation of the periodic Hecke algebra, can be analyzed using the representation theory of the affine Hecke algebra. It is shown how degeneracies in the spectrum of the Hamiltonian arise from the special structure of the non-generic irreducible representations. Finally, a more general relation between translation invariant lattice models and the affine Hecke algebra is proposed.

1. INTRODUCTION

The affine Hecke algebra $AH(t, N)$ of type A_{N-1} is an associative algebra over $\mathbb{C}$. It is generated by the unit element $\mathbf{1}$ and $x_1, g_1, \ldots, g_{N-1}$ satisfying:

$$g_i^2 = (t-1)g_i + t, \tag{1}(a)$$

$$g_i g_{i+1} g_i = g_{i+1} g_i g_{i+1}, \tag{b}$$

$$[g_i, g_j] = 0 \; ; \; \forall \, |i-j| > 1 , \qquad (c)$$
$$[x_1, g_j] = 0 \; ; \; \text{for } j \geq 2 , \qquad (d)$$
$$g_1 x_1 g_1 x_1 = x_1 g_1 x_1 g_1 , \qquad (e)$$

where t is a complex scalar and N a non-negative integer. $1, g_1, \ldots g_{N-1}$ together with relations (a)-(c) give a presentation of $H(t, N)$ - the Hecke algebra of type A_{N-1}.

$AH(t, N)$ has already found several applications. Bernstein and Zelevinskii used it in their study of the representation theory of the p-adic group $GL(N, Q_p)$ [BZ]. The representation theory of other semi-simple p-adic groups is related to other affine Hecke algebras [Lus1,2].

In a different context, Drinfeld has shown that modules of $Y(sl(N))$ (the Yangian based on $sl(N)$) of weight m have the form $M \otimes_{S_m} \mathbb{C}^N$ where M is a degenerate affine Hecke algebra module [Dr]. The degenerate affine Hecke algebra is closely related to $AH(t, N)$. Moreover, Drinfeld conjectured in the same paper an analogous relation between $AH(t, N)$ and quantum affine Kac-Moody algebras of type $A_N^{(1)}$.

Finally, $AH(t, N)$ plays a key role in a generalization of the Knizhnik-Zamolodchikov equation proposed by Cherednik [Che1].

2. REPRESENTATIONS OF THE AFFINE HECKE ALGEBRA

A typical finite lattice model of statistical mechanics can be formulated in a finite dimensional vector space. From this point of view we are interested in finite dimensional representations of $AH(t, N)$. Since $AH(t, N)$ is infinite dimensional, even for finite N, we look for finite dimensional quotients of it. In any finite dimensional quotient the generator x_1 must satisfy some finite order polynomial equation, and conversely, such a condition guarantees that the quotient is finite dimensional. To analyze the representations we introduce the maximal commutative subalgebra of $AH(t, N)$ which was found by Bernstein and Zelevinskii (hereafter the BZ subalgebra). Defining $x_1, \ldots, x_N$ inductively by $x_{i+1} = t^{-1} g_i x_i g_i$, it is easy to prove that

$$[x_i, x_j] = 0 \quad \forall \, i, j \; ; \quad [g_i, x_j] = 0 \; j \neq i, i+1 . \qquad (2)$$

The x_i generate the BZ-subalgebra, and the idea is to look for representations in which the x_i act diagonally (see [Che2]). For finite dimensional quotients defined by the condition

$$\prod_{j=1}^{n} (x_1 - \lambda^{(j)}) = 0 \; , \quad \lambda^{(i)} \neq \lambda^{(j)} \text{ if } i \neq j . \qquad (3)$$

this method gives, for generic $t, \lambda^{(1)}, \ldots, \lambda^{(n)}$ values, the complete list of irreducibles together with an explicit basis [IP]. For non-generic values (i.e. the parameters satisfy some specified special relations), one can define consistent families of semi-simple quotients and get their irreps. In order to show how this works, we will consider a particular quotient with $n = 2$, in which further relations are imposed. This quotient has a relatively simple structure and it also produces all the representations which are relevant to the physical models we wish to discuss later [L1]. We denote it $Y(t, \lambda^{(1)}, \lambda^{(2)}, N)$ and its defining relations are:

$$e_i^2 = e_i, \qquad (4)(a)$$
$$e_i e_{i\pm 1} e_i = \tau e_i, \qquad (b)$$
$$[e_i, e_j] = 0 \qquad \forall\, |i-j| > 1, \qquad (c)$$
$$[x_1, e_j] = 0 \qquad \forall\, j \geq 2, \qquad (d)$$
$$(x_1 - \lambda^{(1)})(x_1 - \lambda^{(2)}) = 0, \qquad (e)$$
$$e_1 x_1 e_1 = \left(\frac{\lambda^{(1)} + \lambda^{(2)}}{1+t}\right) e_1, \qquad (f)$$

where $t \neq -1$, $e_i = \frac{g_i + 1}{1+t}$, $\tau^{-1} = 2 + t + t^{-1}$. Relation (4b) tells us that we have picked a quotient of $H(t, N)$ which is isomorphic to the Temperley-Lieb-Jones (TLJ) algebra [GHJ]. Relations (4e) and (4f) combine to give a particular solution to (1e). We note that $Y(N)$ admits a unique extension of the Markov trace defined on the TLJ algebra. It is defined by $\text{tr}(\mathbf{1}) = 1$, $\text{tr}(x_1) = \frac{\lambda^{(1)} + \lambda^{(2)}}{1+t}$ and $\text{tr}(e_{N-1} w) = \tau \text{tr}(w)$ if $w \in Y(N-1)$.

Since the x_i commute, we can look for an element $V_{\vec{\lambda}(N)}$ in the BZ subalgebra, which is a simultaneous eigenvector of them, namely:

$$x_i V_{\vec{\lambda}(N)} = V_{\vec{\lambda}(N)} x_i = \lambda_i V_{\vec{\lambda}(N)} \qquad ; \qquad i = 1, \ldots, N. \qquad (5)$$

One can prove that the following is a solution to (5):

$$V_{\vec{\lambda}(N)} = \prod_{j=1}^{N} (x_j - b_j) \qquad ; \qquad b_j = \frac{\lambda^{(1)} \lambda^{(2)}}{t^{j-1} \lambda_j}. \qquad (6)$$

We now specify the possible eigenvalues. λ_j has the general form $\lambda_j = \frac{\lambda^{(l)}}{t^{p_j}}$ where l is either 1 (λ_j is an "up spin") or 2 (λ_j is a "down spin"), and p_j is a non-negative integer. Each $\vec{\lambda}(N)$ contains a sub-sequence of k up spins and a sub-sequence of $N-k$ down spins ($0 \leq k \leq N$). The p_j are further constrained to start from zero and to increase monotonically within each sub-sequence. For example, the following is an allowed $N = 6$, $k = 2$ "eigenvalue vector":

$$\vec{\lambda}(6) = \left(\lambda^{(1)}, \lambda^{(2)}, \frac{\lambda^{(2)}}{t}, \frac{\lambda^{(2)}}{t^2}, \frac{\lambda^{(1)}}{t}, \frac{\lambda^{(2)}}{t^3}\right).$$

It is easy to verify that generically there are 2^N distinct $\vec{\lambda}(N)$. The corresponding $V_{\vec{\lambda}(N)}$ provide a basis for the BZ-subalgebra and obey a simple multiplication rule:

$$V_{\vec{\lambda}(N)} V_{\vec{\theta}(N)} = \delta_{\vec{\lambda}(N), \vec{\theta}(N)} N_{\vec{\lambda}(N)} V_{\vec{\lambda}(N)} \qquad ; \qquad N_{\vec{\lambda}(N)} = \prod_{i=1}^{N} (\lambda_i - b_i). \qquad (7)$$

It is also possible to show that all elements of $Y(N)$ of the form $aV_{\vec{\lambda}(N)}b$, where a, b are any elements of $Y(N)$, and $V_{\vec{\lambda}(N)}$ is any eigenvector of the x_i with a fixed number (k) of up spins, form, in the generic case, a minimal two-sided ideal in $Y(N)$

labeled by the integer k. We get that $Y(N)$ is semi-simple in the generic case. Since each minimal ideal gives rise to a unique irreducible representation, $Y(N)$ has $N+1$ inequivalent irreducible representations. The irrep labeled by k acts on a $\binom{N}{k}$ dimensional vector space. Some further simple arguments show how a $Y(N+1)$ irrep decomposes into a direct sum of $Y(N)$ irreps. The structure of a finite dimensional semi-simple associative algebra is conveniently encoded in a Bratteli diagram (see [GHJ]). This is a graph where the vertices are arranged in horizontal rows called floors. The graph has no disconnected vertices and only vertices on neighbouring floors are connected by edges. Each vertex at floor N represents an irrep of $Y(N)$, the number attached to the vertex stands for the dimension of this irrep, and the edges connecting floor N with floor $N+1$ indicate the decomposition discussed above.

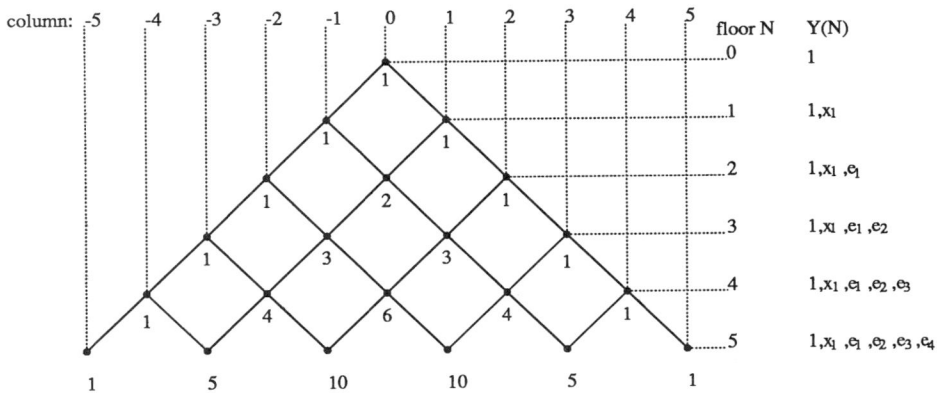

Figure 1. The Bratteli diagram for generic Y with $N \leq 5$.

The Bratteli diagram for the generic Y is just the Pascal triangle (see Figure 1).

A triple $(t, \lambda^{(1)}, \lambda^{(2)})$ is non-generic if it solves one or both of the following equations:

$$(a)\ \lambda^{(1)} - t^{n_1}\lambda^{(2)} = 0 \quad ; \quad (b)\ \lambda^{(2)} - t^{n_2}\lambda^{(1)} = 0 \tag{8}$$

where n_1, n_2 are positive integers. At a non-generic point $Y(N)$ will not be in general semi-simple. A sign of this is the fact that some of the $V_{\bar{\lambda}(N)}$ square to zero if one of the equations above holds. For the physical applications we have in mind, an important class of irreducible representations comes from semi-simple quotients of Y that can be consistently defined only at the non-generic points. To introduce them, assume at first that equation (8a) holds but not (8b). Taking $N = n_1$, there is a unique x_i eigenvector whose trace (defined after eq.(4)) is zero. This is the $V_{\bar{\lambda}(n_1)}$ with all n_1 spins up. Setting this element to zero in $Y(N \geq n_1)$ defines a consistent family of semi-simple algebras. The resulting Bratteli diagram is given by a Pascal triangle cut along column $n_1 - 1$. All the points which are strictly to the right of the cut are deleted. The numbers associated with the remaining points, are adjusted so that a number associated with a point on floor $N+1$ is the sum of the numbers associated with the points on floor N that are connected to it by edges. If both equations are satisfied, one can also set to zero $V_{\bar{\lambda}(n_2)}$ with all n_2 spins down.

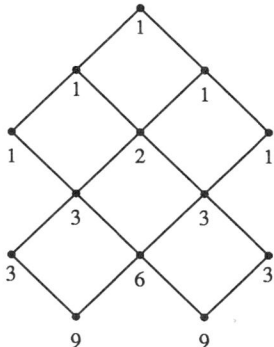

Figure 2. The Bratteli diagram of the $n_1 = 3$, $n_2 = 3$ quotient of Y.

The Bratteli diagram of the resulting semi-simple quotient is given by a Pascal triangle cut on both sides of its symmetry axis where the second cut is along column $-(n_2-1)$ (see Figure 2).

3. LATTICE MODELS REALIZING THE AFFINE HECKE ALGEBRA

We will now give two examples of lattice models which realize representations of Y. The second example will be discussed in detail and we will show how the representation theory described above explains the finite lattice symmetries of the model. For convenience we consider quantum chain models, but the basic ideas apply to the 2D classical statistical mechanics formulation as well. The examples we discuss suggest a natural generalization to other finite dimensional $AH(t,N)$ quotients, which will be mentioned in the summary.

We are looking for quantum chain Hamiltonians which depend on both the g_i and the x_i. An Hamiltonian which depends only on the g_i gives a representation of the $H(t,N)$ subalgebra of $AH(t,N)$ and is not an interesting example from the present point of view. Concerning the dependence on the x_i, they can appear either in bulk terms or in boundary terms or in both. An example of a bulk term which depends on the x_i is provided by the Ising model in an external magnetic field. The quantum chain Hamiltonian is given by:

$$H = \left(\sum_{l=1}^{L}(1-\sigma_l^x) + \lambda \sum_{l=1}^{L-1}(1-\sigma_l^z \sigma_{l+1}^z) + h \sum_{l=1}^{L} \sigma_l^z \right). \qquad (9)$$

In the above, $\sigma_i^a (a = x, y, z)$ acts as a 2×2 σ^a Pauli matrix on the i'th component of the tensor product and as a unit matrix on all other components. L is the length of the chain, λ is the inverse temperature and h the external magnetic field. H is a linear expression in the generators of $AH(i,N)$ in a representation coming from the $n_1 = n_2 = 2$ quotient of $Y(N)$:

$$g_{2j-1} = \tfrac{1}{2}(1+i)(1-\sigma_j^x) - 1 \quad ; \quad g_{2j} = \tfrac{1}{2}(1+i)(1-\sigma_j^z \sigma_{j+1}^z) - 1 \ ;$$
$$x_{2l-1} = -x_{2l} = i^{l-1}\sigma_l^z \ . \qquad (10)$$

An example of dependence through boundary terms which will be considered in some detail is provided by the twisted xxz chain. The Hamiltonian for the N-site chain is given by:

$$H_{xxz} = \sum_{j=1}^{N} V_j(\gamma) \qquad (11)$$
$$= \sum_{j=1}^{N} \frac{1}{2}\left(\sigma_j^x \sigma_{j+1}^x + \sigma_j^y \sigma_{j+1}^y + \cos\gamma(1 - \sigma_j^z \sigma_{j+1}^z) + i\sin\gamma(\sigma_j^z - \sigma_{j+1}^z)\right).$$

The N'th term in the sum is fixed by the following boundary conditions:

$$\sigma_{N+1}^x \pm i\sigma_{N+1}^y = e^{\pm i\phi}(\sigma_1^x \pm i\sigma_1^y) \quad ; \quad \sigma_{N+1}^z = \sigma_1^z, \qquad (12)$$

where ϕ is a parameter. It reads:

$$V_{N,1}(\gamma,\phi) = e^{i\frac{\phi}{2}\sigma_1^z} \frac{1}{2}\left((\sigma_N^x \sigma_1^x + \sigma_N^y \sigma_1^y + \cos\gamma(1 - \sigma_N^z \sigma_1^z) + i\sin\gamma(\sigma_N^z - \sigma_1^z)\right) e^{-i\frac{\phi}{2}\sigma_1^z}. \qquad (13)$$

These boundary conditions are compatible with translation invariance. Translations between two neighbouring sites are generated by:

$$T = e^{i\frac{\phi}{2}\sigma_1^z} s_1 s_2 \cdots s_{N-1}, \qquad (14)$$

where $s_i = V_i(\pi) + 1$. Translation invariance means $[H,T] = 0$. We note that H also commutes with the total spin z operator defined by $\hat{S}^z = \frac{1}{2}\sum_{i=1}^{N} \sigma_i^z$.

The spectrum of (11) was studied by [ABB] and [AGR], using numerical methods, either to diagonalize the finite lattice H directly ([AGR]) or to solve the Bethe ansatz equations [ABB]. The spectrum was found to exhibit an interesting pattern of degeneracies. After a brief review of [AGR] findings, we explain how $Y(N)$ is realized in this model and how does its representation theory account for the degeneracies.

Degeneracies appear when the two parameters of the model, γ, ϕ are rational multiples of π. [AGR] use the following parametrization:

$$\gamma = \pi\left(1 - \frac{m(m+1)}{n^2}\right) \quad ; \quad \phi = 2\pi\frac{k}{n}, \quad k = 0,1,\ldots,n-1, \qquad (15)$$

where n, m and k are integers. Since $\hat{S}^z$ commutes with H we can consider eigenvectors of H which are also eigenvectors of $\hat{S}^z$. Fixing an eigenvalue q of $\hat{S}^z$ (q is an integer or half integer depending on whether N is even or odd), define $H(q,k)$ to be the subset of H eigenvalues which come from the eigenvectors having $\hat{S}^z$ eigenvalue q, and k in (15) fixed to some value. After "choosing a system", namely fixing a relation between n and m, one finds $H(q',k') \subset H(q,k)$ for certain choices of q',k',q,k (other parameters are kept fixed). For example, choosing N even and taking $n = m + 1$, which defines the 2_R systems, one finds that $H(q,k) \subset H(k,q)$ for certain q and k. [AGR] define a procedure ("the projection mechanism") whereby one keeps only that part of $H(q,k)$ which is not contained in $H(q',k')$. The finite size scaling properties of this part are described by a minimal conformal field theory with $c < 1$.

One can compute the number of the eigenvalues which are left after the application of the projection mechanism. A neat formula based on the numerical data appears in [BRS]. The numbers agree with the dimensions of the irreducible representations of the various (n_1, n_2) quotients of $Y(N)$ described earlier. This provides a strong evidence for the relevance of the representation theory of $Y(N)$ to the understanding of the origin of the degeneracies. We will establish the connection in several steps [L2].

1. Let us consider the periodic Hecke algebra generated by $1, g_1, \ldots, g_N$, whose defining relations can be read from the $A_N^{(1)}$ Coxeter graph. This means that the g_i satisfy relations (1a-c), except that instead of $[g_1, g_N] = 0$ we have $g_1 g_N g_1 = g_N g_1 g_N$. We define the periodic TLJ algebra, which is a quotient of periodic Hecke, in an analogous manner. We observe that the xxz Hamiltonian (11) is written in terms of a representation of the periodic TLJ algebra:

$$e_j = (2\cos\gamma)^{-1} V_j(\gamma) , \quad j = 1, \ldots, N-1;$$
$$e_N = (2\cos\gamma)^{-1} V_{N,1}(\gamma, \phi) . \tag{16}$$

2. Consider a representation of the periodic Hecke algebra, acting on a finite dimensional vector space V. Assume that there exists a T acting on V, satisfying $T g_i = g_{i+1} T$ where $i = 1, \ldots, N$ and $N + 1 \equiv 1$. Then we consider:

$$x = g_{N-1}^{-1} \cdots g_1^{-1} T . \tag{17}$$

One can verify that $[x, g_j] = 0$ for $1 \leq j \leq N-2$ and $[x, g_{N-1} x g_{N-1}] = 0$. Comparing with eq. (1), we see that $1, g_1, \ldots, g_{N-1}, x$ satisfy the same relations, up to a simple isomorphism. We therefore have a representation of the affine Hecke algebra, acting on V. If, moreover, T is invertible, the periodic Hecke generators can be expressed in terms of the $AH(t, N)$ generators, where x is identified with x_N^{-1}. In this situation we can use the $AH(t, N)$ representation theory to analyze the periodic Hecke representation.

3. The twisted xxz model provides an example where the conditions specified in 2 are satisfied. The invertible T is just the translation operator (14). Since the xxz representation is finite dimensional, the generators must satisfy some further relations. After some computations one finds that in each $\hat{S}^z$ sector, with a fixed $\hat{S}^z$ eigenvalue q, the defining relations of $Y(N)$ are satisfied with the following values of $(t, \lambda^{(1)}, \lambda^{(2)})$:

$$t = e^{2i\gamma}, \quad \lambda^{(1)} = (-1)^{\frac{N}{2}-q+1} e^{-i(\frac{N}{2}+q)\gamma} e^{-i\frac{\phi}{2}},$$
$$\lambda^{(2)} = (-1)^{\frac{3N}{2}-q-1} e^{-i(\frac{N}{2}-q)\gamma} e^{i\frac{\phi}{2}} . \tag{18}$$

The identification of the parameters allows us to make contact with [AGR] observations. We will not carry the full analysis but just illustrate some key points. We first explain, in a non-rigorous way, why the degeneracies of H are related to non-generic points in the parameter space of the algebra. Degeneracy in the eigenvalues of H is due to the fact that the same irreducible representation appears more than once in the xxz representation. As discussed earlier, for non-generic $(t, \lambda^{(1)}, \lambda^{(2)})$, $Y(N)$ is not, in general, semi-simple. Typically, the dimensions of the irreducible representations are smaller than in the generic case (compare the numbers on the Pascal triangle with the numbers on the two-cut Pascal triangle). On the other hand, the dimension of the xxz representation is fixed and does not depend on the parameters. The multiplicity of irreps contained in the xxz representation can therefore increase, and this will manifest itself in the spectrum as higher degeneracy. We now check how the parameters of the model are constrained by the requirement that the parameters of the algebra take non-generic values. We can solve the system (8) inserting the values (18):

$$\gamma = \frac{\pi p}{n_1 + n_2} \ , \ p = 1, \ldots, n_1 + n_2 - 1;$$

$$\phi = 2\pi \left[l - \frac{p(n_1 + q)}{n_1 + n_2} \right] \ , \ l \text{ integer} .$$
(19)

We see that indeed at non-generic points, γ and ϕ are rational multiples of π. Moreover, since degenerate eigenvalues must come from two representations of the same algebra, q', k' must satisfy equations (19) with the same algebra parameters n_1, n_2. To illustrate the last point consider again the example with N even and $n = m + 1$. After substituting γ and ϕ from (15) we find that q and k appear in the equations only through the combination $q + k$. This is consistent with the choice for q' and k' which satisfies $q' + k' = q + k$.

4. SUMMARY AND GENERALIZATIONS

We have seen that the affine Hecke algebra is realized in the twisted xxz model via its finite dimensional quotient $Y(N)$, and that its representation theory is useful for the study of the degeneracies appearing in the spectrum of the finite lattice Hamiltonian. Reviewing the steps which were taken to establish the connection between $AH(t, N)$ and the xxz model, it should be emphasized that the xxz model is just one particular example of a model satisfying the assumptions made in step 2. The method outlined in step 2 allows us a fairly systematic treatment of periodic Hecke representations of physical interest, using our knowledge of the representation theory of $AH(t, N)$. For example, one can think of higher spin chains which give representations of the periodic Hecke algebra on $(\mathbb{C}^m)^{\otimes N}$, $m > 2$, and are equipped with an invertible T. In fact, it may be possible to consider an even more general setting. Consider an N-site quantum chain with an Hamiltonian H and an invertible T such that:

$$H = \sum_{j=1}^{N} b_j \ ; \ T b_i = b_{i+1} T \ , \ i = 1, \ldots, N \ , \ N + 1 \equiv 1 .$$
(20)

As usual, H acts on some finite dimensional vector space V and therefore the b_i can be thought of as generators of some finite-dimensional associative algebra. If V admits a representation of $H(t, N)$ such that (the same!) T satisfies the following equations:

$$T g_i = g_{i+1} T \ , \ i = 1, \ldots, N - 1 \ ; \ T^2 g_{N-1} = g_1 T^2 ,$$
(21)

then T and $H(t, N)$ generate a representation of $AH(t, N)$. Since $b_i = T^{i-1} b_1 (T^{i-1})^{-1}$, the algebra generated by $AH(t, N)$ and b_1 contains both H and T. If $[b_1, g_j] = 0$ for all $j > 2$, we get a natural inclusion structure and the resulting algebra might be a reasonable starting point for the study of the symmetries of the model. For example, T^N is in the center of this algebra. We can compute T^N using (17) with $x = x_N^{-1}$. One gets:

$$T^N = t^{\frac{1}{2}N(N-1)} x_1^{-1} x_2^{-1} \cdots x_N^{-1} ,$$
(22)

where the x_i are the generators of the BZ subalgebra. The r.h.s. is a central element in $AH(t, N)$ and it remains central after adjoining b_1. This may provide a first clue as to how the $AH(t, N)$ representations organize themselves in the bigger algebra.

Finally, let us mention an example to which such a scheme may be relevant. If one takes the b_i to be the generators of the Birman-Wenzl-Murakami (BMW) algebra (see for instance [BW]) then V automatically admits a representation of the TLJ algebra (and therefore of $H(t,N)$) given by $e_i = 1 - \frac{b_i - b_i^{-1}}{q - q^{-1}}$ where $q \neq 1$ is a parameter. Moreover, it is easy to see from the last formula that the T that translates b_i also translates e_i. The twisted BMW quantum chain was studied in [BRS] and found to exhibit "xxz type" degeneracies at special points of its parameter space.

REFERENCES

[ABB] F.C. Alcaraz, M.N. Barber and M.T. Batchelor, Annals of Physics 182 (1988) 280.

[AGR] F.C. Alcaraz, U. Grimm and V. Rittenberg, Nucl. Phys. B316 (1989) 735.

[BRS] D. Baranowski, V. Rittenberg and G. Schütz, Nucl. Phys. B370 (1992) 551.

[BW] J.S. Birman, H. Wenzl, Trans. Amer. Math. Soc. 313 (1989) 249.

[BZ] I.N. Bernstein, A.V. Zelevinskii, Representations of the group $GL(n,F)$ where F is a local nonarchimedean field. Uspehi Mat. Nauk 31 (1976) 5.

[Che1] I.V. Cherednik, Monodromy representations for generalized KZ equations and Hecke algebras, Kiev preprint, ITP-89-74E.

[Che2] I.V. Cherednik, Duke mathematical Journal, 54 (1987) 563.

[Dr] V.G. Drinfeld, Degenerate affine Hecke algebras and Yangians, Funct. Analysis and Applications, 20 (1986) 69-70.

[IP] In Preparation.

[GHJ] F.M. Goodman, P. de la Harpe, V.F.R. Jones, Coxeter-Dynkin diagrams and towers of algebras, Vol 14, Mathematical Sciences, Research Institute Publications, (1989) Springer Verlag.

[L1] D. Levy, The structure of the affine Hecke quotient underlying the translation invariant xxz chain, Tel-Aviv preprint, (1992) TAUP 1986-92.

[L2] D. Levy, Algebraic Structure of Translation-Invariant Spin-$\frac{1}{2}$ xxz and q-Potts Quantum Chains, Phys. Rev. Lett 67 (1991) 1971.

[Lus1] G. Lustig, Some examples of square integrable representations of semi-simple p-adic groups, Trans. Amer. Math. Soc. 277 (1983) 623.

[Lus2] G. Lustig, Affine Hecke algebras and their graded version, J. Amer. Math. Soc. 2 (1989) 599.

AN EXACT RENORMALISATION IN A VERTEX MODEL

Bruce W. Westbury

Mathematics Institute
University of Warwick
Coventry CV4 7AL

ABSTRACT

In this paper we propose a real-space renormalisation procedure for a class of homogeneous vertex models. This procedure is implemented for the 27-state model associated to a minimal irreducible representation of the quantised enveloping algebra of the exceptional Lie group E_6.

INTRODUCTION

The models we will study in this paper are two-dimensional models on the square lattice with spins associated to edges and interactions associated to vertices. These will be referred to as, simply, vertex models. A general construction of vertex models is to choose a Hopf algebra, H, and a finite-dimensional representation V. The model is constructed by taking the state space of an edge to be V. The Boltzmann weights can be viewed as a linear map $T: V \otimes V \to V \otimes V$ and are chosen so that this map commutes with the action of the Hopf algebra H. If $V \otimes V$ is completely reducible and each irreducible summand has multiplicity 1 then there is a canonical basis of the commutant $\text{End}_H(V \otimes V)$ consisting of primitive orthogonal idempotents. The Boltzmann weights are defined by taking an arbitrary linear combination of these idempotents. The coefficients are interaction coefficients.

A real-space renormalisation procedure is constructed by first blocking the edges and then giving a blocking move on the resulting state spaces on the edges. The blocking is defined by choosing a positive integer n and then collecting together n parallel lines at a time. The effect of this is to replace the state space of an edge V by $\otimes^n V$. The blocking move is defined by choosing an idempotent $e \in \text{End}_H(\otimes^n V)$. Let the image of e be W. Then the blocking move consists in applying e to the state space of each edge. The result of this procedure is to reduce the size of the square lattice by a factor n and to replace the state space V by the state space W. Typically W is not isomorphic to V, or to its dual, and so this procedure has changed the model. This construction is used in knot theory to construct new link invariants from old ones and is known as cabling. Exceptionally, W is isomorphic to V and so the model has not changed and we have a renormalisation procedure.

In detail then, let $T\colon V\otimes V \to V\otimes V$ be the Boltzmann weights and write T_i for the usual linear map $\otimes^n V \to \otimes^n V$ which is T on the i-th and $(i+1)$-th factors and the identity on the others. Then the renormalisation procedure changes T to

$$(e\otimes e)\left(\prod_{i=1}^{n-1}\prod_{j=0}^{i} T_{n-i+2j+1}\right)\left(\prod_{i=1}^{n} T_{2i-1}\right)\left(\prod_{i=n-1}^{1}\prod_{j=0}^{i} T_{n-i+2j+1}\right)(e\otimes e)$$

This defines a non-linear map of the vector space $\mathrm{End}_H(V\otimes V)$. However the model is unaltered if all the Boltzmann weights are multiplied by a scalar and not all Boltzmann weights are zero, so this map should be regarded as a non-linear map, Φ, of the projective space associated to $\mathrm{End}_H(V\otimes V)$. This map has degree n^2. If $\mathrm{End}_H(V\otimes V)$ has dimension r and a basis has been chosen, then this renormalisation operator acts on $\mathbb{C}P^{r-1}$ and is given by r homogeneous polynomials each of degree n^2.

The general theory of real-space renormalisation methods in lattice models is described in (Binney et al. [1992, Chapters 5]). The main physical predictions which can be obtained from studying the dynamics of the renormalisation operator, Φ, are as follows. A particular model will define a path in the space of interaction parameters which is parametrised by a temperature-like variable. A critical point is a point p on this path such that, for some $k \geqslant 1$, the sequence of points $\{\Phi^{ik}(p) : i \geqslant 0\}$ converges to a fixed point of Φ^k. The derivative of Φ^k is a linear endomorphism of the tangent space to the fixed point. Let λ be the eigenvalue of this endomorphism with largest magnitude and assume that $|\lambda| > 1$. Then the formula for the critical exponent, ν, is

$$\nu = \frac{\log \lambda}{k \log n}.$$

Since the blocking move is linear, the argument in (Niemeijer and van Leeuwen [1976]) shows that the critical exponents (but not the critical temperatures) predicted by this method will be exact.

The simplest case to consider is to take $n = 2$ and $r = 2$. This means we are looking for representations V such that $V \otimes V$ has two composition factors one of which is either V or its dual. The only non-trivial representations such that $V \otimes V$ has two composition factors are the fundamental representations of $SL(n)$ for all $n \geqslant 2$. The only one of these which has V or its dual as a factor is the case $n = 3$. The case we consider in this paper is the case $n = 2$ and $r = 3$.

The main example in this paper is the homogeneous vertex model associated to either of the two irreducible representations of the exceptional Lie group E_6 of dimension 27. These two representations are a dual pair. Although this seems like an exotic example to look at this representation was used to construct E_6 before the classification of simple Lie algebras. There are 27 lines on a non-singular cubic hypersurface and the configuration of these lines is also independent of the choice of hypersurface. This configuration gives a symmetric trilinear form on the vector space, V, with basis the set of 27 lines. The identity component of the group of automorphisms of V which preserve this trilinear form is E_6. A recent reference which discusses this construction is (Aschbacher [1987]).

LOCAL RELATIONS

For any semi-simple Lie algebra $\mathfrak{g}$ there is a quasi-triangular Hopf algebra $U_q\mathfrak{g}$, defined in (Jimbo [1989]), which is a deformation of the enveloping algebra of $\mathfrak{g}$, $U\mathfrak{g}$. If q is not a root of unity then the algebra $U_q\mathfrak{g}$ is isomorphic to the algebra $U\mathfrak{g}$ and, in particular, every finite dimensional representation is completely reducible and the irreducible finite dimensional representations of $U_q\mathfrak{g}$ are labelled by the dominant weights of $\mathfrak{g}$, see (Lusztig [1988]). The irreducible representation associated to a dominant weight λ will be denoted by V_λ.

Definition 0.1. For each n and each λ the Hopf algebra structure gives an action of $U_q\mathfrak{g}$ on $\otimes^n V_\lambda$. The quasi-triangular structure of $U_q\mathfrak{g}$ gives, for each n and each λ, a homomorphism

$$\beta_n \to \mathrm{End}_{U_q\mathfrak{g}}(\otimes^n V_\lambda)$$

where β_n is the braid group on n strings. The linear span of the image is denoted by $C_n(\mathfrak{g}; \lambda)$. In the language of lattice models, these are the algebras generated by the single vertex operators.

The following is list of representations V such that $V \otimes V$ has at most three composition factors. The minimal polynomial of the image of σ_i in $C_n(\mathfrak{g}; V)$ is independent of both n and i. The degree of this polynomial is (typically) the number of composition factors of $V \otimes V$ and so in each of these cases the image of σ_i has at most three distinct eigenvalues and satisfies a cubic equation.

(1) $\mathfrak{g} = \mathfrak{su}(N)$ and V is the fundamental representation.
(2) $\mathfrak{g} = \mathfrak{so}(N)$ and V is the fundamental representation.
(3) $\mathfrak{g} = \mathfrak{sp}(N)$ and V is the fundamental representation.
(4) $\mathfrak{g} = \mathfrak{su}(N)$ and V is the symmetric square of the fundamental representation.
(5) $\mathfrak{g} = \mathfrak{su}(N)$ and V is the exterior square of the fundamental representation.
(6) $\mathfrak{g} = \mathfrak{so}(8)$ or $\mathfrak{g} = \mathfrak{so}(10)$ and V is one of the two spin representations of $\mathfrak{g}$.
(7) $\mathfrak{g} = E_6$ and V is either of the two irreducible representations of dimension 27.

For any V there is a decomposition of $V \otimes V$ into the symmetric and exterior squares. This shows that if $\otimes V$ has at most three factors then the symmetric square of V or the exterior square is irreducible. Of these examples the only ones where the symmetric square of V is irreducible are (1) and (3). In all these examples except (3) the exterior square of V is irreducible.

The algebras $C_n(\mathfrak{g}; \lambda)$ for (1) are quotients of the Hecke algebras, see (Jones [1987]). The algebras $C_n(\mathfrak{g}; \lambda)$ for (2) and (3) are quotients of specialisations of the Birman-Wenzl algebras, see (Birman and Wenzl [1989]) and (Wenzl [1990]). Equivalently, in the language of lattice models, the single vertex operators, $\{\sigma_i\}$, in (1) satisfy the Hecke relations and in (2) and (3) they satisfy the Birman-Wenzl relations.

In these examples the natural inclusion

$$C_n(\mathfrak{g}; \lambda)@ >>> \mathrm{End}_{U_q\mathfrak{g}}(\otimes^n V_\lambda)$$

is an isomorphism. However, in general, the algebra $C_n(\mathfrak{g}; \lambda)$ is a proper subalgebra of the commutant. For the applications in this paper it is essential to take these subalgebras and not the commutants. For the applications to knot theory these subalgebras are more convenient since link invariants can be constructed by defining Markov traces on these subalgebras. For the applications to lattice models these subalgebras are more efficient since the row-to-row transfer matrix is an element of this subalgebra and so this gives a more refined temperature-independent block diagonalisation of the row-to-row transfer matrix.

The aim of this section is to find a set of local relations which are satisfied by the single vertex operators, $\{\sigma_i\}$, in all the cases listed above.

Notation 0.2. Let V be a representation of $H_n(2)$. Then the characteristic polynomial of σ_i is independent of i and is

$$(\lambda - x)^p (\lambda - y)^q (\lambda - z)^r$$

for some ordered triple of non-negative integers (p, q, r). In particular the dimension of V is $p + q + r$. In this paper we will label the representations by these ordered triples. This notation does not determine n which has to be understood from the context.

Notation 0.3. The convention for ordering the vertices of the Dynkin diagrams of the Lie algebras D_5 and E_6 is given in Figure 1.

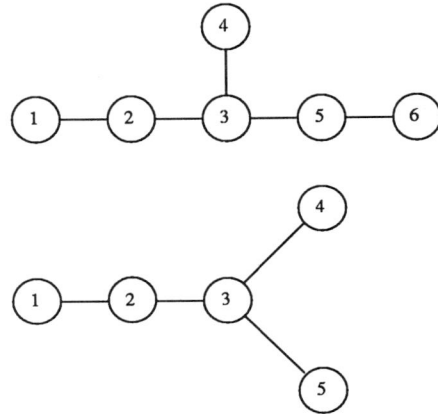

Figure 1. The Dynkin diagrams of the Lie algebras D_5 and E_6

The correspondence between the Bratteli diagram in Figure 3 and the highest weights of the irreducible representations is given in Figure 2.

Definition 0.4. Let a and b be any two constants satisfying $ab + xy = 0$. Then let $\rho(x, y)$ be the representation of β_3 defined by

$$\sigma_1 \mapsto \begin{pmatrix} x & 0 \\ a & y \end{pmatrix}, \quad \sigma_2 \mapsto \begin{pmatrix} y & b \\ 0 & x \end{pmatrix}.$$

Up to equivalence, this representation is independent of the choice of constants.

Definition 0.5. There is a representation of β_3 defined by

$$\sigma_1 \mapsto \begin{pmatrix} x & 0 & 0 \\ -1 & y & 0 \\ -y & (xz + y^2) & z \end{pmatrix}, \quad \sigma_2 \mapsto \begin{pmatrix} z & (xz + y^2) & -y \\ 0 & y & -1 \\ 0 & 0 & x \end{pmatrix}.$$

	$A_3 = SL(4)$	$A_6 = SL(7)$	$D_5 = SO(10)$	E_6
(1,0,0)	[0, 2, 0]	[0, 0, 0, 1, 0, 0]	[0, 0, 0, 0, 2]	[2, 0, 0, 0, 0, 0]
(0,1,0)	[2, 1, 0]	[1, 0, 1, 0, 0, 0]	[0, 0, 1, 0, 0]	[0, 0, 1, 0, 0, 0]
(0,0,1)	[4, 0, 0]	[0, 2, 0, 0, 0, 0]	[1, 0, 0, 0, 0]	[0, 0, 0, 0, 0, 1]

	$A_3 = SL(4)$	$A_6 = SL(7)$	$D_5 = SO(10)$	E_6
(1,0,0)	[0, 0, 2]	[0, 0, 0, 0, 0, 1]	[0, 0, 0, 0, 3]	[3, 0, 0, 0, 0, 0]
(1,1,0)	[1, 1, 1]	[1, 0, 0, 1, 0, 0]	[0, 0, 1, 0, 1]	[1, 0, 1, 0, 0, 0]
(1,1,1)	[2, 2, 0]	[0, 1, 0, 1, 0, 0]	[1, 0, 0, 0, 1]	[0, 0, 0, 1, 0, 0]
(0,1,0)	[3, 0, 1]	[]	[0, 1, 0, 1, 0]	[1, 0, 0, 0, 0, 1]
(0,1,1)	[4, 1, 0]	[1, 1, 1, 0, 0, 0]	[0, 0, 0, 1, 0]	[0, 1, 0, 0, 0, 0]
(0,0,1)	[6, 0, 0]	[0, 3, 0, 0, 0, 0]	–	[0, 0, 0, 0, 0, 0]

Figure 2. The correspondence between highest weights and the Bratteli diagram

The matrix $\sigma_1\sigma_2\sigma_1 = \sigma_2\sigma_1\sigma_2$ is

$$-xyz \begin{pmatrix} 0 & 0 & 1 \\ 0 & 1 & 0 \\ 1 & 0 & 0 \end{pmatrix}.$$

The algebra $H_3(2)$ is the semi-simple algebra of dimension twenty whose irreducible representations have dimensions

$$1 \quad 2 \quad 1 \quad 3 \quad 2 \quad 1.$$

The algebra $H_n(2)$ is generically a direct sum of matrix algebras. A list of inequivalent irreducible representations of $H_n(2)$ can be obtained from irreducible representations of C_n by adding the representation (r, q, p) whenever (p, q, r) corresponds to a representation of C_n and (r, q, p) does not. This shows that the dimensions of the algebras $H_n(2)$, for $2 \leqslant n \leqslant 6$ are at least

$$\begin{array}{ccccc} 2 & 3 & 4 & 5 & 6 \\ 3 & 20 & 200 & 680 & 5042 \end{array}$$

The Bratteli diagram for the sequence of algebras $H_n(2)$ has the Bratteli diagram of C_n as a subdiagram and has an involution which interchanges (p, q, r) and (r, q, p). The start of the Bratteli diagram of the sequence $H_n(2)$ is given in Figure 3.

These homomorphisms can be combined in the following commutative diagram.

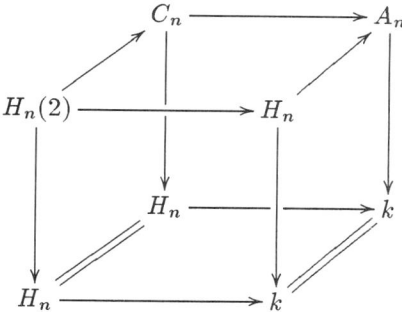

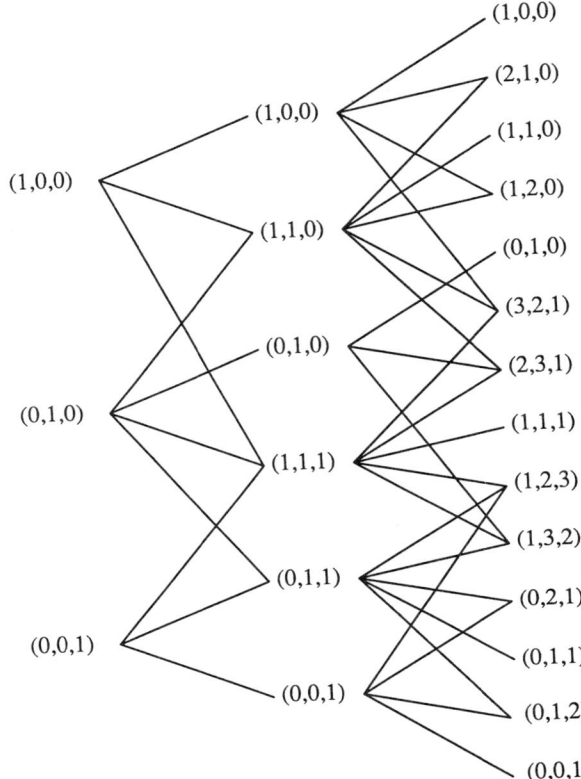

Figure 3. The start of the Bratteli diagram of $H_n(2)$

RENORMALISATION

The only representations in the previous list which have V or V^* as one of the irreducible factors of $V \otimes V$ are

(1) $\mathfrak{g} = \mathfrak{su}(2) \cong \mathfrak{so}(3)$ and $V = \mathbb{C}^3$,
(2) $\mathfrak{g} = \mathfrak{su}(3)$ and $V = \mathbb{C}^3$,
(3) $\mathfrak{g} = \mathfrak{su}(3)$ and $V = \mathbb{C}^6$,
(4) $\mathfrak{g} = E_6$ and $V = \mathbb{C}^{27}$.

In this list of examples, in (1) there is a copy of V and not V^* in $V \otimes V$. Also, in (2) the single vertex operators satisfy the Hecke relations and so are a different case. This leaves the remaining two cases.

The irreducible representation $(1,1,1)$ of $H_4(2)$ restricted to $H_3(2)$ is the irreducible representation $(1,1,1)$ and the actions of u_3 and v_3 are given by $u_3 = u_1$ and $v_3 = v_1$.

Lemma 0.6. *The irreducible representation $(1,3,2)$ of $H_4(2)$ is given by the following matrices and the irreducible representation $(1,2,3)$ of $H_4(2)$ is given by interchanging y and z in these matrices.*

$$g_1 = \begin{pmatrix} x & 0 & yz(y+z) & 0 & 0 & y^2z(y+z) \\ 0 & 0 & -yz & 0 & 0 & 0 \\ 0 & 1 & (y+z) & 0 & 0 & 0 \\ 0 & 0 & 0 & 0 & 0 & -yz \\ 0 & 0 & 0 & 0 & y & 0 \\ 0 & 0 & 0 & 1 & 0 & (y+z) \end{pmatrix}$$

$$g_2 = \begin{pmatrix} 0 & -yz & 0 & 0 & 0 & 0 \\ 1 & (y+z) & 0 & 0 & 0 & 0 \\ 0 & x(y+z)/yz & x & 0 & z(y+z) & 0 \\ 0 & 0 & 0 & y & 0 & 0 \\ 0 & 0 & 0 & 0 & (y+z) & y \\ 0 & 0 & 0 & 0 & -z & 0 \end{pmatrix}$$

$$g_3 = \begin{pmatrix} y & 0 & 0 & 0 & 0 & 0 \\ 0 & 0 & 0 & -yz & 0 & 0 \\ 0 & 0 & 0 & 0 & 0 & -yz \\ 0 & 1 & 0 & (y+z) & 0 & 0 \\ 0 & 0 & 0 & x(y+z)/yz & x & x(y+z)/z \\ 0 & 0 & 1 & 0 & 0 & (y+z) \end{pmatrix}$$

The change of model parameters can be calculated as follows. The algebra $H'_2(2)$ has dimension 3. The algebra $H_4(2)$ has three distinct irreducible representations in which $v_1 v_3 \neq 0$. These are the representations $(1,1,1)$, $(1,3,2)$ and $(1,2,3)$. In each of these representations $v_1 v_3$ is a rank one idempotent. This shows that, in each of these three representations, $v_1 v_3 \sigma_2 \sigma_1 \sigma_3 \sigma_1 v_1 v_3$ is proportional to $v_1 v_3$ and the transformation is given by these constants of proportionality.

Explicit matrices for these representations have been given and a direct calculation gives the following result.

$(1,1,1)$: $\quad e\sigma_2\sigma_1\sigma_3\sigma_1 e = y^2 z^2 e,$

$(1,3,2)$: $\quad e\sigma_2\sigma_1\sigma_3\sigma_1 e = xy^3 e,$

$(1,2,3)$: $\quad e\sigma_2\sigma_1\sigma_3\sigma_1 e = xy^2 z e.$

The transformation
$$\phi_R \colon (A, B) \mapsto (A^2 B^{-1}, B)$$
has been obtained using $v_1 v_3$ as the projection operator. By symmetry, the projection operator $u_1 u_3$ gives the transformation
$$\phi_L \colon (A, B) \mapsto (A, B^2 A^{-1}).$$
Each of these two transformations is an involution but these two transformations do not commute. The action of the group generated by these two involutions is completely described by
$$(\phi_R \phi_L)^n \colon (A, B) \mapsto \left(\frac{B^{2n}}{A^{2n-1}}, \frac{B^{2n+1}}{A^{2n}} \right)$$
$$(\phi_L \phi_R)^n \colon (A, B) \mapsto \left(\frac{A^{2n+1}}{B^{2n}}, \frac{A^{2n}}{B^{2n-1}} \right)$$
and
$$\phi_L (\phi_R \phi_L)^n \colon (A, B) \mapsto \left(\frac{A^{2n+1}}{B^{2n}}, \frac{A^{2n+2}}{B^{2n+1}} \right)$$
$$\phi_R (\phi_L \phi_R)^n \colon (A, B) \mapsto \left(\frac{B^{2n}}{A^{2n-1}}, \frac{B^{2n-1}}{A^{2n-2}} \right).$$

The point (A, B) is fixed by either of the last two transformations if and only if $A^2 = B^2$. The point (A, B) is fixed by either of the first two transformations if and only if $A^{2n} = B^{2n}$.

Definition 0.7. Define a one parameter family of maps
$$\phi_L(x) \colon \mathbb{C}P^2 \to \mathbb{C}P^2$$
$$\phi_L(x) \colon [p : q : r] \mapsto [P : Q : R]$$
where the homogeneous polynomials P, Q and R are:
$$P = \frac{p^2 (p^2 - 4pq + 6q^2)}{16} + \frac{p(p^2 - 2pq - 2q^2) r}{8}$$
$$+ \frac{(p^2 + 2pq + 3q^2) r^2}{8} + \frac{(p + 2q) r^3}{8} + \frac{r^4}{16},$$
$$Q = q^2 r^2 + \frac{-(p^2 q^2) - p^3 r + 2p^2 qr - 4pqr^2 + q^2 r^2 + pr^3 + 2qr^3}{4x}$$
$$+ \frac{pq^3 - 3pq^2 r - q^3 r + 3pqr^2 + 3q^2 r^2 - pr^3 - 3qr^3 + r^4}{(1+x)^2}$$
$$+ \frac{2(q^3 r - 2q^2 r^2 + qr^3)}{1+x},$$
$$R = \frac{q^2 (-p^2 + 2pq - 2qr + 5r^2)}{4}$$
$$+ \frac{q(-p^3 + 2p^2 q - 2pq^2 - 2p^2 r + 4pqr - 2q^2 r - 3pr^2 + 2qr^2 + 2r^3)}{4x}$$
$$+ \frac{(p-r)(q-r)^3}{2(1+x)^2} + \frac{qr(-q+r)^2}{1+x}.$$

Definition 0.8. Define a one parameter family of maps

$$\phi_R(x)\colon \mathbb{C}P^2 \to \mathbb{C}P^2$$
$$\phi_R(x)([p:q:r]) = \phi_L(x)([r:q:p])$$

and, finally, define a one parameter family of maps

$$\Phi(x)\colon \mathbb{C}P^2 \to \mathbb{C}P^2$$

by

$$\Phi(x)([p:q:r]) = \phi_R(\phi_L(x)([p:q:r])) \ .$$

This map has fixed points

$$[1:1:1],$$
$$[2^{4/3}:1:1],$$
$$[x:1:x],$$
$$[x^{-1}:1:x^{-1}].$$

The Lefschetz number of a smooth map of degree d acting on $\mathbb{C}P^2$ is $(2d+1)$. Assuming the fixed points are isolated then, by the Lefshetz fixed point theorem, this is a lower bound for the number of fixed points as some fixed points may contribute -1. However for a holomorphic map all isolated fixed points contribute $+1$ and so, if the fixed points are isolated then there are precisely $(2d+1)$ fixed points.

REFERENCES

M. Aschbacher [1987], The 27-dimensional module for E_6, Invent. Math. **89**, 159.

J.J. Binney, N.J. Dowrick, A.J.Fisher and M.E.J. Newman [1992], "The theory of critical phenomena", Oxford Univ. Press, New York.

J.S. Birman and H. Wenzl [1989] Braids, link polynomials and a new algebra, *Trans. Am. Math. Soc.* **313**, 249.

M. Jimbo [1989], Introduction to the Yang-Baxter equation, *in* "Braid Group, Knot Theory and Statistical Mechanics", C.N. Yang and M.L. Ge eds., Advanced Series in Mathematical Physics, vol. **9**, World Scientific.

V.F.R. Jones [1987], Hecke algebra representations of braid groups and link polynomials, *Ann. of Math* (2) **126**, 335.

G. Lusztig [1988], Quantum deformations of certain simple modules over enveloping algebras, *Adv. in Math.* **70**, 237.

T. Niemeijer and J.M.J. van Leeuwen [1976], Renormalization theory for Ising-like spin systems, *in* "Phase Transitions and critical phenomena", C. Domb and M.S. Green, eds., vol. 6, Academic Press, New York.

H. Wenzl [1990], Quantum groups and subfactors of type B,C,D, *Comm. Math. Phys.* **133**, 383.

NEW REPRESENTATIONS OF THE TEMPERLEY-LIEB ALGEBRA WITH APPLICATIONS

H.N.V. Temperley

Emeritus Professor of Applied Mathematics
University of Wales, Swansea[1]

ABSTRACT

The Temperley-Lieb algebra, using a new representation in terms of Young tableaux and in conjunction with Bruria Kaufman's "eigenoperator method", is applied to obtain new exact results for the Potts model. Cases in which q, the number of colours, is a "Beraha number" can apparently be solved completely. The Young tableaux representation of the Temperley-Lieb algebra is related to the Schensted construction. The connections with B. Kaufman's "eigenoperator" approach are also discussed.

1. INTRODUCTION

The Potts model [1] involves colouring the points of a graph or lattice with q colours. Each colouring is weighted with a factor $1 + f$ for each horizontal nearest neighbour pair that are coloured alike and with a factor $1 + F$ for a like vertical nearest neighbour pair. (All unlike neighbour pairs are given a weight unity). The problem is to calculate the generating function as a function of q, f and F. This generating function is also known as the Whitney-Tutte polynomial, but connections with braid topology and with knot polynomials have recently emerged. Various particular cases are of special interest. $q = 0 + \delta$ is the tree problem, known for any graph. $q = 1 + \delta$ is the percolation problem, partially solved by Temperley and Lieb [2] by reduction to Lieb's six vertex F model. $q = 2$ is the Onsager-Ising problem, solved by Onsager in 1944 for the plane square lattice and since extended to many other planar lattices. Baxter [3] solved the critical cases $q = Ff$ of the Potts model using the Temperley-Lieb formalism. f and F both put equal to -1 means that like pairs of nearest neighbours are not allowed. The generating function then becomes the chromatic polynomial, known for many graphs and obtained by Baxter for a large triangular lattice. If we formally put $q = -1$ in the chromatic polynomial R.P. Stanley showed in 1973 [4] that we get the number of acyclic orientations of the bonds of the lattice or graph.

---

[1] Present address, Thorney House, Thorney, Langport, Somerset TA10 0DW, England.

The four-colour theorem on any planar graph is equivalent to the statement that the chromatic polynomial is finite for $q = 4$ for any subgraph formed by deleting and contracting lines on the planar graph.

2. METHODS OF ATTACK

We want the generating function for all possible colourings of the points on a graph when like neighbouring pairs are weighted with factors $1 + f$, $1 + F$ as explained above.

$$W(q, F, f) = q^N + AFq^{N-1} + Bfq^{N-1} + \ldots , \qquad (1)$$

where A is the number of vertical nearest neighbour pairs and B is the number of horizontal nearest neighbour pairs in the graph.

The deletion-contraction algorithm has long been known. Let G be our graph, G' the graph obtained from it by omitting one vertical line and G'' the graph obtained by contracting the same line and identifying its ends. The we have easily, for a vertical line of G,

$$W(G) = W(G') + F\, W(G'') , \qquad (2)$$

with a similar equation obtained by removing a horizontal line from G. In principle this is enough for any finite graph, but in physics we are interested in the properties of a large periodic lattice, usually a plane square, and we have to systematise the process. One method is to set up a row-transfer operator as in other problems in lattice statistical mechanics. This expresses the effect of adding a row of points to the lattice, colouring them in q colours and introducing extra f's and F's as appropriate.

One general result seems to have been due to Whitney [5]. It is fairly easy to deduce from (1) and (2) by induction on the number of lines that

$$W(q, F, f) = \sum_g F^l f^m q^c , \qquad (3)$$

where the sum is taken over all subgraphs g of G, where the g's are obtained by omitting lines but not points from G and l, m are the numbers of vertical and horizontal lines in g and c is the number of disconnected components (counting isolated points) of g.

The row-transfer operators used by Temperley and Lieb [2] were as follows

$$\text{Products of} \quad 1 + f\frac{U_{AB}}{q^{\frac{1}{2}}} \quad \text{for pair } AB \text{ in the bottom row ,}$$

$$1 + f\frac{U_{BC}}{q^{\frac{1}{2}}} \quad \text{for pair } BC \text{ in the bottom row , } \qquad (4)$$

$$1 + f\frac{U_{CD}}{q^{\frac{1}{2}}} \quad \text{for pair } CD \text{ in the bottom row ,}$$

$$F\left(1 + \frac{q^{\frac{1}{2}}}{F}U_A\right) \quad \text{for vertical pair } A'A \text{ in two successive rows ,}$$

$$F\left(1 + \frac{q^{\frac{1}{2}}}{F}U_B\right) \quad \text{for vertical pair } B'B \text{ in two successive rows , } \qquad (5)$$

$$F\left(1 + \frac{q^{\frac{1}{2}}}{F}U_C\right) \quad \text{for vertical pair } C'C \text{ in two successive rows ,}$$

where the U operators have the following representation used by Potts [1]

$$q^{\frac{1}{2}}U_A : \quad R_A \to R_{A'} + Y_{A'} + G_{A'} + \ldots ,$$

$$Y_A \to R_{A'} + Y_{A'} + G_{A'} + \ldots , \tag{6}$$
$$G_A \to R_{A'} + Y_{A'} + G_{A'} + \ldots ,$$

$$\frac{U_{AB}}{q^{\frac{1}{2}}} : \quad R_A R_B \to R_A R_B ,$$

$$R_A Y_B \to 0 ,$$
$$R_A G_B \to 0 , \tag{7}$$
$$Y_A Y_B \to Y_A Y_B ,$$
$$Y_A R_B \to 0 , \text{ etc },$$

which select like pairs of neighbours. It is easily verified that these operators satisfy the von Neumann relations.

$$U_A^2 = q^{\frac{1}{2}} U_A , \quad U_{AB}^2 = q^{\frac{1}{2}} U_{AB} , \quad U_B^2 = q^{\frac{1}{2}} U_B , \ldots \tag{8}$$

$$U_A U_{AB} U_A = U_A , \quad U_{AB} U_A U_{AB} = U_{AB} , \tag{9}$$
$$U_B U_{AB} U_B = U_B , \quad U_{AB} U_B U_{AB} = U_{AB} , \ldots .$$

Operators with no common suffix commute as do operators like U_{AB}, U_{BC} (U_{23} and U_{45}). (These relations are not in the original Temperley-Lieb paper but we discovered them subsequently). Unfortunately, representations (6) and (7) only make sense if q, the number of colours, is a positive integer. Equation (3) defines the generating function/Whitney polynomial for any q.

3. THE TEMPERLEY-LIEB REPRESENTATION

We arrived at the Temperley-Lieb representation [2] by a systematic study of Heisenberg like operators. These act, not on the actual sites of the lattice, but on the two-valued spin variables μ_i associated with lines of the medial lattice in Figure 1.

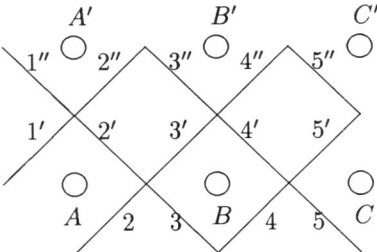

Figure 1. $A, B, C, \ldots A', B', \ldots$ points on the physical lattice, $1, 2, 3, \ldots 1', 2', \ldots$ lines on the medial lattice.

The representation is, Temperley and Lieb [2] expression (20),

$$U_A = U_{12} = C_1 C_2 \tfrac{1}{2}(1 - S_1 S_2) + e^{\theta} \tfrac{1}{4}(1 + S_1)(1 - S_2) + e^{-\theta} \tfrac{1}{4}(1 - S_1)(1 + S_2) , \tag{10}$$

where $q^{\frac{1}{2}} = 2 \cosh \theta$. C_i and S_i are Pauli operators acting on the spin variable μ_i with

$$C_i : \quad C_i 1 = 1 ,$$
$$C_i \mu_i = -\mu_i ,$$
$$S_i : \quad S_i 1 = \mu_i ,$$
$$S_i \mu_i = 1 .$$

The expression for $U_{AB} = U_{23}$ is just (10) with 1 replaced by 2 and 2 by 3 and so on for $U_B = U_{34}$, $U_{BC} = U_{45}$, etc. They satisfy (8) and (9) for any q. This representation leads to a six-vertex problem on the medial lattice, of the type previously solved by Lieb, when applied to the percolation and Potts problems. Unfortunately the "generalised Bethe Ansatz" method works only for the critical cases of these two problems where the numerical coefficients of the U_A and U_{AB} operators in (4) and (5) are the same i.e. $Ff = q$ which is known to be the critical condition for the Potts model. Since repeated efforts to generalise the Bethe Ansatz method have failed, we worked with three further representations.

4. THE NON-INTERSECTING STRING MODEL (NIS)

We generalise the six-vertex model on the medial lattice by allowing the line variables to take several colours instead of just two. However, we only allow three types of vertex as shown in Figure 2.

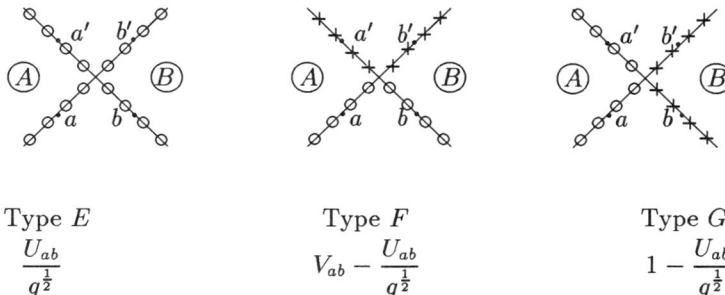

Type E
$\dfrac{U_{ab}}{q^{\frac{1}{2}}}$

Type F
$V_{ab} - \dfrac{U_{ab}}{q^{\frac{1}{2}}}$

Type G
$1 - \dfrac{U_{ab}}{q^{\frac{1}{2}}}$

Figure 2. $A, B, \ldots$ points on the physical lattice, $a, b, \ldots a', b', \ldots$ midpoints of lines on the medial lattice.

All four lines in a type E vertex are coloured with the same colour in q ways while neighbouring pairs of lines in vertices of types F and G are coloured with two different colours each in $q(q-1)$ ways. This was considered as a lattice model in its own right by Stroganov [6] and more generally by Perk et al. in a number of papers [7]. It will be evident that, if we build up a lattice using only these vertices, we shall finish up with non-intersecting domains whose boundaries are each coloured with one colour and possibly some open strings each of one colour. Two strings of different colours may have vertices of type F or G in common, but they cannot intersect, hence the name non-intersecting string model.

We can also regard this non-intersecting string lattice as the medial lattice of a plane square lattice. Label the mid points of the lines of the medial lattice a, b, a', b' etc. as shown in Figure 2 and let U_{ab}, U_a, U_b etc be a set of U operators associated with this lattice ("the medial lattice of the medial lattice"). Referring to Figure 2, the operator $U_{ab}/q^{\frac{1}{2}}$ implies that a and b and a' and b' are all coloured alike, while $1 - U_{ab}/q^{\frac{1}{2}}$ implies that a and b are coloured differently but a and a' are alike, as are b and b'. Now define a new operator V_{ab} as $U_{ab}U_aU_bU_{ab}$. $U_{ab}/q^{\frac{1}{2}}$ implies that a and b are alike, $q^{\frac{1}{2}}U_a$ that a' is coloured in all q colours, $q^{\frac{1}{2}}U_b$ that b' is coloured in all q colours while the final operator $U_{ab}/q^{\frac{1}{2}}$ ensures that a' and b' are coloured alike. This also includes the cases where all four lines are alike as in a type E vertex so we must subtract these out and the operator $V_{ab} - U_{ab}/q^{\frac{1}{2}}$ is associated with a vertex of type F.

We can now express the deletion contraction algorithm for the physical lattice in the form given by L.H. Kauffman [8]. Let A and B in Figure 2 be two horizontal neighbours in the physical lattice. Express the deletion contraction algorithm in Kauffman's "skein-relation" form as in Figure 3.

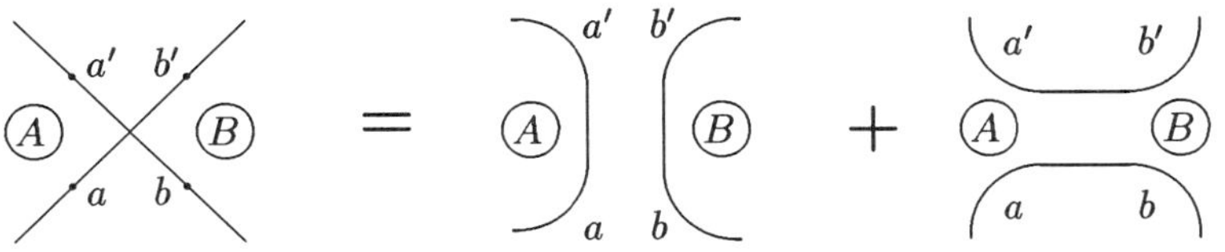

$$AB \text{ operator} \quad = \quad \begin{pmatrix} 1 - U/q^{\frac{1}{2}} \\ +U/q^{\frac{1}{2}} \end{pmatrix} \quad + \quad \frac{f}{q}\begin{pmatrix} V - U/q^{\frac{1}{2}} \\ +U/q^{\frac{1}{2}} \end{pmatrix} . \quad (11)$$

Figure 3

Lines such as aa', ab are part of non-intersecting strings of uniform colour. However, we must consider separately the cases where strings aa' and bb' are coloured alike or coloured differently, so we have $G + E$ as the first term and $\frac{f}{q}(F + E)$ as the second term in Figure 3. The first term on the right of figure 3 corresponds to the line AB being deleted while the second term corresponds to its being contracted so that A and B on the physical lattice are coloured alike.

In other words, the AB interaction operator is just $V_{ab} = U_{ab}U_aU_bU_{ab}$. Similarly it can be shown that $V_{bc} = U_{bc}U_bU_cU_{bc}$ describe the interaction between B and B' on the physical lattice. It is, in fact, quite easy to show from (8) and (9) that V_{ab} and V_{bc} obey the relations $V_{ab}V_{bc}V_{ab} = V_{ab}$ etc and $V_{ab}^2 = qV_{ab}$ etc, that is to say that operators U_{ab}, U_b etc acting on the q-colour variables on the medial lattice generate a set of operators V_{ab}, V_{bc} etc that act as Temperley-Lieb operators on the physical lattice but with $q^{\frac{1}{2}}$ replaced by q. This result was deduced by Perk et al. [7], but by quite other means.

5. THE YOUNG TABLEAU REPRESENTATION

The close resemblance of (10) to a Heisenberg exchange operator, $(1 + U_{12}$ in fact becomes the exchange operator if we put $\theta = \pi i$ in (10)), led Dr. Colin Vout and myself to look for a Young tableau representation of the U operators. We found it and it has been reported at various conferences though not formally published. (Very similar results have been reported by V. Pasquier and by P.P. Martin.) We found that one does not need Young tableaux with more than two rows. This is because our operators are less general than full permutation operators since they only act on two-valued symbols, the μ's. We are concerned with the Heisenberg group rather than the full symmetric group. The one-row operators are defined recursively as follows,

$$\boxed{2\,3} = 1 - \frac{U_{23}}{A_1}, \quad \boxed{2\,3\,4} = \boxed{2\,3}\left(1 - \frac{U_{34}}{A_2}\right)\boxed{2\,3},$$

$$\boxed{2\,3\,4\,5} = \boxed{2\,3\,4}\left(1 - \frac{U_{45}}{A_3}\right)\boxed{2\,3\,4} \quad \text{etc.} \quad (12)$$

where $A_1 = q^{\frac{1}{2}}$, $A_2 = (q-1)/q^{\frac{1}{2}}$, $A_3 = q^{\frac{1}{2}}(q-2)/(q-1)$ and generally

$$A_r = \frac{\sinh(r+1)\theta}{\sinh r\theta}, \quad q^{\frac{1}{2}} = 2\cosh\theta .$$

These operators are idempotent and also invariants in the sense that e.g. $\young(2345)$ vanishes when multiplied by U_{23}, U_{34} or U_{45}.

Two-row operators are made as follows. We begin with a product like $\dfrac{U_{23}}{q^{\frac{1}{2}}}\dfrac{U_{45}}{q^{\frac{1}{2}}}$ which corresponds to $\young(24,35)$ and we may add on the right a single row of boxes like $\young(678)$ defined analogously to $\young(234)$. We use operators $1 - A_1 U_{34}$ to exchange 3 and 4 in $\young(24,35)$ thus $\young(23,45)$ is $(1 - A_1 U_{34})\,\young(24,35)\,\dfrac{1 - A_1 U_{34}}{A_1 A_2}$. We use an operator like $1 - A_2 U_{45}$ to exchange 5 for 4 at a "Knight's move" away, e.g.

$$\young(234,5) = (1 - A_2 U_{45})\,\young(235,4)\,\left(\dfrac{1 - A_2 U_{45}}{A_2 A_3}\right) \tag{13}$$

and so on for larger distances. We can, of course, undo the effect of an operator such as $1 - A_2 U_{45}$ by multiplying in front or behind by its inverse $1 - U_{45}/A_3$. So far we have found idempotents. To obtain non-diagonal operators it is sufficient simply to omit some factors like $1 - A_1 U_{34}$, $1 - A_2 U_{45}$.

These operators were found to have many of the properties of the operators corresponding to standard Young tableaux. Operators corresponding to different shapes of tableau are orthogonal, so that the U algebra is split into sub-algebras, one sub-algebra for each shape of tableau. The most important differences are, first that we only need two rows, secondly if q, the number of colours, is a "Beraha number" $4\cos^2(\pi/k)$ with k an integer ≥ 3. This situation corresponds to θ being imaginary in (10) and (11) and e^θ being a root of unity. This means that some of the A_r as defined in (11) vanish so that we have a limit both to the number of boxes in a 1-row tableau and to the distance apart two successive numbers can be if we can exchange them in the manner illustrated in (13). However, the early Beraha numbers are of great interest. $k = 3$ corresponds to $q = 1$ which is almost trivial, $k = 4$ corresponds to $q = 2$ which is the Onsager problem solved by him in 1944, $k = 5$ corresponds to $q = \frac{1}{2}(3 + \sqrt{5})$ and this is closely connected with Baxter et al.'s extensive work [9] on the generalised hard hexagon models as will become clearer when we look at the "walks" representation of the algebra. $k = 6$ corresponds to $q = 3$ which has been extensively studied by series expansion and renormalisation methods.

Using the Potts representation (6) and (7) of the U operators we have verified explicitly that, for $q = 2$ the operator corresponding to a line of three boxes vanishes identically. The same is true for $q = 3$ and a line of five boxes. We have no Potts representation for the case of $q = \frac{1}{2}(3 + \sqrt{5})$ but indirect arguments show that the operator for four boxes in a single line vanishes. Thus, if there is only one box in the lowest row, we can only have three standard tableau $\young(2456,3)$, $\young(2356,3)$ and $\young(2346,5)$, the operator $\young(2345,6)$ dropping out.

For the case $q = 2$ we can only have two standard tableau of the type $\young(245,3)$ and $\young(235,4)$. For $q = 3$ we can have four standard tableaux in this sub-algebra.

6. THE "WALKS" REPRESENTATION

It is well known that 2-row tableaux are in one to one correspondence with 1-step walks up or down that never go below the starting point. If we exhibit the walks on a diagonal lattice the algorithm is, if in the tableau representation r is in the top row the step ending in r is up, if r is in the bottom row the step is down. This implies that, if q is a Beraha number $4\cos^2 \pi/k$, the walks corresponding to permissible tableaux never go outside a lattice of depth $k-1$. Thus, for $q=2$, $k=4$, all the permissible walks are included in the lattice shown in Figure 4.

We can describe the walks in a great many ways by means of spin variables, for example the rth step may be up or down, or point $2r+1$ may be at the top or bottom corresponding to $-$ or $+$. If we constrain the spin on point A to be positive that on point B can be positive or negative corresponding to the sign of μ_3. Looking at the operators we see that $U_{23}/q^{\frac{1}{2}}$ and $1-U_{23}/q^{\frac{1}{2}}$ correspond to A and B being alike or different while $q^{\frac{1}{2}}U_{34}-1$ is just Onsager's operator C_B which changes the sign of spin B or of the spin at station 3 in Figure 4. Thus we are back to Onsager's treatment which uses operators $S_A S_B$ etc. and C_B, C_C etc. so this treatment give nothing new. However, it is otherwise with the cases $k=5$ and $k=6$. We show the relevant portion (of depth 4 for $k=5$) of the lattice in Figure 5.

Again we can label all possible walks by associating a spin variable with each vertical pair of points. If we choose the signs as shown the possible walks are subject to the constraint that no positives can come together but we can have $1+2-3+$, $1+2-3-4+$, $1+2-3-4-5+$ and so on, as many minuses as we like between each pair of pluses. Each permissible walk corresponds to a permissible row of pluses and minuses in Baxter's "hard hexagon" problem. We have verified in detail that the critical Potts model, which is the only one we can solve at present, corresponds, as it should, to the generalised hard hexagon problem on one of its transition lines found by Baxter et al. [9].

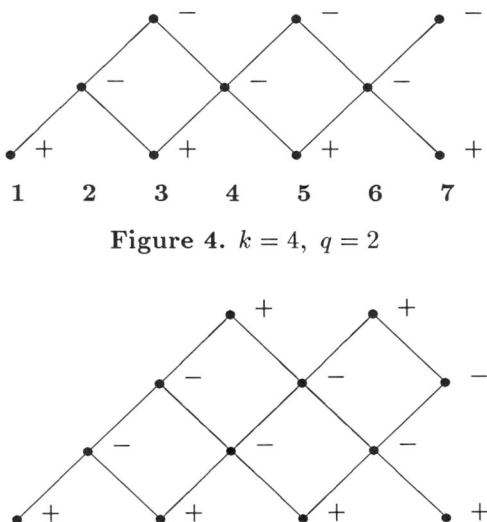

Figure 4. $k=4$, $q=2$

Figure 5. $k=5$, $q=\frac{1}{2}(3+\sqrt{5})$, outside points positive, inside points negative

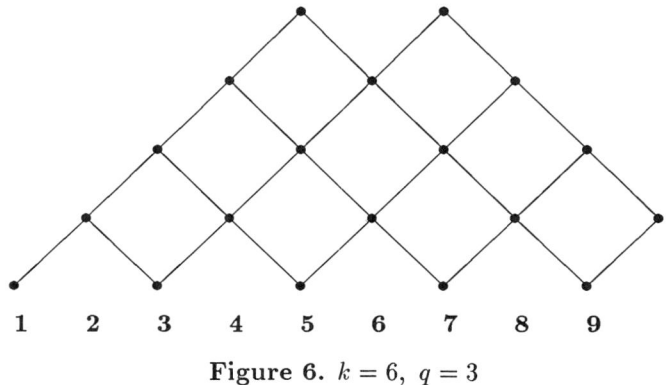

Figure 6. $k = 6$, $q = 3$

The case $k = 6$, $q = 3$ has a lattice that is obtained by putting another set of diagonal squares on top of Figure 5. (Fig. 6).

We can no longer describe the walks by two-valued spin variables, we need three-valued lines for columns 5, 7, 9 etc. in Figure 6. However the fact that the walks lattice is finite for q a Beraha number suggests that, for these cases, it may be possible to generalise Bruria Kaufman's eigenoperators process which appears to be very powerful and was used with considerable effect on the Onsager problem ($q = 2$) [10]. It has been successfully applied to a generalised Ising model by Aguilar and Braun [11]. For applications to various under-determined physical problems see Bevensee [12].

The basic idea is simple. Instead of looking at the problem of finding the top eigenvalue in the equation

$$V\Psi = \lambda \Psi , \qquad (14)$$

where V is the row-transfer operator, we look for an operator O which, when transformed by V satisfies the equation

$$VOV^{-1} = \lambda' O . \qquad (15)$$

Such operators clearly exist, for let ψ and χ be right and left eigenfunctions of V with eigenvalues λ_1 and λ_2. Then the dyadic product $\psi \cdot \chi$ satisfies (15), with $\lambda' = \lambda_1/\lambda_2$. Kaufman and Onsager [13] only used spin operators C_i and S_i but the method is clearly valid for a much wider class of algebras. The great advantage is that the product of the eigenoperators is another eigenoperator but, on the other side of the coin, all eigenoperators with $\lambda' \neq 1$ are necessarily singular. (Take the determinants of both sides of (15)).

It is easy enough to find eigenoperators when q is a Beraha number, and to form product expressions for the eigenvalues, but we have to be careful about our choice of algebra for the operators O. One possible approach is to use the sub-algebra corresponding to the tableau shapes with only one box in the bottom row, but this is not sufficiently general. It seems that we must use the sub-algebras corresponding to equal numbers of boxes in the top and bottom rows.

For $q = 2$ this is simply the spinor algebra, reproducing B. Kaufman's results [10]. The method seems to give sensible results for $q = \frac{1}{2}(3 + \sqrt{5})$ and $q = 3$ but they have still to be checked, I have been "had" so many times by this problem! We need a criterion for a given algebra to produce at least the maximum eigenvalue of V.

7. THE SCHENSTED CONSTRUCTION

This is a method [14] of associating a permutation with a pair of standard tableaux. It does not, of itself, lead to a representation of the Temperley-Lieb algebra, but is so like it that we record it here, in the hope that it may help other workers. The prescription is simple:

(1) Write the permutation e.g. (243) (5) in the standard way

$$\begin{array}{cccc} 2 & 3 & 4 & 5 \\ 4 & 2 & 3 & 5 \end{array}.$$

(2) Write the numbers in the bottom line in the left hand tableau in succession, beginning with the top left hand box. Put them into the top row in succession, but a number that is smaller than one previously added displaces it into the next row.

(3) Fill up the right hand tableau with the numbers 2, 3, 4, 5,... in succession filling the boxes in the order in which they were added to form the left hand tableau.

Thus, with the above permutation, fill up the boxes as shown

$$4\ 2\ 3\ 5 \longrightarrow \boxed{4}\ ,\ \begin{array}{|c|}\hline 2 \\ \hline 4 \\ \hline\end{array}\ ,\ \begin{array}{|c|c|}\hline 2 & 3 \\ \hline 4 & \\ \hline\end{array}\ ,\ \begin{array}{|c|c|c|}\hline 2 & 3 & 5 \\ \hline 4 & & \\ \hline\end{array}\ \Bigg|\ \begin{array}{|c|c|c|}\hline 2 & 4 & 5 \\ \hline 3 & & \\ \hline\end{array}\ . \qquad (16)$$

A permutation that is just a simple exchange, of successive numbers like (23) (4) (5) produces an idempotent, (both tableaux the same)

$$3\ 2\ 4\ 5 \longrightarrow \boxed{3}\ ,\ \begin{array}{|c|}\hline 2 \\ \hline 3 \\ \hline\end{array}\ ,\ \begin{array}{|c|c|}\hline 2 & 4 \\ \hline 3 & \\ \hline\end{array}\ ,\ \begin{array}{|c|c|c|}\hline 2 & 4 & 5 \\ \hline 3 & & \\ \hline\end{array}\ \Bigg|\ \begin{array}{|c|c|c|}\hline 2 & 4 & 5 \\ \hline 3 & & \\ \hline\end{array}\ . \qquad (17)$$

For pairs of tableaux with only one box at the bottom the only types of permutation that we need are:

(a) the identity permutation which produces a single row of boxes giving invariant operators,
(b) the simple exchange of two successive numbers, which produces idempotents,
(c) an ascending or descending cycle of successive numbers, which produces non-diagonal operators.

As they stand, they are not representations of the sub-algebra, because the "product rule" for two successive operations does not always work.

Acknowledgements

I have benefited from discussions with Dr. R. Muse, my research assistant at Swansea, who convinced me that a proposed generalisation of the Bethe Ansatz method did not work. Also from Dr. C. Vout, also my research assistant at Swansea, who developed the tableau representation jointly with me. Professor R.J. Baxter pointed out in 1973 that the Temperley-Lieb formalism originally developed for the percolation problem could solve the critical Potts model. I thank him for this and also for many helpful discussions over the years. I also thank Drs. V. Pasquier and P.P. Martin for helpful correspondence. (They have done work very closely related to the tableau representation). I also thank Prof. B. Evans and Professor V.F.R. Jones for discussion and correspondence. I thank the Science and Engineering Research Council for research grants at Swansea and support at Cambridge during 1975-6. I thank the Leverhulme Foundation for an Emeritus Fellowship, also the Royal Society for awarding me the Rumford Medal. Lastly, I thank the organisers for inviting me to take part in this workshop.

REFERENCES

[1] R.B. Potts, Proc. Camb. Phil. Soc. **48**, 106-9 (1952).

[2] H.N.V. Temperley & E. Lieb, Proc. Roy. Soc. A **322**, 251-80 (1971).

[3] R.J. Baxter, Journ. Phys. C **6**, L445-8 (1973); R.J. Baxter & I.G. Enting, Journ. Phys. A **9**, L149-52 (1976).

[4] R.P. Stanley, Discrete Mathematics **5**, 171-8 (1973).

[5] H. Whitney, Ann. Math. **33**, 688-718 (1932).

[6] Y.G. Stroganov, Phys. Lett. **74A**, 116-18 (1979).

[7] H. Au-Yang, B.M. McCoy, J.H.H. Perk, S. Tang & M.L. Yang, Phys. Lett., **123A**, 219-23 (1987);
B.M. McCoy, J.H.H. Perk, S. Tang & C.H. Sah, Phys. Lett., **125A**, 9-14 (1987);
R.J. Baxter, J.H.H. Perk & H. Au-Yang, Phys. Lett., **128A**, 138-42 (1988);
G. Albertini, B.M. McCoy, J.H.H. Perk & S. Tang, Nucl. Phys. **B314**, 741-63 (1989);
G. Albertini, B.M. McCoy & J.H.H. Perk, Phys. Lett., **139A**, 204-12 (1989).

[8] L.H. Kauffman, Topology **26**, 395-407 (1987); AMS Contemporary Math. Series **78**, 263-97 (1989).

[9] R.J. Baxter & S.K. Tsang, Journ. Phys. A **13**, 1023-30 (1980); R.J. Baxter, Journ. Phys. A **13**, L61-70 (1980).

[10] B. Kaufman, Phys. Rev. **76**, 1232-43 (1949).

[11] A. Aguilar & E. Braun, Physica A **170**, 643-62 (1991).

[12] R.M. Bevensee, Maximum Entropy Solutions to Scientific Problems, Prentice-Hall, New Jersey (1992).

[13] B. Kaufman and L. Onsager, Phys. Rev. **76**, 1244-52 (1949).

[14] C. Schensted, Can. Jour. Math **13**, 179-91 (1961).

ORDER–DISORDER QUANTUM SYMMETRY IN G-SPIN MODELS

K. Szlachányi

Central Research Institute for Physics
H-1525 Budapest 114, P.O.B. 49, Hungary

ABSTRACT

Generalizing the Ising model from $Z(2)$ to an arbitrary finite group G we find that the double Hopf algebra $\mathcal{D}(G)$ plays the role of the $Z(2) \times Z(2)$ symmetry group. Non–Abelian 'parafermion' fields are constructed that are irreducible tensor operators with respect to $\mathcal{D}(G)$ and that generate amplifying homomorphisms of the observable algebra. The quantum symmetry and braid group statistics of the model are analysed in spirit of the Doplicher–Haag–Roberts programme.

1. INTRODUCTION

This talk is based on work done in collaboration with P. Vecsernyés. In a recent paper [1] a non–Abelian generalization of the Ising and of the $Z(N)$ clock models has been proposed in order to demonstrate a new kind of quantum symmetry originating in the interplay between the order and disorder parameters. It is well known that in $Z(N)$ spin models at intermediate values of the coupling constant where neither the original $Z(N)$ nor the dual $Z(N)$ is broken (a critical point for $N \leq 4$ and an interval of critical points for $N > 4$) there exist sectors possessing $Z(N)$ and dual $Z(N)$ charges at the same time. Such composite charges are parafermions: Interchanging two identical particles results in a phase factor $\lambda_{q\tilde{q}} = \exp i2\pi q\tilde{q}/N$ where the mod N integers q and $\tilde{q}$ label the $Z(N) \times Z(N)$ charge of the parafermion. In order to have this strange statistics it is important to have both q and $\tilde{q}$ non–zero. In other words parafermion fields are products of order and disorder fields.

Similarly in G-spin models, in which $Z(N)$ is replaced by a finite non–Abelian group G, one can construct non–Abelian parafermion fields of the form *order* × *disorder*. These fields obey non–Abelian braid commutation relations and transform as irreducible representations of the Hopf algebra $\mathcal{D}(G)$, the double of G. The appearence of $\mathcal{D}(G)$ as a symmetry algebra can be understood as follows. In G-spin models the space of states is spanned by vectors $|\sigma\rangle$ where σ is a one dimensional spin configuration $\sigma\colon \mathbf{Z} \to G$. Symmetries of the model are on the one hand the operators $Q(g)$, $g \in G$ multiplying all spins on the chain by g from the left,

$$Q(g)|\sigma\rangle = |\ldots, g\sigma_x, g\sigma_{x+1}, \ldots\rangle, \tag{1.1}$$

and on the other hand the operators $P(h)$, $h \in G$ projecting to those states that have twists in the boundary condition equal to h,

$$P(h)|\sigma\rangle = \delta_{\sigma_\infty, h\sigma_{-\infty}}|\sigma\rangle. \tag{1.2}$$

These operators obey the relations

$$\begin{aligned} Q(g_1)Q(g_2) &= Q(g_1 g_2), \\ Q(g)P(h) &= P(ghg^{-1})Q(g), \\ P(h_1)P(h_2) &= \delta_{h_1, h_2} \cdot P(h_2), \end{aligned} \tag{1.3}$$

that are just the defining relations of $\mathcal{D}(G)$ as an algebra. The coalgebra structure of $\mathcal{D}(G)$ will be discerned when we study the action of $\mathcal{D}(G)$ on the field operators. For G Abelian $\mathcal{D}(G)$ is the group algebra of $G \times G^*$, where G^* is the Pontryagin dual of G, and we recover the usual symmetry group of Abelian spin systems. If G is non–Abelian $\mathcal{D}(G)$ is a non–commutative, non–cocommutative quasitriangular Hopf algebra, hence a 'quantum group'. It has been studied earlier in the contexts of orbifold construction in conformal field theory [2] and invariants of 3–manifolds [3]. Its representation theory was worked out in [4]. Though not Lorentz covariant, G-spin models are the simplest local and non–chiral QFT models in which $\mathcal{D}(G)$ can be shown to act as a symmetry. Nevertheless, this quantum symmetry might be related to conformal invariance because one expects — in analogy with $Z(N)$ models — that $\mathcal{D}(G)$ is realized only for special values of the couplings where neither the order nor the disorder symmetry is broken and the system is critical.

In Section 2 we define the field algebra $\mathcal{F}$ of G-spin models and an action of the symmetry algebra $\mathcal{D}(G)$ on $\mathcal{F}$. Non–Abelian parafermion fields are constructed and shown to obey the F-algebra relations. In Section 3 we analyse the model from the point of view of the Doplicher–Haag–Roberts (DHR) theory [5]. The point is to derive the emergence of the $\mathcal{D}(G)$ quantum symmetry solely from the structure of the observable algebra $\mathcal{A}$, which is the $\mathcal{D}(G)$-invariant subalgebra of $\mathcal{F}$. To carry out this programme it will be necessary to generalize the DHR–endomorphisms and use amplifying homomorphisms $\mu\colon \mathcal{A} \to M_n(\mathcal{A})$ instead. The F-algebra relations of the parafermion fields help to construct concrete examples of such amplimorphisms μ. Our main results can be summarized in the following two theorems.

Theorem 1. There is a full subcategory $\mathbf{Amp_0}\,\mathcal{A}$ of the category $\mathbf{Amp}\,\mathcal{A}$ of localized, transportable amplifying homomorphisms such that $\mathbf{Amp_0}\,\mathcal{A}$ is isomorphic to the category $\mathbf{Rep}\,\mathcal{D}(G)$ of representations of $\mathcal{D}(G)$. The isomorphism is to be understood between strict monoidal braided C^*-categories with direct sums, subobjects, and conjugates.

This theorem is of purely kinematical nature. Dynamics enters through the assumption on a faithful irreducible representation π_0 of $\mathcal{A}$, called the vacuum representation.

Theorem 2. Assume that the vacuum representation π_0 satisfies Haag duality, that is for any interval Λ

$$\pi_0(\mathcal{A}(\Lambda^c))' = \pi_0(\mathcal{A}(\Lambda)) \tag{1.4}$$

where $\mathcal{A}(\Lambda)$ denotes the algebra of observables localized in Λ and Λ^c is the 'complement' of Λ (See Subsection 3.1.).

Let **Rep** $\mathcal{A}$ be the category of those representations π of $\mathcal{A}$ that satisfy the following selection criterion. There exists a finite interval Λ such that — when restricted to $\mathcal{A}(\Lambda^c)$ — π is unitarily equivalent to a finite multiple of π_0.

Then **Rep** $\mathcal{A}$ and **Amp** $\mathcal{A}$ are isomorphic as C^*-categories with subobjects and direct sums.

These theorems together imply that $\mathcal{D}(G)$ describes the symmetry of (a subset of) the superselection sectors of $\mathcal{A}$. If we could prove that actually $\mathbf{Amp_0}\,\mathcal{A}$ is equivalent to $\mathbf{Amp}\,\mathcal{A}$ then we would be able to claim that $\mathcal{D}(G)$ is the quantum symmetry of all superselection sectors satisfying the above selection criterion.

2. $\mathcal{D}(G)$ QUANTUM SYMMETRY IN THE FIELD ALGEBRA

2.1. The field algebra $\mathcal{F}$

The field algebra of G-spin models is generated by order parameters $\delta_g(x)$, $x \in \mathbf{Z}$, $g \in G$ and disorder parameters $\rho_g(l)$, $l \in \mathbf{Z} + \frac{1}{2}$, $g \in G$ satisfying the following defining relations.

$$\delta_g(x)\delta_h(x) = \delta_{g,h} \cdot \delta_h(x),$$
$$\rho_g(l)\rho_h(l) = \rho_{gh}(l),$$

$$\delta_g(x)\delta_h(x') = \delta_h(x')\delta_g(x),$$

$$\rho_g(l)\delta_h(x) = \begin{cases} \delta_{gh}(x)\rho_g(l), & l < x, \\ \delta_h(x)\rho_g(l), & l > x, \end{cases}$$

$$\rho_g(l)\rho_h(l') = \begin{cases} \rho_h(l')\rho_{h^{-1}gh}(l), & l > l', \\ \rho_{ghg^{-1}}(l')\rho_g(l), & l < l', \end{cases}$$

and

$$\rho_e(l) = 1, \qquad \sum_{g \in G} \delta_g(x) = 1.$$

The meaning of $\delta_g(x)$ is to project on those states $|\sigma\rangle$ that have $\sigma_x = g$ and of $\rho_g(l)$ is to multiply all spins σ_y with $y > l$ by g from the left. $\delta_g(x)$ is defined to be Hermitean and $\rho_g(l)$ to be unitary. The C^*-algebra generated by the δ-s and ρ-s is denoted by $\mathcal{F}$.

2.2. The symmetry algebra $\mathcal{D}(G)$

As an algebra $\mathcal{D}(G)$ is the crossed product of $\mathbf{C}(G)$ and $\mathbf{C}G$ with respect to the adjoint action of the latter on the former. The coproduct Δ, the counit ε, the antipode S, and the *-operation are defined on the basis elements $(g,h) \equiv P(g)Q(h)$ $g,h \in G$ as follows.

$$\Delta(g,h) = \sum_{f \in G}(f,h) \otimes (f^{-1}g,h), \tag{2.1}$$

$$\varepsilon(g,h) = \delta_{g,e}, \tag{2.2}$$

$$S(g,h) = (h^{-1}g^{-1}h, h^{-1}), \tag{2.3}$$

$$(g,h)^* = (h^{-1}gh, h^{-1}). \tag{2.4}$$

$\mathcal{D}(G)$ is a C^*–Hopf algebra, that is Δ, ε, and S commute with $*$. For the coproduct this means that if $a \in \mathcal{D}(G)$, $\Delta(a) = a^{(1)} \otimes a^{(2)}$ then $\Delta(a^*) = \Delta(a)^* = a^{(1)*} \otimes a^{(2)*}$. $\mathcal{D}(G)$ is quasitriangular with universal R-matrix

$$R = \sum_{g \in G}(g,e) \otimes (E,g) \tag{2.5}$$

where e is the unit of $\mathbf{C}G$ and E is the unit of $\mathbf{C}(G)$.

$\mathcal{D}(G)$ is semisimple. For $r \in \widehat{\mathcal{D}(G)} \equiv \{$equivalence classes of irreps of $\mathcal{D}(G)\}$ let D_r be concrete matrix representation from the class r, n_r be its dimension, and M_r be the corresponding minimal central idempotent in $\mathcal{D}(G)$. For the trivial class $r = 0$ D_0 is the counit ε and M_0 is the integral z defined by

$$z = \frac{1}{|G|}\sum_{g \in G}(e,g). \tag{2.6}$$

Each $r \in \widehat{\mathcal{D}(G)}$ is a pair $r = (A, \tau)$ where A is a conjugacy class of G and $\tau \in \hat{C}_A$ where C_A is the centralizer subgroup of an element $g_A \in A$.

The set of finite dimensional matrix representations of $\mathcal{D}(G)$ are the objects of a category $\mathbf{Rep}\,\mathcal{D}(G)$ in which the set of morphisms from D to D' is the space

$$(D'|D) = \{t \in \mathrm{Mat}(n' \times n, \mathbf{C}) \,|\, D'(a)t = tD(a), a \in \mathcal{D}(G)\} \tag{2.7}$$

of intertwiners. This category is a strict monoidal braided C^*-category with direct sums, subobjects and conjugates [6,7]. The star operation and the conjugation are contravariant functors from $\mathbf{Rep}\,\mathcal{D}(G)$ to itself acting on the objects and morphisms respectively as follows: $*: D \mapsto D$, $t \mapsto t^*$, $\bar{}: D \mapsto \bar{D} \equiv D^T \circ S$, $t \mapsto t^T$. The strict monoidal structure is given by the covariant functor $\times: \mathbf{Rep}\,\mathcal{D}(G) \times \mathbf{Rep}\,\mathcal{D}(G) \to \mathbf{Rep}\,\mathcal{D}(G)$, $D \times D' = (D \otimes D') \circ \Delta$, $t_1 \times t_2 = t_1 \otimes t_2$. The functors $*$ and $\bar{}$ are monoidal: $* \circ \times = \times \circ (*,*)$, $\bar{} \circ \times \sim \times \circ (\bar{},\bar{})$, where the natural equivalence $\sim$ is given by $(D, D') \mapsto (\bar{D} \otimes \bar{D}')(R)$, with the universal R-matrix R. The braiding structure is given by the natural equivalence $(D, D') \mapsto B(D, D') = P_{12} \cdot (D \otimes D')(R)$ between the functors $\times$ and $\times^{op}$, where P_{12} interchanges the factors in the tensor product of the representation spaces of D and D', and $\times^{op}$ is the product in the opposite order.

2.3. The action of $\mathcal{D}(G)$ on $\mathcal{F}$

The action of the symmetry algebra $\mathcal{D}(G)$ on the field algebra $\mathcal{F}$ is a bilinear map $\gamma: \mathcal{D}(G) \times \mathcal{F} \to \mathcal{F}$ satisfying the the following properties.

(i) $\quad\gamma_a(\gamma_b(F)) = \gamma_{ab}(F)$

(ii) $\quad\gamma_a(F_1 F_2) = \gamma_{a^{(1)}}(F_1)\gamma_{a^{(2)}}(F_2)$

(iii) $\quad\gamma_a(F^*) = \gamma_{\bar{a}}(F)^*$

for all $a, b \in \mathcal{D}(G)$, $F, F_1, F_2 \in \mathcal{F}$, and for $\bar{a} \equiv S(a^*)$. (i—iii) are natural generalizations of the properties of the action of a group algebra to the case of Hopf algebras. Notice also that (ii) is incompatible with a quasi Hopf algebra unless $\mathcal{F}$ is non–associative [8].

As for the G-spin model we define the action γ by

$$\gamma_{(g,h)}(\delta_f(x)) = \delta_{g,e} \cdot \delta_{hf}(x), \qquad x \in \mathbb{Z},$$
$$\gamma_{(g,h)}(\rho_f(l)) = \delta_{g,hfh^{-1}} \cdot \rho_{hfh^{-1}}(l), \qquad l \in \mathbb{Z} + \tfrac{1}{2}. \tag{2.8}$$

For Abelian groups G this action reproduces the usual action of G and G^* on the order and disorder parameters respectively.

Using the central idempotents M_r one can decompose the field algebra into linear subspaces $\mathcal{F} = \oplus_r \mathcal{F}_r$ with $\mathcal{F}_r = \gamma_{M_r}(\mathcal{F})$. Especially $\mathcal{A} = \gamma_z(\mathcal{F})$ is a C^*–subalgebra of $\mathcal{F}$ and will be interpreted as the algebra of observables of G-spin models.

2.4. The non–Abelian parafermion fields

For $x \in \mathbb{Z}$ and $l \in \mathbb{Z} + \tfrac{1}{2}$ being fixed the fields $\rho_g(l)\delta_h(x)$ $g, h \in G$ transform under γ as a left regular $\mathcal{D}(G)$-multiplet. Therefore each row of the matrix of fields

$$F_r^{ij}(l, x) = \sum_{g, h \in G} D_r^{ij}((g, h)^*) \cdot \rho_g(l) \delta_h(x) \tag{2.9}$$

is a D_r–multiplet for $r \in \widehat{\mathcal{D}(G)}$. They obey braided commutation relations

$$F_{r_1}^{i_1 j_1}(l_1, x_1) \, F_{r_2}^{i_2 j_2}(l_2, x_2) = F_{r_2}^{i_2 j_2'}(l_2, x_2) \, F_{r_1}^{i_1 j_1'}(l_1, x_1) \cdot B_{r_1 r_2}^{j_2' j_1'\ j_1 j_2}(\lessgtr) \tag{2.10}$$

if $\{l_1, x_1\} \lessgtr \{l_2, x_2\}$ where the braiding matrices are

$$B_{r_1 r_2}(<) = P_{12} \cdot D_{r_1} \otimes D_{r_2}(R),$$
$$B_{r_1 r_2}(>) = D_{r_2} \otimes D_{r_1}(R^{-1}) \cdot P_{12}.$$

Furthermore the fields $F^{ij} \equiv F_r^{ij}(l, x)$ satisfy the F–algebra relations

$$F_r^{ij*} F_r^{ik} = \mathbf{1}\,\delta^{jk}, \tag{2.11}$$
$$F_r^{ij} F_r^{kj*} = \mathbf{1}\,\delta^{ik}. \tag{2.12}$$

In general an $m \times n$ matrix F with entries in $\mathcal{F}$ is called a D–multiplet matrix if each row of F is a D–multiplet and it satisfies equation (2.11), i.e. $F^* F = \mathbf{1} \otimes I_n$. If F satisfies (2.12) too, it is called non–degenerate. An associative product of multiplet matrices can be given by $(F \times F')^{ii'\ jj'} = F^{ij} F'^{i'j'}$.

The F–algebra (2.11–12) seems to be the proper generalization (at least for semisimple Hopf algebra symmetries) of the Cuntz algebra [9] which played an important role in the reconstruction of the symmetry group [6] in the DHR theory of superselection sectors.

3. $\mathcal{D}(G)$ QUANTUM SYMMETRY OF THE SUPERSELECTION SECTORS

3.1. Amplifying homomorphisms

The observable algebra $\mathcal{A}$ is generated by operators $v_g(x), w_g(l)$ with relations

$$v_{g_1}(x)v_{g_2}(x) = v_{g_1 g_2}(x),$$
$$w_{h_1}(l)w_{h_2}(l) = \delta_{h_1,h_2} w_{h_2}(l),$$
$$v_g(x)w_h(x + \tfrac{1}{2}) = w_{gh}(x + \tfrac{1}{2})v_g(x),$$
$$v_g(x)w_h(x - \tfrac{1}{2}) = w_{hg^{-1}}(x - \tfrac{1}{2})v_g(x).$$

The local algebras $\mathcal{A}(\Lambda) = \langle v_g(x), w_g(l) \, | \, x, l \in \Lambda, \, g \in G \rangle$, for $\Lambda \subset \tfrac{1}{2}\mathbf{Z}$ finite, satisfy isotony, $\mathcal{A}(\Lambda_1) \subset \mathcal{A}(\Lambda_2)$ if $\Lambda_1 \subset \Lambda_2$, locality, $\mathcal{A}(\Lambda_1) \subset \mathcal{A}(\Lambda_2)'$ if $\operatorname{dist}(\Lambda_1, \Lambda_2) \geq 1$, and Haag duality, $\mathcal{A}(\Lambda^c)' = \mathcal{A}(\Lambda)$ if Λ is an interval and $\Lambda^c = \{\xi \in \tfrac{1}{2}\mathbf{Z} \, | \, \operatorname{dist}(\xi, \Lambda) \geq 1\}$.

In the DHR theory of superselection sectors charges are created by localized, transportable endomorphisms $\rho: \mathcal{A} \to \mathcal{A}$. In G-spin models, however, all such endomorphisms are automorphisms because the local algebras $\mathcal{A}(\Lambda)$ are finite dimensional. Therefore in order to recover the fairly non–Abelian structure of superselection sectors in G-spin models one is forced to investigate more general notions of morphisms.

We studied amplifying homomorphisms $\mu: \mathcal{A} \to M_m(\mathcal{A}) \equiv \operatorname{Mat}(m, \mathcal{A})$ that are C^*–maps (possibly not unit preserving) from $\mathcal{A}$ to one of its finite dimensional matrix amplification $M_m(\mathcal{A})$. μ is called localized in Λ if

$$\mu(A) = \mu(1) \cdot (A \otimes I_m) \qquad A \in \mathcal{A}(\Lambda^c) \,. \tag{3.1}$$

The space of intertwiners from $\nu: \mathcal{A} \to M_n(\mathcal{A})$ to $\mu: \mathcal{A} \to M_m(\mathcal{A})$ is

$$(\mu|\nu) = \{T \in \operatorname{Mat}(m \times n, \mathcal{A}) \, | \, \mu(A)T = T\nu(A) \, A \in \mathcal{A}, \, \mu(1)T = T = T\nu(1) \} \,. \tag{3.2}$$

μ and ν are called equivalent, $\mu \sim \nu$, if $\exists U \in (\mu|\nu)$ such that $UU^* = \mu(1)$ and $U^*U = \nu(1)$. μ is called transportable if its translate $\alpha_x \circ \mu \circ \alpha_{-x} \sim \mu$ for all $x \in \mathbf{Z}$.

The objects of the category $\mathbf{Amp}\,\mathcal{A}$ are the localized, transportable amplimorphisms μ and the morphisms from ν to μ are the intertwiners $T \in (\mu|\nu)$. There is a natural notion of a product of two amplimorphisms, $\mu \times \nu = (\mu \otimes \operatorname{id}_{M_n(\mathbf{C})}) \circ \nu: \mathcal{A} \to M_{mn}(\mathcal{A})$, and of the intertwiners $T_1 \in (\mu_1|\nu_1)$ and $T_2 \in (\mu_2|\nu_2)$, $T_1 \times T_2 = \mu_1(T_2)(T_1 \otimes I_{n_2}) \in (\mu_1 \times \mu_2|\nu_1 \times \nu_2)$. The product $\times$ makes $\mathbf{Amp}\,\mathcal{A}$ to be a strict monoidal category. The physical interpretation of $\times$ is composition of charges.

The amplimorphisms μ_F generated by some multiplet matrix $F \in \operatorname{Mat}(m \times n, \mathcal{F})$

$$\mu_F(A) = F(A \otimes I_n)F^* \,, \tag{3.3}$$

are the objects of the full subcategory $\mathbf{Amp}_0\,\mathcal{A}$ of $\mathbf{Amp}\,\mathcal{A}$. $\mu_{F_1} \sim \mu_{F_2}$ iff F_i is a D_i–multiplet matrix, $i = 1, 2$, and $D_1 \sim D_2$. One checks easily that $\mu_{F_1} \times \mu_{F_2} = \mu_{F_1 \times F_2}$. For all μ there exists a unit preserving amplimorphism ν such that $\nu \sim \mu$. (The dimension of this ν is uniquely determined by the equivalence class of μ and is called the essential dimension of μ.) Hence the inclusion of non–unital morphisms μ into the discussion eventually did not enlarge the available set of superselection sectors.

3.2. The statistics operator

The braiding structure on $\mathbf{Amp}\,\mathcal{A}$ is given by the statistics operators $\epsilon(\mu_1, \mu_2) \in (\mu_2 \times \mu_1 | \mu_1 \times \mu_2)$ defined as follows. Choose $\tilde{\mu}_i \sim \mu_i$ $i = 1, 2$ and equivalences $U_i \in (\tilde{\mu}_i | \mu_i)$ such that the localization region of $\tilde{\mu}_1$ stands left from that of $\tilde{\mu}_2$. Then

$$\epsilon(\mu_1, \mu_2) = \mu_2(U_1^*)(U_2^* \otimes I_1) P_{12}(U_1 \otimes I_2) \mu_1(U_2) \,. \tag{3.4}$$

These operators obey naturalness relations, hexagonal identities, and coloured braid relations that are 'amplified' generalizations of the ones one finds in the endomorphism case [5]. As a consequence of these $g_i \mapsto \mu^{\times(i-1)}(\epsilon(\mu, \mu))$ $i = 1, 2, \ldots$ defines a representation of the braid group in the infinite amplification $\hat{\mathcal{A}} = \mathcal{A} \otimes M_m \otimes M_m \otimes \ldots$ of the observable algebra, the equivalence class of which is uniquely determined by that of μ.

In order to get numbers characterizing the statistics one applies a left inverse of μ on $\epsilon(\mu, \mu)$ to get a c–number λ_μ, the statistical parameter. For $\mu = \mu_F$ with a D_r–multiplet matrix we found the following result. If $r = (A, \tau)$ then $\lambda_\mu = \omega_r / d_r$ where the statistics phase ω_r and the statistical dimension d_r are given by

$$\omega_r = \operatorname{tr} \tau(g_A)/\dim \tau \,, \qquad d_r = n_r \,. \tag{3.5}$$

We mention that in proving Theorem 1 the multiplet matrices play the role of a functor realizing the equivalence of $\mathbf{Amp}\,\mathcal{A}$ and $\mathbf{Rep}\,\mathcal{D}(G)$. Especially for the braiding structures the equivalence is to be read off from the relation

$$\epsilon(\mu_{F_1}, \mu_{F_2}) = (F_2 \times F_1) \cdot B(D_1, D_2) \cdot (F_1 \times F_2)^* \,, \tag{3.6}$$

which can be thought as an amplified and non–Abelian generalization of the group theoretical commutator of two unitary charged fields, let us say, in the $Z(N)$ model: $\epsilon(\mu_{F_1}, \mu_{F_2}) = F_2 F_1 F_2^* F_1^* = \lambda_{q\bar{q}} 1$.

3.3. Charged sectors of $\mathcal{A}$

The tacit assumption in the above discussion was that the superselection sectors of $\mathcal{A}$ are in one–to–one correspondence with the equivalence classes of amplimorphisms μ in $\mathbf{Amp}\,\mathcal{A}$. Only in this case can we claim that the $\mathcal{D}(G)$ quantum symmetry is realized in the quantum theory. The required equivalence between the representation theory of $\mathcal{A}$ and $\mathbf{Amp}\,\mathcal{A}$ can hold only for special coupling constants. The requirement on the dynamics is formulated by assuming that the vacuum representation π_0 of $\mathcal{A}$ satisfies Haag duality (1.4). This can be interpreted as assuming the absence of symmetry breaking. Once such a π_0 is chosen, charged representations π_μ can be defined for each object μ of $\mathbf{Amp}\,\mathcal{A}$. Namely, let $\pi_\mu = (\pi_0 \otimes \mathrm{id}) \circ \mu$ which is a representation on $\mathcal{H}_0 \otimes \mathbf{C}^m$ where $\mathcal{H}_0$ is the representation space of π_0. Now one can prove, using Haag duality, that the category of representations of $\mathcal{A}$ of the form π_μ is isomorphic to $\mathbf{Amp}\,\mathcal{A}$. Then Theorem 2 follows by showing that the selection criterion is the necessary and sufficient condition for a representation π of $\mathcal{A}$ to have the form $\pi = \pi_\mu$.

3.4. Miscellany

For each irreducible object μ one can define a Markov trace on the braid group using the iterated left inverse [10]. The Markov trace in turn determines link invariants

$I_{G,r}$ for each $r \in \widehat{\mathcal{D}(G)}$ and for each finite group G. These are the obvious ones one obtains from the R–matrix (2.5).

For DHR endomorphisms ρ Longo proved [11] that the index of the inclusion $\rho(\mathcal{A}) \subset \mathcal{A}$ is equal to the square of the statistical dimension of ρ. For amplimorphisms $\mu: \mathcal{A} \to M_m(\mathcal{A})$ this has the following generalization. μ has an extension $\hat{\mu}$ to the infinite amplification $\hat{\mathcal{A}}$ as an endomorphism. For μ irreducible the index $[\hat{\mathcal{A}}: \hat{\mu}(\hat{\mathcal{A}})]$ is equal to the square of the statistical dimension d_μ. d_μ in turn is equal to the dimension n_r of the corresponding $\mathcal{D}(G)$ representation and also to the essential dimension of μ.

Representations of the modular group can be obtained, on the one hand, on the space of characters of $\mathcal{D}(G)$ [12], and on the other hand, on formal linear combinations of superselection sectors [13]. In the G-spin model these representations turn out to be equivalent.

3.5. The integrable S_3–spin model

A candidate for a vacuum representation π_0 satisfying Haag duality and therefore leading to $\mathcal{D}(G)$ symmetry in the superselection sectors can be found in the S_3-spin model. Let χ_1 and χ_2 denote respectively the characters of the non-trivial 1-dimensional and of the 2-dimensional representation of S_3. The model is defined by the action $S(\sigma) = -\sum_{<x,y>}[\beta_1 \chi_1(\sigma_x^{-1}\sigma_y) + \beta_2 \chi_2(\sigma_x^{-1}\sigma_y)]$ where $\sigma: \mathbf{Z}^2 \to G$. Monte Carlo simulations and exact correlation inequalities show 3 phases in the ferromagnetic part of the (β_1, β_2)-plane. A one–parameter solution of the QYBE was found in [14] which is of genus 0, depends on an anisotropy parameter, and intersects the (β_1, β_2)-plane in the point $\beta_1^* = \frac{1}{8}\ln 2 + \frac{1}{6}\ln(1 + \frac{3}{\sqrt{2}})$, $\beta_2^* = \frac{1}{3}\ln(1 + \frac{3}{\sqrt{2}})$. This point is conjectured to be a tricritical point where the three phase boundaries meet and where none of the symmetries are broken. Therefore $\mathcal{D}(S_3)$ quantum symmetry is expected to be realized there.

REFERENCES

[1] K. Szlachányi, P. Vecsernyés, Quantum symmetry and braid group statistics in G-spin models, KFKI–1992–08/A, preprint 1992.

[2] R. Dijkgraaf, C. Vafa, E. Verlinde and H. Verlinde, Commun. Math. Phys. **123**, 485 (1989).

[3] D. Altschuler and A. Coste, Invariants of 3-manifolds from finite groups, CERN-TH.6204/91, preprint 1991.

[4] R. Dijkgraaf, V. Pasquier and P. Roche, Talk presented at Intern. Coll. on Modern Quantum Field Theory, Tata Institute, 8-14 January 1990.

[5] S. Doplicher, R. Haag and J.E. Roberts, Commun. Math. Phys. **13**, 1 (1969); ibid **15**, 173 (1969); ibid **23**, 199 (1971); ibid **35**, 49 (1974).

[6] S. Doplicher and J.E. Roberts, Bull. Am. Math. Soc. **11**, 333 (1984); Ann. Math. **130**, 75 (1989); Invent. Math. **98**, 157 (1989).

[7] S. Majid, Int. J. Mod. Phys. **A5**, 1 (1990).

[8] G. Mack and V. Schomerus, Nucl. Phys. **B370**, 185 (1992).

[9] J. Cuntz, Commun. Math. Phys. **57**, 173 (1977).

[10] K. Fredenhagen, K.-H. Rehren and B. Schroer, Commun. Math. Phys. **125**, 201 (1989).

[11] R. Longo, Commun. Math. Phys. **126**, 217 (1989); ibid **130**, 285 (1990).

[12] P. Bántay, Lett. Math. Phys. **22**, 187 (1991).

[13] K.-H. Rehren, Braid Group Statistics and their Superselection Rules, in: Algebraic Theory of Superselection Sectors, ed. D. Kastler, p. 333; World Scientific 1990.

[14] K. Szlachányi and P. Vecsernyés, Phys. Lett. **273B**, 273 (1991).

QUANTUM GROUPS, QUANTUM SPACETIME, AND DIRAC EQUATION

Arne Schirrmacher

Max–Planck–Institut für Physik
Föhringer Ring 6, W-8000
München 40, Germany

ABSTRACT

A simple modification of the Dirac algebra yields the q-deformed version corresponding to the framework of quantized Minkowski space. The method of deriving $\hat{R}$-matrix relations for Minkowski coordinates, differentials, derivatives, and γ-matrices is explained. Taking a suitable conjugation structure on the derivatives and q-Dirac matrices allows for a q-analog of the Dirac equation that iterates to a Klein-Gordon equation. Its covariance is shown.

1 INTRODUCTION

Quantum groups are widely recognized by now in conformal field theory, integrable models, knot theory, low dimensional topology and elsewhere. The common aim of most of these areas is a better understanding of quantum field theory as the foundation of physics on the whole. On first sight the intriguing idea of the quantum deformation of spacetime may seem to be a rather different project, the goal, however, is the same. The q-deformation brings the theory on a lattice in such a way that the spacetime symmetries are still remembered and a regularization is built in from scratch. Once the concept of q-deformed Minkowski space is introduced its algebraic, analytical and topological structures open a wide field — just to mention some key words as Hopf algebra, q-difference equations, q-special functions, braid statistics etc.[1-3]

In this paper we will approach some of these problems in an introductory style which is of physicist's conception. Hence if we mention "consistency" for a set of relations this can be read by mathematicians as shorthand for the matching of the Poincaré series of the (differential) graded q-algebra with that of the undeformed one.

This "consistency" is guaranteed by $\hat{R}$-matrix formulation of the algebraic structures that frees us from exhaustive battles with the diamond lemma and similar.*

The q-Dirac algebra serves as introduction to the structure of q-deformed algebras. A kind of change of basis gives a hardly anticipated generalized Clifford structure with a new grading.

As projectors on Lorentz invariant subspaces also exist for the q-deformed Lorentz group, they can be used to explain how the $\hat{R}$-matrix relations for coordinates, differentials, derivatives and Dirac matrices are constructed.

After adopting an appropriate conjugation structure on derivatives and γ-matrices the formulation of a q-analog of the Dirac equation is straightforward. The analysis of the covariance reveals the well-known q-spinor structure with slight modifications. The q-Klein-Gordon equation is derived from the q-Dirac equation. Its covariance is shown and the action of the Lorentz generators on the q-Dirac spinor discussed. We do not solve these q-relativistic wave equations here. This will be done in a more general context of Hilbert space reresentation in forthcoming publications[6].

2 THE q-DIRAC ALGEBRA

We can use the well-known γ-matrices

$$(\gamma^0)^2 = -1, \quad (\gamma^{1,2,3})^2 = +1, \quad \gamma^i\gamma^j + \gamma^j\gamma^i = 0, \quad (i \neq j), \tag{1}$$

to introduce the q-deformed Dirac algebra without assuming any knowledge on quantum groups. Consider the following four combinations† depending on a parameter q,

$$\begin{aligned}
\gamma^A &= \tfrac{1}{2q}(\gamma^1 - i\gamma^2), & \gamma^C &= \tfrac{1}{2}(q^2\gamma^0 + \gamma^3) + \tfrac{i}{2}(q^2 - 1)\gamma^1\gamma^2\gamma^3, \\
\gamma^B &= \tfrac{q}{2}(\gamma^1 + i\gamma^2), & \gamma^D &= \tfrac{1}{2}(\gamma^0 - \gamma^3).
\end{aligned} \tag{2}$$

For $q = 1$ these are just 'lightcone coordinates', for $q \neq 1$ γ^C has an additional piece of third order. We find the relations, using $\lambda \equiv q - q^{-1}$,

$$\begin{aligned}
(\gamma^{A,B,D})^2 &= 0 & (\gamma^C)^2 &= -q\lambda\gamma^B\gamma^A \\
\gamma^A\gamma^B + \gamma^B\gamma^A &= 1 & \gamma^A\gamma^C + q^{-2}\gamma^C\gamma^A &= -q\lambda\gamma^D\gamma^A \\
\gamma^A\gamma^D + \gamma^D\gamma^A &= 0 & \gamma^B\gamma^C + q^2\gamma^C\gamma^B &= q^3\lambda\gamma^D\gamma^B \\
\gamma^B\gamma^D + \gamma^D\gamma^B &= 0 & \gamma^C\gamma^D + \gamma^D\gamma^C &= q\lambda\gamma^B\gamma^A - q^2.
\end{aligned} \tag{3}$$

This is the Clifford algebra related to the q-deformation of Minkowski space[1,2,9]. We may call these γ-matrices the *Dirac quantum plane*. There is an identification of this quantum algebra with the undeformed one since we have a finite algebra were the 16 ordered monomials form a basis. The q-deformation is just a kind of change of basis, the grading of the two algebras is, however, different thus making the deformation nontrivial. It is clear from this presentation that relations (3) form a consistent set which is in other cases a hard task to prove. In the theory of quantum groups this is ensured by the existence of a $\hat{R}$-matrix. We will show that relations (3) can be written as already proposed in ref. 7 (cf. also refs. 1,8; note, however, that the summation order is different)

$$\gamma^J\gamma^I + \hat{R}^{IJ}{}_{KL}\gamma^L\gamma^K = q^2 g^{IJ}, \tag{4}$$

---

*For an example how this makes life easier see ref. 4 in contrast to ref. 5.

†A, B, C, D are used here as labels like $0, 1, 2, 3$ in (1) conforming with notation in ref. 9; variables $I, J, ...$ take values A,B,C,D and summation is understood accordingly.

using the metric g here in the (A,B,C,D) basis

$$g^{IJ} = \begin{pmatrix} 0 & 1 & 0 & 0 \\ q^{-2} & 0 & 0 & 0 \\ 0 & 0 & -q\lambda & -1 \\ 0 & 0 & -1 & 0 \end{pmatrix}^{IJ}. \tag{5}$$

3 $\hat{R}$-MATRIX RELATIONS FOR QUANTUM PLANES

In order to derive $\hat{R}$-matrix relations we use the projector decomposition. The spaces of symmetric and antisymmetric coordinates are invariant subspaces under Lorentz transformations. The orthogonality allows to split the symmetric projection into one related to the invariant length, the trace projector P_T, and the symmetric traceless part P_S. In four dimensions also the antisymmetric projection can be split due to the existence of a totally antisymmetric ϵ-tensor allowing to define selfdual and anti-selfdual subspaces, i.e. P_+ and P_-, respectively.

In ref. 9 it has been shown that this structure remains under q-deformation. Without reference to the explicit structure of the projectors, we can understand them as follows: All four projectors add up to unity $\mathbf{1}$ (i.e. $\mathbf{1}^{ij}_{kl} = \delta^i_k \delta^j_l$). Then there is the analog of the interchanging operator $V = P_{\text{symm}} - P_{\text{anti}}$ ($V^{ij}_{kl} \longrightarrow \delta^i_l \delta^j_k$ as $q \to 1$). The trace projector is built from the metric g and the selfduality is related to the ϵ-tensor.

$$\mathbf{1} = \begin{matrix} P_{\text{symm}} \\ + \\ P_{\text{anti}} \end{matrix} = \begin{matrix} P_T & = & \omega^{-2}\, g \otimes g^{-1} \\ + & & \\ P_S & = & \frac{1}{2}(\mathbf{1}+V) - \omega^{-2}\, g \otimes g^{-1} \\ + & & \\ P_+ & = & \frac{1}{4}(\mathbf{1}-V) + \frac{i}{2q\omega}\,\epsilon \\ + & & \\ P_- & = & \frac{1}{4}(\mathbf{1}-V) - \frac{i}{2q\omega}\,\epsilon. \end{matrix} \tag{6}$$

Here, $\omega \equiv q + q^{-1}$, g^{-1} is the inverse of (5). The generalized entities have still many of the old properties:

$$\begin{array}{ll} V^2 = \mathbf{1} & \text{(interchanging property)}, \\ V\epsilon = -\epsilon = \epsilon V & \text{(antisymmetry of } \epsilon\text{-tensor)}, \\ \epsilon^2 = -q^2\omega(\mathbf{1}-V) & \text{(contraction over two indices)}. \end{array} \tag{7}$$

It has been shown[9] that two combinations of these projectors satisfy the Yang-Baxter equation thus giving $\hat{R}$-matrices generating consistent quantum algebras:

$$\begin{aligned} \hat{R}_I &= P_S + P_T - q^2 P_+ - q^{-2} P_-, \\ \hat{R}_{I\!I} &= q^{-2} P_S + q^2 P_T - P_+ - P_-. \end{aligned} \tag{8}$$

To find the $\hat{R}$-matrix relation for the symmetric Lorentz coordinates we require the antisymmetric projectors to vanish [‡§]

$$\mathbf{1} X_1 X_2 = (P_S + P_T) X_1 X_2 = \hat{R}_I\, X_1 X_2, \tag{9}$$

[‡] We use leg numbering notation i.e. $(\hat{R} X_1 X_2)^{i_1 i_2} = \hat{R}^{i_1 i_2}_{k_1 k_2} X^{k_1} X^{k_2}$.
[§] For an alternative derivation of such relations using only one $\hat{R}$-matrix see refs. 7, 1.

For the derivatives we just get the same relation. We will, however, need them in the form where they have lower indices $\partial_I = g_{IK}\partial^K$. Due to the properties of the $\hat{R}$-matrix that gives

$$\partial_2\partial_1 = \partial_2\partial_1 \, \hat{R}_I \, . \tag{10}$$

On the antisymmetric differentials the symmetric projectors annihilate, yielding

$$1 dX_1 dX_2 = (P_+ + P_-)dX_1 dX_2 = -\hat{R}_{I\!I} \, dX_1 dX_2 \, . \tag{11}$$

The γ's are clearly not symmetric, but the trace does not vanish

$$1\gamma_2\gamma_1 = (P_T + P_+ + P_-)\gamma_2\gamma_1 = -\hat{R}_{I\!I} \, \gamma_2\gamma_1 + q^2 \, g \, , \tag{12}$$

where we normalized $g_{IJ}\gamma^J\gamma^I = q\omega$, giving the relation proposed in the previous section. The matrix representation of the γ-matrices is given in the appendix. There we also discuss briefly the construction of the analog of γ^5 and its relations[10].

For completeness let us write down relations for vectors, i.e. X^I and ∂^I. It is the lightcone coordinates

$$\begin{array}{llll} A & \sim & X^1 - iX^2 & \qquad B \sim X^1 + iX^2 \\ C & \sim & X^0 + X^3 & \qquad D \sim X^0 - X^3 \end{array} \tag{13}$$

to be deformed:

$$\begin{array}{rclcrcl} AB & = & BA - q^{-1}\lambda CD + q\lambda D^2 & \quad & BC & = & CB - q^{-1}\lambda BD \\ AC & = & CA + q\lambda AD & & BD & = & q^2 DB \\ AD & = & q^{-2}DA & & CD & = & DC \, . \end{array} \tag{14}$$

Interestingly $X^0 \sim C + D$, i.e. the time coordinate (or the energy in case of the ∂), is central.

4 CONJUGATION AND DIRAC OPERATOR

Using $\gamma^{0+} = -\gamma^0$, $\gamma^{k+} = \gamma^k$ the hermitian adjoint of $\gamma^A \ldots \gamma^D$ can be calculated. They are not anymore linear combinations of $\gamma^A \ldots \gamma^D$. Note also that γ^0 is a rather complicated element of the deformed Clifford algebra, cf. appendix. More important for relativistic wave equations are, however, the relations for the Dirac adjoint $(\gamma^0\gamma^I)^+$:

$$\begin{array}{rclcrcl} (\gamma^0\gamma^A)^+ & = & \dfrac{1}{q^2}\gamma^0\gamma^B & \quad & (\gamma^0\gamma^B)^+ & = & q^2 \, \gamma^0\gamma^A \\ (\gamma^0\gamma^C)^+ & = & \gamma^0\gamma^C & & (\gamma^0\gamma^D)^+ & = & \gamma^0\gamma^D \, . \end{array} \tag{15}$$

We define the Dirac operator as $\not{\partial} \equiv \gamma^I\partial_I$. In ref. 9 it was shown that in the q-deformed case it is not possible to have a linear conjugation structure for both the coordinates and the derivatives. For the latter relations of type $\bar{\partial} \sim \partial + \lambda X\square$ were found. Since linearity is a fundamental principle in quantum physics¶ that we would like to maintain in the q-deformed theory, we have to insist on the linear conjugation structure on the derivatives and may allow for a complicated one on the Minkowski

---

¶As Wigner[11] puts it in the beginning of his famous 1937 work: "It is perhaps the most fundamental principle of Quantum Mechanics that the system of states forms a *linear manifold*, in which a unitary *scalar product* is defined...".

space[12] (which does then not anymore inherits its conjugation properties from a spinor calculus). Thus motivated by (13) we use for the derivatives

$$\begin{aligned}\overline{\partial^A} &= \partial^B \\ \overline{\partial^B} &= \partial^A \\ \overline{\partial^C} &= \partial^C \\ \overline{\partial^D} &= \partial^D\end{aligned} \quad \text{or} \quad \begin{aligned}\overline{\partial_A} &= q^2\, \partial_B \\ \overline{\partial_B} &= q^{-2}\, \partial_A \\ \overline{\partial_C} &= \partial_C \\ \overline{\partial_D} &= \partial_D\end{aligned} \tag{16}$$

and find‖

$$\begin{aligned}(\gamma^0 \partial\!\!\!/)^+ &= (\gamma^0 \gamma^I)^+ \overline{\partial_I} \\ &= \frac{1}{q^2}\, \gamma^0 \gamma^B\, q^2 \partial_B + q^2\, \gamma^0 \gamma^A\, \frac{1}{q^2}\, \partial_A + \gamma^0 \gamma^C \partial_C + \gamma^0 \gamma^D \partial_D \\ &= \gamma^0 \partial\!\!\!/ .\end{aligned} \tag{17}$$

We calculate the square of $\partial\!\!\!/$ with help of (10), (8) and (4)

$$\begin{aligned}\partial\!\!\!/\,\partial\!\!\!/ &= \gamma^I \partial_I\, \gamma^J \partial_J \\ &= \partial_I \partial_J\, \gamma^I \gamma^J \\ &= \partial_I \partial_J\, \hat{R}^{JI}_{I\,LK} \gamma^K \gamma^L \\ &= \partial_I \partial_J\, (q^2 \hat{R}_{I\!I} - q^2 \lambda \omega P_T + q\lambda P_-)^{JI}_{LK} \gamma^K \gamma^L \\ &= \partial_I \partial_J\, (q^2(-\gamma^I \gamma^J + q^2\, g^{JI}) - q^3 \lambda g^{JI}) \\ &= -q^2 \partial_I \partial_J \gamma^I \gamma^J + q^2\, g^{JI} \partial_I \partial_J ,\end{aligned} \tag{18}$$

i.e.

$$\partial\!\!\!/\,\partial\!\!\!/ = \frac{q}{\omega} g^{JI} \partial_I \partial_J . \tag{19}$$

In a more suitable normalization $g_{IJ} \gamma^J \gamma^I = -\omega^2$ (instead of $q\omega$, i.e. $\gamma \to i\sqrt{\omega/q}\, \gamma$) we can write down the Dirac equation

$$(i\partial\!\!\!/ - m)\psi = 0 . \tag{20}$$

The square of the Dirac operator now gives

$$(i\partial\!\!\!/)(i\partial\!\!\!/) = g^{JI} \partial_I \partial_J \equiv -\Box \quad \text{or} \quad p\!\!\!/ \cdot p\!\!\!/ = -p^2 = m^2 . \tag{21}$$

Thus we find that solutions of the Dirac equation also satisfy the Klein-Gorden equation for each component

$$-\Box \psi = m^2 \psi . \tag{22}$$

Using (16) it is easy to prove that the Klein-Gordon operator is real.

Understanding ∂ now as operator ($\partial = \overrightarrow{\partial} = -\overleftarrow{\partial}$) we have for the Dirac adjoint spinor $\overline{\psi} \equiv (i\gamma^0 \psi)^+ = \psi^+ i\gamma^0$

$$\overline{\psi}(i\partial\!\!\!/ - m) = 0 . \tag{23}$$

Hence the solutions of the q-Dirac equation are the kernel of the hermitean operator $i\gamma^0(i\partial\!\!\!/ - m)$.

‖ We should remark that in the definition of the q-deformed γ-matrices was a freedom of rescaling $\gamma^A \to \alpha \gamma^A$ while $\gamma^B \to \alpha^{-1} \gamma^B$ that now has been fixed.

5 COVARIANCE OF THE q-DIRAC EQUATION

In this section we seek to determine the transformation properties of the Dirac spinor such that the Dirac equation is form invariant under Lorentz transformations. It has to transform linearly:

$$\psi \to \psi' = S\psi . \tag{24}$$

The derivatives transform with the quantum matrix T of the quantum Lorentz group

$$\partial_A = g_{AI}\partial^I \to \partial'_A = g_{AI}T^I{}_K g^{KB}\partial_B = \partial_B T^{tB}{}_A . \tag{25}$$

For the transformed Dirac equation we find

$$m\psi' = i\gamma^I \partial'_I \psi' \iff mS\psi = i\partial_I T^{tI}{}_J \gamma^J S\psi , \tag{26}$$

i.e.

$$m\psi = i\partial_I(T^{tI}{}_J S^{-1}\gamma^J S)\psi . \tag{27}$$

Thus covariance amounts to the relation

$$T^J{}_I \gamma^I = S^{-1}\gamma^J S \tag{28}$$

In the chiral representation the q-deformed γ-matrices have the form (see appendix)

$$\gamma^I = \begin{pmatrix} & \bar{\sigma}^I \\ \sigma^I & \end{pmatrix} \tag{29}$$

where the bar does not mean conjugation. Setting

$$\psi = \begin{pmatrix} \phi \\ \chi \end{pmatrix} \qquad S = \begin{pmatrix} M & \\ & M^* \end{pmatrix} \tag{30}$$

the Dirac equation separates into spinor equations for ϕ and χ. M and M^* have to be two-dimensional representations of the Lorentz group. If we choose M to be a $SL(2)$ quantum matrix it is easy to show that $M^* = M^{+-1}$ also is. Relation (28) becomes

$$T^I{}_J \sigma^J = M^+ \sigma^I M \qquad \text{and} \qquad T^I{}_J \bar{\sigma}^J = M^{-1}\bar{\sigma}^I M^* . \tag{31}$$

In order to verify these relations we need the following equalities

$$\text{Tr}_q(\sigma^I \bar{\sigma}_J) = \delta^I{}_J = \text{Tr}_q(\bar{\sigma}^I \sigma_J) , \tag{32}$$

where $\text{Tr}_q A = \text{tr}\left[\begin{pmatrix} q^{-2} & \\ & 1 \end{pmatrix} A\right]$. In addition $\{\sigma^I\}$ and $\{\bar{\sigma}^I\}$ span the space of 2×2 matrices. Hence we can identify

$$X \equiv x_I \sigma^I \qquad \text{for any vector } x_I \tag{33}$$

and we find

$$\tfrac{1}{2}\text{Tr}_q(X\bar{\sigma}_I) = x_I \tag{34}$$

and

$$\tfrac{1}{2}\text{Tr}_q(X\bar{\sigma}_I)\sigma^I = X \qquad \text{for any matrix } X . \tag{35}$$

Relation (31) for σ is valid if

$$T^I{}_J \equiv \tfrac{1}{2}\text{Tr}_q(M^+ \sigma^I M \, \bar{\sigma}_J) \tag{36}$$

since

$$T^I{}_J \sigma^J = \tfrac{1}{2}\text{Tr}_q([M^+ \sigma^I M]\bar{\sigma}_J)\sigma^J = M^+ \sigma^I M . \tag{37}$$

This is true because (36) was just the way the $\widehat{R}$-matrix for the Lorentz group was constructed[1,9,13]:

$$T \equiv M^* \otimes M, \qquad X^I = x_a^* \, \sigma^{Ia}{}_b \, x^b \tag{38}$$

$$A = q^{-1} x^* y, \quad B = q \, y^* x, \quad C = y^* y, \quad D = x^* x \, . \tag{39}$$

There are, however, two minor changes: The reality structure we introduced implies that x^* cannot be identified with $\bar{x}$ as in the undeformed case. Also q-factors are new but the scaling $A \to qA$, $B \to q^{-1}B$ leaves commutation relations (14) invariant.

To prove the relation (31) for $\bar{\sigma}$ we must understand how σ and $\bar{\sigma}$ are related. Define

$$\epsilon = \begin{pmatrix} & \sqrt{q} \\ -1/\sqrt{q} & \end{pmatrix} \tag{40}$$

the q-spinor metric. For M a $SU_q(2)$ matrix we have

$$\epsilon^{-1} M \epsilon = M^{-1t} \, , \qquad \epsilon^+ M^+ \epsilon^{+-1} = M^{-1+t} = M^{*t} \, . \tag{41}$$

Omitting the Lorentz index for σ, rewrite

$$\sigma' = M^+ \sigma M \tag{42}$$

as

$$\epsilon^+ \sigma' \epsilon = M^{*t}(\epsilon^+ \sigma \epsilon) M^{-1t} \tag{43}$$

or

$$(\epsilon^+ \sigma' \epsilon)^a{}_b = M^{*t\,a}{}_m (\epsilon^+ \sigma \epsilon)^m{}_n M^{-1t\,n}{}_b \tag{44}$$

then the transpose reads

$$(\epsilon^+ \sigma' \epsilon)^{tb}{}_a = M^{*\,m}{}_a (\epsilon^+ \sigma \epsilon)^{tn}{}_m M^{-1\,b}{}_n \, . \tag{45}$$

M^* and M^{-1} have the wrong order when compared with (31). From ref. 1

$$\widehat{R} M_1^* M_2 = M_1 M_2^* \widehat{R} \quad \Longrightarrow \quad M_2^* \widehat{R} M_2^{-1} = M_1^{-1} \widehat{R} M_1^* \tag{46}$$

(or $\widehat{R} \leftrightarrow \widehat{R}^{-1}$), $\widehat{R}$ the $\widehat{R}$-matrix of $SU_q(2)$, and identifying

$$\bar{\sigma} \equiv \mathrm{tr}_2(\widehat{R}_{12}\,(\epsilon^+ \sigma \epsilon)_2^t) \quad \text{i.e.} \quad \bar{\sigma}^c{}_d \equiv \widehat{R}^{ca}{}_{db} (\epsilon^+ \sigma \epsilon)^{tb}{}_a \tag{47}$$

we find

$$\begin{aligned}
\bar{\sigma}'^c{}_d &= \widehat{R}^{ca}{}_{db}(\epsilon^+ \sigma' \epsilon)^{tb}{}_a \\
&= (\epsilon^+ \sigma' \epsilon)^{tn}{}_m M^{*\,m}{}_a \widehat{R}^{ca}{}_{db} M^{-1\,b}{}_n \\
&= (\epsilon^+ \sigma' \epsilon)^{tn}{}_m M^{-1\,c}{}_k \widehat{R}^{km}{}_{ln} M^{*\,l}{}_d \\
&= M^{-1\,c}{}_k \bar{\sigma}^k{}_l M^{*\,l}{}_d
\end{aligned} \tag{48}$$

or

$$\bar{\sigma}' = M^{-1} \bar{\sigma} M^* \, . \tag{49}$$

This shows that the identification was correct. From (47) it also follows $\bar{\sigma}' = T \bar{\sigma}$.
As a consequence of this q-covariance $\bar{\psi}\psi$ is invariant. The other bilinear covariants are discussed elsewhere.

6 REPRESENTATION OF THE LORENTZ ALGEBRA

We complete the analysis of the q-deformed Dirac theory with the discussion of the infinitesimal Lorentz transformations on the Dirac spinor.

By constructing generators acting on two-dimensional spinors and their conjugates consistent with their commutation relations from an initially seven generator algebra a convincingly simple chiral algebra (that has, however, a complicated comultiplication) was derived in ref. 14:

$$\begin{aligned}
q^2 M^3 M^+ - q^{-2} M^+ M^3 &= \omega M^+ & q^{-2} N^3 N^+ - q^2 N^+ N^3 &= \omega N^- \\
q^{-1} M^+ M^- - q M^- M^+ &= M^3 & q N^+ N^- - q^{-1} N^- N^+ &= N^3 \\
q^{-2} M^3 M^- - q^2 M^- M^3 &= -\omega M^- & q^2 N^3 N^- - q^{-2} N^- N^3 &= -\omega N^+ .
\end{aligned} \qquad (50)$$

The M's and N's commute up to factors in q depending on some freedom in definition.

We introduce q-deformed $\sigma^{\mu\nu}$ using $(1+V)\gamma_2\gamma_1 = 2q\omega^{-1} g$ from (8) with γ's in the initial normalization as

$$\sigma^{IJ} = \frac{1}{2}[\gamma^J, \gamma^I]_V \equiv \frac{1}{2}(\gamma^J \gamma^I - V^{IJ}{}_{KL} \gamma^L \gamma^K) = \gamma^J \gamma^I - \frac{q}{\omega} g^{IJ} . \qquad (51)$$

Hence we find $V\sigma = -\sigma$ and $P_T\sigma = 0$. With the quantum Dirac algebra it turns out that the following σ's are the spinor representation of the chiral Lorentz generators

$$M^3 \sim q\, \sigma^{AB} - q^{-1}\sigma^{CD} , \qquad M^+ \sim -q^{-1}\sigma^{BD} , \qquad M^- \sim -q^{-1}\sigma^{AC} ,$$

$$N^3 \sim -q^{-1}(\sigma^{AB} + \sigma^{CD}) , \qquad N^+ \sim -q\, \sigma^{AD} , \qquad N^- \sim q^{-3}\, \sigma^{CB} . \qquad (52)$$

Thus the infinitesimal transformation of the Dirac spinor under the Lorentz group is determined. The explicit matrix representation is given in the appendix.**

7 APPENDIX

7.1 Inverse identifications

The relations (2) can be inverted:

$$\begin{aligned}
\gamma^0 &= q^{-2}\gamma^C + \gamma^D + q^{-1}\lambda\gamma^C\gamma^B\gamma^A - q\lambda\gamma^D\gamma^B\gamma^A \\
\gamma^1 &= q\gamma^A + q^{-1}\gamma^B \\
\gamma^2 &= i(q\gamma^A - q^{-1}\gamma^B) \\
\gamma^3 &= q^{-2}\gamma^C - \gamma^D + q^{-1}\lambda\gamma^C\gamma^B\gamma^A - q\lambda\gamma^D\gamma^B\gamma^A .
\end{aligned} \qquad (53)$$

7.2 γ^5 matrix

Defining[10]

$$\gamma^5 \equiv \epsilon_{IJKL}\, \gamma^I \gamma^J \gamma^K \gamma^L \cdot const \qquad (54)$$

(in the initial normalization $const = q^{-3}\omega/(\omega^2+2) \to 2^3/4!$) we calculate

$$\gamma^5 = -\frac{4}{q^2}\gamma^D\gamma^C\gamma^B\gamma^A - \frac{\omega}{q}\gamma^B\gamma^A + \frac{2}{q^2}\gamma^D\gamma^C + 1 , \qquad (55)$$

**For calculations the seven generator Lorentz algebra is sometimes of advantage[9,13]; with the matrix representation (61) that for the seven generator algebra is also provided.

with the following simple relations

$$\gamma^A \gamma^5 + q^{-2} \gamma^5 \gamma^A = 0 \qquad \gamma^C \gamma^5 + \gamma^5 \gamma^C = 0$$
$$\gamma^B \gamma^5 + q^2 \gamma^5 \gamma^B = 0 \qquad \gamma^D \gamma^5 + \gamma^5 \gamma^D = 0 .\qquad (56)$$

It is easy to check that these relations are consistent with (3). Calculating the square $(\gamma^5)^2 = 1 - \lambda\omega q^{-2} \gamma^B \gamma^A$ reveals that in the deformed case there are rather two such objets coinciding in the limit, γ^5 and

$$(\gamma^5)^{-1} = -2\,\frac{q^4+1}{q^2}\, \gamma^D \gamma^C \gamma^B \gamma^A - q\omega\, \gamma^B \gamma^A + \frac{2}{q^2}\, \gamma^D \gamma^C + 1 . \qquad (57)$$

Interestingly enough, one finds $q\omega^{-1}\,\gamma^5 + q^{-1}\omega^{-1}\,(\gamma^5)^{-1} = \gamma \equiv i\gamma^0\gamma^1\gamma^2\gamma^3$ the pseudoscalar of the undeformed Clifford algebra. It is, however, not possible to write commutation relations with γ alone thus in the quantum group case the appropriate objects will be γ^5 and its inverse.

7.3 Matrix representation

The q-deformed γ-matrices can be calculated easily using the identification (2). In the chiral representation[††] we take ($\sigma^1 = \begin{pmatrix} 0 & 1 \\ 1 & 0 \end{pmatrix}$, $\sigma^2 = \begin{pmatrix} 0 & -i \\ i & 0 \end{pmatrix}$, $\sigma^3 = \begin{pmatrix} 1 & 0 \\ 0 & -1 \end{pmatrix}$)

$$\gamma^0 = i\begin{pmatrix} 0 & -1 \\ -1 & 0 \end{pmatrix} \quad \gamma^1 = i\begin{pmatrix} 0 & \sigma^1 \\ -\sigma^1 & 0 \end{pmatrix} \quad \gamma^2 = \frac{1}{i}\begin{pmatrix} 0 & \sigma^2 \\ -\sigma^2 & 0 \end{pmatrix} \quad \gamma^3 = \frac{1}{i}\begin{pmatrix} 0 & \sigma^3 \\ -\sigma^3 & 0 \end{pmatrix} \qquad (58)$$

giving

$$\gamma^A = \begin{pmatrix} \cdot & \cdot & \cdot & i/q \\ \cdot & \cdot & \cdot & \cdot \\ \cdot & -i/q & \cdot & \cdot \\ \cdot & \cdot & \cdot & \cdot \end{pmatrix} \qquad \gamma^B = \begin{pmatrix} \cdot & \cdot & \cdot & \cdot \\ \cdot & \cdot & iq & \cdot \\ \cdot & \cdot & \cdot & \cdot \\ -iq & \cdot & \cdot & \cdot \end{pmatrix}$$

$$\gamma^C = \begin{pmatrix} \cdot & \cdot & -iq^2 & \cdot \\ \cdot & \cdot & \cdot & -iq\lambda \\ \cdot & \cdot & \cdot & \cdot \\ \cdot & -i & \cdot & \cdot \end{pmatrix} \qquad \gamma^D = \begin{pmatrix} \cdot & \cdot & \cdot & \cdot \\ \cdot & \cdot & \cdot & -i \\ -i & \cdot & \cdot & \cdot \\ \cdot & \cdot & \cdot & \cdot \end{pmatrix} \qquad (59)$$

In this representation the γ^5 matrix is diagonal

$$\gamma^5 = \begin{pmatrix} 1 & \cdot & \cdot & \cdot \\ \cdot & q^{-2} & \cdot & \cdot \\ \cdot & \cdot & -1 & \cdot \\ \cdot & \cdot & \cdot & -q^{-2} \end{pmatrix} . \qquad (60)$$

The chiral Lorentz generators have the following representation

$$M^+ = \begin{pmatrix} \cdot & \cdot & \cdot & \cdot \\ 1 & \cdot & \cdot & \cdot \\ \cdot & \cdot & \cdot & \cdot \\ \cdot & \cdot & \cdot & \cdot \end{pmatrix} \qquad M^- = \begin{pmatrix} \cdot & 1 & \cdot & \cdot \\ \cdot & \cdot & \cdot & \cdot \\ \cdot & \cdot & \cdot & \cdot \\ \cdot & \cdot & \cdot & \cdot \end{pmatrix} \qquad M^3 = \begin{pmatrix} -q & \cdot & \cdot & \cdot \\ \cdot & 1/q & \cdot & \cdot \\ \cdot & \cdot & \cdot & \cdot \\ \cdot & \cdot & \cdot & \cdot \end{pmatrix}$$

[††]The choice is made such that (2) and (13) is the same change of basis in the limit and the generators (52) are given in the 'spinor basis' $(x, y, \bar{x}, \bar{y})$, cf. ref. 9; using $(y, x, \bar{y}, \bar{x})$ instead gives a uniform factor i in (58).

$$N^+ = \begin{pmatrix} . & . & . & . \\ . & . & . & . \\ . & . & . & -1 \\ . & . & . & . \end{pmatrix} \quad N^- = \begin{pmatrix} . & . & . & . \\ . & . & . & . \\ . & . & . & . \\ . & . & -1 & . \end{pmatrix} \quad N^3 = \begin{pmatrix} . & . & . & . \\ . & . & . & . \\ . & . & q & . \\ . & . & . & -1/q \end{pmatrix}.$$

(61)

REFERENCES

[1] U. Carow-Watamura, M. Schlieker, M. Scholl, S. Watamura, *Z. Phys. C* **48**, 159 (1990); *Int. J. Mod. Phys.* **A6**, (1991) 3081.

[2] W.B. Schmidke, J. Wess, B. Zumino, *Z. Phys. C* **52**, (1991) 471.

[3] J. Schwenk, J. Wess, *Phys. Lett.* **B291**, (1992) 273.

[4] A. Schirrmacher, Remarks on the use of $\hat{R}$-matrices, *in* "Quantum Groups and related topics", 55, Kluwer (1992) (Proceedings of 1st Max Born Symp., Sept. 27–29, 1991, Wroclaw, Poland), R. Gielerak *et al.* eds.

[5] Yu.I. Manin, Quantum groups and non-commutative de Rham complexes, Bonn preprint MPI 91/47 (Arbeitstagung).

[6] Forthcoming publications by M. Pillin, W.B. Schmidke, J. Wess.

[7] N.Yu. Reshetikhin, L.A. Takhtadzhyan, L.D. Faddeev, *Leningrad Math. J.* **1**, (1990) 193.

[8] B. Zumino, *Mod. Phys. Lett.* **A13**, (1991) 1225.

[9] O. Ogievetsky, W.B. Schmidke, J. Wess, B. Zumino, *Commun. Math. Phys.* **150**, (1992) 495.

[10] A. Schirrmacher, The algebra of q-deformed γ-matrices, *in* "Proceedings of XIX Colloquium on group theoretical methods in physics", Salamanca, Spain, 1992.

[11] E. Wigner, *Ann. Math.* **40**, (1939) 39.

[12] J. Wess, private communication.

[13] A. Schirrmacher, Aspects of quantising Lorentz symmetry, *in* "New Symmetry Principles in Quantum Field Theory", (Proceedings of the ASI Cargèse, July 16-27 1991), J. Frölich *et al.* eds.

[14] O. Ogievetsky, W.B. Schmidke, J. Wess, B. Zumino, *Lett. Math. Phys.* **23**, (1991) 233.

HAMILTONIAN STRUCTURE OF EQUATIONS APPEARING IN RANDOM MATRICES

J. Harnad[†,1], C. A. Tracy[2], and H. Widom[3]

[1] Department of Mathematics and Statistics, Concordia University
7141 Sherbrooke W., Montréal, Canada H4B 1R6, and
Centre de recherches mathématiques, Université de Montréal
C. P. 6128-A, Montréal, Canada H3C 3J7

[2] Department of Mathematics and Institute of Theoretical Dynamics
University of California, Davis, CA 95616, USA

[3] Department of Mathematics
University of California, Santa Cruz, CA 95064, USA

ABSTRACT

The level spacing distributions in the Gaussian Unitary Ensemble, both in the "bulk of the spectrum," given by the Fredholm determinant of the operator with the *sine kernel* $\frac{\sin \pi(x-y)}{\pi(x-y)}$ and on the "edge of the spectrum," given by the *Airy kernel* $\frac{\text{Ai}(x)\text{Ai}'(y)-\text{Ai}(y)\text{Ai}'(x)}{(x-y)}$, are determined by compatible systems of nonautonomous Hamiltonian equations. These may be viewed as special cases of isomonodromic deformation equations for first order 2×2 matrix differential operators with regular singularities at finite points and irregular ones of Riemann index 1 or 2 at ∞. Their Hamiltonian structure is explained within the *classical R-matrix* framework as the

---

[†] Research supported in part by the Natural Sciences and Engineering Research Council of Canada and by the National Science Foundation, U.S.A., DMS–9001794 and DMS–9216203.

equations induced by spectral invariants on the loop algebra $\widetilde{sl}(2)$, restricted to a Poisson subspace of its dual space $\widetilde{sl}_R^*(2)$, consisting of elements that are rational in the loop parameter.

0. INTRODUCTION

It has long been known for a class of exactly solvable models in statistical mechanics that the correlation functions are expressible in terms of Painlevé transcendents or more generally solutions to completely integrable differential equations possessing the Painlevé property [WMTB], [MTW], [SMJ]. (For a review see [Mc].) It was the great insight of the Kyoto School to invent the notion of a τ-function associated to a system of completely integrable differential equations [SMJ], [JMMS], [JMU], [JM] and to realize that in this class of solvable models the correlation functions are τ-functions.

In the theory of random matrices it has also long been known (see, e.g., [Po], [Me1]) that the level spacing distributions are expressible in terms of the Fredholm determinant of an integral operator. In a particular scaling limit the kernel of this integral operator is the *sine kernel*. For this case it was shown in [JMMS] that the Fredholm determinant was also a τ-function. This then leads in the simplest case to formulas for the level spacing distributions in terms of a particular Painlevé V transcendent. More generally, the deformation equations consist of a compatible set of nonautonomous Hamiltonian equations.

Since the original work of [JMMS] on random matrices and impenetrable bosons, the methods for deducing the relevant deformation equations and properties of the τ-functions have been simplified and generalized, both in the study of random matrices [BTW], [Me2], [Dy], [MM], [TW1], [TW2], [W] and quantum correlation functions [IIKS], [IIKV]. The Hamiltonian structure of the underlying deformation equations can most simply be understood as a nonautonomous version of the equations induced by spectral invariants on a loop algebra with respect to a suitably chosen *classical R-matrix* Poisson bracket structure [H].

In the following section, the problem of computing level spacing distributions for the Gaussian Unitary Ensemble (GUE), both in the "bulk of the spectrum" and at the "edge of the spectrum" is formulated in these terms—the two cases differing only in the kernels resulting from the different scaling limits. For the first case, this is the above mentioned *sine kernel*

$$\frac{1}{\pi} \frac{\sin \pi(x-y)}{x-y},$$

while for the second it is the *Airy kernel*

$$\frac{\mathrm{Ai}(x)\mathrm{Ai}'(y) - \mathrm{Ai}(y)\mathrm{Ai}'(x)}{x-y}.$$

In §2, the resulting Hamiltonian equations are explained in terms of spectral invariants on the dual space $\widetilde{sl}^*(2)$ of the loop algebra $\widetilde{sl}(2)$, with respect to a standard classical R-matrix Poisson bracket structure, restricted to Poisson subspaces consisting of rational coadjoint orbits.

1. SPECTRAL DISTRIBUTIONS FOR RANDOM MATRICES

1a. Fredholm Determinant Representation

The GUE measure on the spectral space of $N \times N$ hermitian matrices with eigenvalues $(X_1, \ldots, X_N)$ is [Me1]

$$P_N(X_1, \ldots, X_N) \, dX_1 \cdots dX_N = C_N \prod_{j<k} |X_j - X_k|^2 \exp\left(-\sum_j X_j^2\right) dX_1 \cdots dX_N. \tag{1.1}$$

Averaging over all but n of the eigenvalues gives the n-point correlation function [Me1]:

$$R_{n,N}(X_1, \ldots, X_n) = \frac{N!}{(N-n)!} \int_{-\infty}^{\infty} \cdots \int_{-\infty}^{\infty} P_N(X_1, \ldots, X_N) \, dX_{n+1} \cdots dX_N$$
$$= \det\left(K_N(X_j, X_k)\Big|_{j,k=1}^n\right), \tag{1.2}$$

where

$$K_N(X, Y) := \sum_{k=0}^{N-1} \varphi_k(X) \varphi_k(Y). \tag{1.3}$$

Here $\{\varphi_k(X)\}$ is the sequence obtained by orthonormalizing the sequence $\{X^k \exp(-X^2/2)\}$ over $(-\infty, \infty)$. In particular $R_{1,N}(X)$, the density of eigenvalues at X, equals $K_N(X, X)$.

For fixed X,

$$R_{1,N}(X) \sim \frac{1}{\pi}\sqrt{2N} \text{ as } N \to \infty. \tag{1.4}$$

The scaling limit in the bulk of the spectrum at the point X_0 is the limit

$$N \to \infty, \; X_j \to X_0, \text{ such that } x_j = (X_j - X_0) R_{1,N}(X_0) \text{ is fixed.} \tag{1.5}$$

In this limit the scaled n-point correlation functions $R_n(x_1, \ldots, x_n)$, defined by

$$R_n(x_1, \ldots, x_n) \, dx_1 \cdots dx_n = \lim_{\substack{N \to \infty, X_j \to X_0 \\ x_j \text{ fixed}}} R_{n,N}(X_1, \ldots, X_n) \, dX_1 \cdots dX_n, \tag{1.6}$$

are given by [Me1]

$$R_n(x_1, \ldots, x_n) = \det\left(K(x_j, x_k)\Big|_{j,k=1}^n\right) \tag{1.7}$$

where

$$K(x, y) := \frac{1}{\pi} \frac{\sin \pi(x-y)}{x-y}. \tag{1.8}$$

Now consider a set of "spectral windows"

$$J = J_1 \cup \cdots \cup J_m$$

where

$$\{J_k = [a_{2k-1}, a_{2k}] \subset \mathbf{R}\}_{k=1,\ldots m}, \qquad a_1 < a_2 < \ldots < a_{2m} \tag{1.9}$$

is an ordered set of disjoint intervals on the real axis. The probability that a randomly chosen GUE matrix has precisely $\{n_1, n_2, \ldots, n_m\}$ scaled eigenvalues in the respective intervals $\{J_1, J_2, \ldots, J_m\}$ is given by the *level spacing distribution*:

$$E(n_1, \ldots, n_m, J) = \frac{(-1)^n}{n_1! \cdots n_m!} \frac{\partial^n D(J; \lambda)}{\partial \lambda_1^{n_1} \cdots \partial \lambda_m^{n_m}} \bigg|_{\lambda_1 = \cdots = \lambda_m = 1} \tag{1.10}$$

with $n = n_1 + \cdots + n_m$. Here $D(J; \lambda)$ is the Fredholm determinant

$$D(J; \lambda) := \det(I - \widehat{K}) \tag{1.11}$$

of the integral operator $\widehat{K}$ with kernel

$$K_J(x, y) := \sum_{k=1}^{m} \lambda_k K(x, y) \chi_{J_k}(y), \tag{1.12}$$

where $\chi_{J_k}(y)$ is the characteristic function for the interval J_k, and

$$\lambda := (\lambda_1, \lambda_2, \ldots, \lambda_m) \tag{1.13}$$

denotes the set of generating function parameters. The case of one interval is in [**Me1**]. The more general formula (1.10) was derived in [**TW1**].

In the following, we shall see that $D(J; \lambda)$ is determined by a system of partial differential equations consisting of a compatible set of nonautonomous Hamiltonian equations defined on a suitable phase space. The quantity of central interest will be the total differential

$$\omega(a) := d_a \log D(J; \lambda) = \sum_{i=1}^{2m} \frac{\partial \log D(J; \lambda)}{\partial a_i} da_i \tag{1.14}$$

of the logarithm of the generating function $D(J; \lambda)$, taken with respect to the set of parameters

$$a := (a_1, \ldots, a_{2m}) \tag{1.15}$$

giving the boundaries of the spectral domains. The dynamical equations of Jimbo, Miwa, Môri and Sato [**JMMS**], which determine the dependence of this quantity on the parameters $(a_1, \ldots, a_{2m})$, are given below, following the approach developed in [**TW1**]. We note that in [**JMMS**], [**TW1**], and [**TW2**] the case of equal λ_j's only was considered in the derivation of the equations.

1b. Dynamics of Distribution Functions

The key object in what follows is the resolvent operator

$$\widehat{R} := (1 - \widehat{K})^{-1} \widehat{K}. \tag{1.16}$$

Introducing the functions

$$Q(x, a) := (1 - \widehat{K})^{-1} A, \tag{1.17a}$$
$$P(x, a) := (1 - \widehat{K})^{-1} A', \tag{1.17b}$$

where
$$A(x) := \frac{1}{\pi}\sin\pi x, \tag{1.18}$$

the kernel of the resolvent operator can be shown to be given by

$$R(x,y) = \sum_{k=1}^{m} \lambda_k \frac{Q(x,a)P(y,a) - P(x,a)Q(y,a)}{x-y} \chi_{J_k}(y), \quad x \neq y, \tag{1.19a}$$

$$R(x,x) = \sum_{k=1}^{m} \lambda_k \left(\frac{dQ}{dx}(x,a)P(x,a) - \frac{dP}{dx}(x,a)Q(x,a)\right)\chi_{J_k}(x). \tag{1.19b}$$

Using the identity

$$d_a \log\left(\det(I-\widehat{K})\right) = -\operatorname{tr}\left((I-\widehat{K})^{-1} d_a \widehat{K}\right), \tag{1.20}$$

it is easy to show that

$$\omega(a) = \sum_{j=1}^{2m}(-1)^{j+1} R(a_j, a_j)\, da_j, \tag{1.21}$$

where $R(a_j, a_k)$ is obtained by taking the limits $(x,y) \to (a_j, a_k)$ inside J. Let

$$q_j := \lim_{\substack{x \to a_j \\ x \in J}} \sqrt{\lambda_j} Q(x,a), \tag{1.22a}$$

$$p_j := \lim_{\substack{x \to a_j \\ x \in J}} \sqrt{\lambda_j} P(x,a), \quad j=1,\ldots,2m. \tag{1.22b}$$

Then we have

$$R(a_j, a_k) = \frac{q_j p_k - p_j q_k}{a_j - a_k}, \quad j \neq k, \tag{1.23a}$$

$$R(a_j, a_j) = \pi^2 q_j^2 + p_j^2 + \sum_{k=1}^{2m}(-1)^k R(a_j, a_k) R(a_k, a_j)(a_j - a_k). \tag{1.23b}$$

Thus, the parameter dependence of $\det(I-\widehat{K})$ is determined through the quantities $\{q_j, p_j\}_{j=1,\ldots 2m}$. Differentiating equations (1.17a,b) with respect to the parameters a_j and taking the appropriate limits, these may be shown to satisfy the following system of equations:

$$\begin{aligned}
\frac{\partial q_j}{\partial a_k} &= (-1)^k R(a_j, a_k) q_k, \quad j \neq k, \\
\frac{\partial p_j}{\partial a_k} &= (-1)^k R(a_j, a_k) p_k, \quad j \neq k, \\
\frac{\partial q_j}{\partial a_j} &= p_j - \sum_{k \neq j}(-1)^k R(a_j, a_k) q_k, \\
\frac{\partial p_j}{\partial a_j} &= -\pi^2 q_j - \sum_{k \neq j}(-1)^k R(a_j, a_k) p_k.
\end{aligned} \tag{1.24}$$

This overdetermined system, which we refer to as the JMMS equations, is in fact Frobenius integrable, and has a Hamiltonian structure which we summarize in the next subsection.

1c. Hamiltonian Structure of the JMMS Equations

For convenience, define a new normalization of the coordinates

$$q_{2j} := -\tfrac{1}{2} i x_{2j}, \quad q_{2j+1} := \tfrac{1}{2} x_{2j+1},$$
$$p_{2j} := -i y_{2j}, \quad p_{2j+1} := y_{2j+1}, \tag{1.25}$$

and let

$$G_j(x,y) := \frac{\pi^2}{4} x_j^2 + y_j^2 - \frac{1}{4} \sum_{\substack{k=1 \\ k \neq j}}^{2m} \frac{(x_j y_k - x_k y_j)^2}{a_j - a_k}. \tag{1.26}$$

In terms of these quantities, we have

$$\omega(a) = \sum_{j=1}^{2m} G_j(x,y) \, da_j. \tag{1.27}$$

Defining canonical Poisson brackets:

$$\{x_j, x_k\} = \{y_j, y_k\} = 0, \quad \{x_j, y_k\} = \delta_{jk}, \quad j,k = 1,\ldots,2m, \tag{1.28}$$

the system of equations (1.24) is seen to have the nonautonomous Hamiltonian form

$$d_a x_j = \{x_j, \omega(a)\} \quad \text{and} \quad d_a y_j = \{y_j, \omega(a)\}. \tag{1.29}$$

Moreover, the set of Hamiltonians $\{G_j\}_{j=1,\ldots 2m}$ are easily verified to be generically functionally independent and in involution:

$$\{G_j, G_k\} = 0 \quad \text{for all} \quad j,k = 1,\ldots,2m. \tag{1.30}$$

From this follows that the system (1.29), or equivalently (1.24), is Frobenius integrable, and that ω is locally an exact differential:

$$\omega = d_a \log \tau, \tag{1.31}$$

where the value of the tau-function τ along the integral curves coincides with $\log(\det(I - \widehat{K}))$.

The significance of this Hamiltonian structure in terms of spectral invariants on loop algebras will be explained in §2. First, another similar system occurring in the computation of level spacing distribution functions of scaled random matrices at the "edge of the spectrum" will be described. For the details of what follows see [**TW2**]. (Here also only the case of equal λ_j's was considered, the generalization to the case of unequal λ_j's being straightforward.)

1d. The Airy Kernel System and Distributions at the Edge of the Spectrum:

Scaling at the "edge of the spectrum" [Mo], [Fo], [TW2], corresponds to choosing $X_0 \sim \pm\sqrt{2N}$ and gives rise to a different Fredholm determinant, in which the sine kernel (1.8) is replaced by the *Airy kernel*

$$K(x,y) := \frac{A(x)A'(y) - A'(x)A(y)}{x-y}$$
$$= \lim_{N\to\infty} \frac{1}{2^{1/2}N^{1/6}} K_N\left(\sqrt{2N} + \frac{x}{2^{1/2}N^{1/6}}, \sqrt{2N} + \frac{y}{2^{1/2}N^{1/6}}\right), \quad (1.32)$$

where now, $A(x)$ is an Airy function

$$A(x) = \text{Ai}(x). \quad (1.33)$$

The logarithmic differential of the Fredholm determinant (1.14) in this case is also given by formula (1.21) where, using the same notation as above, the resolvent kernel $R(x,y)$ is still of the form (1.19a,b), with the functions $Q(x,a), P(x,a)$ obtained by replacing $A(x)$ in eq. (1.17a,b), (1.18) by (1.33). Define $\{q_j, p_j\}_{j=1,\ldots 2m}$ as in (1.22a,b), and introduce two further quantities:

$$u := \sum_{j=1}^{m} \lambda_j \int_{a_{2j-1}}^{a_{2j}} A(x)Q(x,a)\,dx, \quad (1.34a)$$

$$v := \sum_{j=1}^{m} \lambda_j \int_{a_{2j-1}}^{a_{2j}} A(x)P(x,a)\,dx. \quad (1.34b)$$

Then similarly to the previous case, we have,

$$R(a_j, a_k) = \frac{q_j p_k - p_j q_k}{a_j - a_k}, \quad j \neq k, \quad (1.35a)$$

$$R(a_j, a_j) = \sum_{k \neq j}(-1)^k \frac{(q_j p_k - p_j q_k)^2}{a_j - a_k} + p_j^2 - a_j q_j^2 - 2p_j q_j u + 2q_j^2 v, \quad (1.35b)$$

where all the limits are again taken within the intervals J_j, J_k. Differentiating with respect to the parameters $\{a_j\}_{j=1,\ldots 2m}$ gives the system

$$\frac{\partial q_j}{\partial a_k} = (-1)^k \frac{q_j p_k - p_j q_k}{a_j - a_k} q_k, \quad (j \neq k),$$

$$\frac{\partial p_j}{\partial a_k} = (-1)^k \frac{q_j p_k - p_j q_k}{a_j - a_k} p_k, \quad (j \neq k),$$

$$\frac{\partial q_j}{\partial a_j} = -\sum_{k \neq j}(-1)^k \frac{q_j p_k - p_j q_k}{a_j - a_k} q_k + p_j - q_j u,$$

$$\frac{\partial p_j}{\partial a_j} = -\sum_{k \neq j}(-1)^k \frac{q_j p_k - p_j q_k}{a_j - a_k} p_k + a_j q_j + p_j u - 2q_j v, \quad (1.36)$$

$$\frac{\partial u}{\partial a_j} = (-1)^j q_j^2,$$

$$\frac{\partial v}{\partial a_j} = (-1)^j p_j q_j.$$

Again, introducing new coordinates $\{x_j, y_j\}_{j=0,\ldots 2m}$ by:

$$q_{2j} := -\tfrac{1}{2}ix_{2j}, \quad q_{2j+1} := \tfrac{1}{2}x_{2j+1},$$
$$p_{2j} := -iy_{2j}, \quad p_{2j+1} := y_{2j+1}, \quad (1.37)$$
$$u := y_0, \quad v := \tfrac{1}{2}x_0.$$

and defining canonical Poisson brackets

$$\{x_j, x_k\} = \{y_j, y_k\} = 0, \quad \{x_j, y_k\} = \delta_{jk}, \quad j,k = 0,\ldots,2m, \quad (1.38)$$

eqs. (1.36) have the Hamiltonian form

$$d_a x_j = \{x_j, \omega(a)\} \quad \text{and} \quad d_a y_j = \{y_j, \omega(a)\}, \quad j=0,\ldots,2m, \quad (1.39)$$

where

$$\omega(a) = d_a \log(\det(I - \widehat{K})) = \sum_{j=1}^{2m} G_j(x,y)\, da_j \quad (1.40)$$

and

$$G_j(x,y) = y_j^2 - \tfrac{1}{4}a_j x_j^2 - x_j y_j y_0 + \tfrac{1}{4}x_j^2 x_0 - \frac{1}{4}\sum_{\substack{k=1\\k\neq j}}^{2m} \frac{(x_j y_k - x_k y_j)^2}{a_j - a_k}. \quad (1.41)$$

It is straightforward to verify that these G_j's are also in involution, implying the Frobenius integrability of the system (1.39) and the existence of a τ-function as in eq. (1.31). However, in this case, the phase space is $2(m+1)$–dimensional, so the G_j's alone do not form a complete set of commuting integrals. It is easily verified, though, that the following additional independent function

$$G_0(x,y) := y_0^2 - x_0 - \frac{1}{4}\sum_{i=1}^{2m} x_i^2 \quad (1.42)$$

is also in involution with the G_j's, thereby forming a complete set $\{G_j\}_{j=0,\ldots 2m}$ of commuting integrals. Introducing an additional flow parameter a_0 corresponding to the Hamiltonian G_0, and letting

$$\tilde{\omega}(\tilde{a}) := \omega(a) + G_0 da_0, \quad (1.43)$$

where

$$\tilde{a} := (a_0, \ldots, a_{2m}), \quad (1.44)$$

we have the following extended system, which is also Frobenius integrable:

$$d_{\tilde{a}} x_j = \{x_j, \tilde{\omega}(\tilde{a})\} \quad \text{and} \quad d_{\tilde{a}} y_j = \{y_j, \tilde{\omega}(\tilde{a})\}, \quad j=0,\ldots,2m. \quad (1.45)$$

Moreover, since G_0 is not explicitly dependent on the parameters a_j and Poisson commutes with all the $\{G_j\}_{j=1,\ldots 2m}$, it is in fact a constant. Since all the quantities $u, v, q_j, p_j \to 0$ as the $a_j \to \infty$, this constant value is just 0.

2. CLASSICAL R-MATRIX AND ISOMONODROMY FORMULATION

Hamiltonian systems of the type encountered above can be understood within a general Lie algebraic setting (cf. [AHP], [H], [RS], [S]) which we now summarize. Let $\tilde{g}$ be the loop algebra of smooth (real or complex), $sl(r)$-valued functions $X(\lambda)$, defined on a circle S^1 centered at the origin in the complex λ-plane. Define the Ad-invariant scalar product $\langle\,,\,\rangle$ on $\tilde{g}$ by:

$$\langle X, Y \rangle = \frac{1}{2\pi i} \oint_{S^1} \operatorname{tr}(X(\lambda)Y(\lambda))\, d\lambda. \tag{2.1}$$

Interpreting this as the evaluation of a linear form X on elements $Y \in \tilde{g}$ allows us to identify $\tilde{g}$ with an open, dense subspace of the dual space, which will be denoted $\tilde{g}^*$. Under this identification, both the adjoint and coadjoint actions of the corresponding loop group $\tilde{\mathcal{G}}$ are given by conjugation:

$$\operatorname{Ad}_g : X \longmapsto gXg^{-1}, \qquad \operatorname{Ad}_g^* : X \longmapsto gXg^{-1}, \qquad g \in \tilde{\mathcal{G}}, \quad X \in \tilde{g}. \tag{2.2}$$

As a linear space, $\tilde{g}$ may be decomposed into a direct sum

$$\tilde{g} = \tilde{g}_+ + \tilde{g}_-, \tag{2.3}$$

where $\tilde{g}_+, \tilde{g}_-$ are the subalgebras consisting of elements $X(\lambda)$ that admit, respectively, holomorphic extensions to the interior and exterior of S^1, with the elements $X \in \tilde{g}_-$ satisfying $X(\infty) = 0$. Under the identification $\tilde{g} \sim \tilde{g}^*$, we have

$$\tilde{g}_+^* \sim \tilde{g}_-, \qquad \tilde{g}_-^* \sim \tilde{g}_+, \tag{2.4}$$

where $\tilde{g}_\pm^* \subset \tilde{g}^*$ are identified as the annihilators of $\tilde{g}_\mp$.

Let

$$\begin{aligned} P_\pm &: \tilde{g} \longrightarrow \tilde{g}_\pm, \\ P_\pm &: X \longmapsto X_\pm, \end{aligned} \tag{2.5}$$

be the projections to these two subspaces relative to the decomposition (2.3) (determined, e.g., by splitting the Fourier series on S^1 into positive and negative parts) and define the endomorphism $R : \tilde{g} \to \tilde{g}$ as the difference:

$$R := P_+ - P_-. \tag{2.6}$$

Then R is a *classical R-matrix* [S], in the sense that the bracket $[\,,\,]_R$ defined on $\tilde{g}$ by:

$$[X,Y]_R := \tfrac{1}{2}[RX,Y] + \tfrac{1}{2}[X,RY] \tag{2.7}$$

is skew symmetric and satisfies the Jacobi identity, determining a new Lie algebra structure on the same space as $\tilde{g}$, which we denote $\tilde{g}_R$. The Lie Poisson bracket on $\tilde{g}_R^* \sim \tilde{g}^*$ associated to the Lie bracket $[\,,\,]_R$ is:

$$\{f,g\}|_X = \langle [df, dg]_R, X \rangle \tag{2.8}$$

for smooth functions f, g on $\tilde{g}_R^*$ (with the usual identifications, $df|_X, dg|_X \in \tilde{g}^* \sim \tilde{g}$). The corresponding group $\tilde{\mathcal{G}}_R$ is identified with the direct product $\tilde{\mathcal{G}}_+ \times \tilde{\mathcal{G}}_-$, where $\tilde{\mathcal{G}}_\pm$ are the loop groups associated to $\tilde{g}_\pm$. The adjoint and coadjoint actions are given by:

$$\begin{aligned} \operatorname{Ad}_R(g) &: (X_+ + X_-) \longmapsto g_+ X_+ g_+^{-1} + g_- X_- g_-^{-1}, \\ \operatorname{Ad}_R^*(g) &: (X_+ + X_-) \longmapsto (g_- X_+ g_-^{-1})_+ + (g_+ X_- g_+^{-1})_-, \\ X_\pm &\in \tilde{g}_\pm, \qquad g_\pm \in \tilde{\mathcal{G}}_\pm. \end{aligned} \tag{2.9}$$

Let $a := \{a_i\}_{i=1}^n$ be a set of n distinct real (or complex) numbers and $\kappa := \{k, k_i\}_{i=1}^n$ a set of $n + 1$ nonnegative integers. Define the subspace $\mathbf{g}_{a,\kappa} \subset \tilde{\mathbf{g}}_R^*$ to consist of elements $X = X_+ + X_-$ with $X_+ \in \tilde{\mathbf{g}}_+$ a polynomial in λ of degree k and $X_- \in \tilde{\mathbf{g}}_-$ a rational function of λ, with poles of degree $\{k_1, \ldots, k_n\}$ at the points $\{a_1, \ldots, a_n\}$. Then $\mathbf{g}_{a,\kappa} \subset \tilde{\mathbf{g}}_R^*$ is Ad_R^*-invariant, and hence a Poisson subspace. The coefficient Y of the leading term in the degree k polynomial X_+ is Ad_R^* invariant, and hence constant under all Hamiltonian flows. Let $\mathcal{I}_{a,\kappa} := I(\tilde{\mathbf{g}}^*)|_{\mathbf{g}_{a,\kappa}}$ be the ring of Ad^*-invariant polynomials on $\tilde{\mathbf{g}}^* \sim \tilde{\mathbf{g}}_R^*$, restricted to $\mathbf{g}_{a,\kappa}$, and define, for $X \in \mathbf{g}_{a,\kappa}$, the $sl(r)$-valued polynomial function:

$$L(\lambda) := a(\lambda)X(\lambda), \qquad (2.10)$$

where

$$a(\lambda) := \prod_{i=1}^n (\lambda - a_i)^{k_i}. \qquad (2.11)$$

Then the ring $\mathcal{I}_{a,\kappa}$ is generated by the coefficients of the characteristic polynomial:

$$\mathcal{P}(\lambda, z) := \det(L(\lambda) - zI), \qquad (2.12)$$

and the characteristic equation

$$\mathcal{P}(\lambda, z) = 0 \qquad (2.13)$$

defines, after suitable compactification (and possible desingularization), an algebraic curve Γ, called the "spectral curve", which is generically an r-fold branched cover of $\mathbb{CP}^1$.

It follows from the Adler-Kostant-Symes (AKS) theorem that:
1) The ring $\mathcal{I}_{a,\kappa}$ is Poisson commutative (with respect to the $\{\ ,\ \}_R$ Poisson bracket).
2) For a Hamiltonian $G \in \mathcal{I}_{a,\kappa}$, Hamilton's equations are given by:

$$\frac{dX}{dt} = \{X, G\} = [dG_+, X] = -[dG_-, X] \qquad (2.14)$$

(The last equality following from the fact that $[dG, X] = 0$, which is the infinitesimal form of Ad^*-invariance).

This means that the spectral curve Γ is invariant under the flows generated by the elements of $\mathcal{I}_{a,\kappa}$, since any two coefficients of the spectral polynomial $\mathcal{P}(\lambda, z)$ Poisson commute. It may also be shown [AHP], [AHH1] that, for sufficiently generic initial conditions and coadjoint orbits $\mathcal{O}$ in $\mathbf{g}_{a,\kappa}$, the elements of $\mathcal{I}_{a,\kappa}$, restricted to $\mathcal{O} \subset \mathbf{g}_{a,\kappa}$, define completely integrable Hamiltonian systems; that is, the number of independent generators in $\mathcal{I}_{a,\kappa}$ equals one half the dimension of the orbit. Through a standard construction [AvM], [RS], [AHH1], [AHH2], the flows they generate are shown to linearize on the Jacobi variety associated to Γ.

We now specialize to the case $\{k = 1, k_i = 1\}_{i=1,\ldots n}$. Then $X(\lambda)$ has the form

$$X(\lambda) = \lambda Y + N_0 + \sum_{i=1}^n \frac{N_i}{\lambda - a_i} \qquad (2.15)$$

and the Ad_R^*-action is:

$$Ad_R^*(g_+, g_-) : X(\lambda) \longmapsto \lambda Y + N_0 + [\gamma_0, Y] + \sum_{i=1}^n \frac{g_i N_i g_i^{-1}}{\lambda - a_i}, \qquad (2.16)$$

where
$$\gamma_0 := \frac{1}{2\pi i} \oint_{S^1} g_-(\lambda) \, d\lambda, \qquad g_i := g_+(a_i), \qquad i = 1, \ldots n. \qquad (2.17)$$

Hence, each coadjoint orbit in $\mathbf{g}_{a,\kappa}$ consists of elements of the form:
$$X(\lambda) = \lambda Y + C + [\gamma_0, Y] + \sum_{i=1}^{n} \frac{N_i}{\lambda - a_i}, \qquad (2.18)$$

where
$$N_0 := C + [\gamma_0, Y], \qquad (2.19)$$

with Y and C constant matrices, and the N_i's have fixed Jordan normal form.

Let
$$\tfrac{1}{2}\mathrm{tr}X^2 = y\lambda^2 + c\lambda + G_0 + \sum_{i=1}^{n} \frac{G_i}{\lambda - a_i} + \sum_{i=1}^{n} \frac{K_i}{(\lambda - a_i)^2}, \qquad (2.20)$$

where
$$y := \tfrac{1}{2}\mathrm{tr}Y^2, \qquad c := \mathrm{tr}(YN_0) = \mathrm{tr}(YC) \qquad (2.21)$$

and
$$G_0 := \frac{1}{4\pi i} \oint_{S^1} \mathrm{tr}X^2(\lambda) \frac{d\lambda}{\lambda} = \mathrm{tr}\left(Y \sum_{i=1}^{n} N_i\right) + \tfrac{1}{2}\mathrm{tr}N_0^2,$$

$$G_i := \frac{1}{4\pi i} \oint_{\lambda = a_i} \mathrm{tr}X^2(\lambda) \, d\lambda = a_i \mathrm{tr}(YN_i) + \mathrm{tr}(N_0 N_i) + \sum_{j \neq i}^{n} \frac{\mathrm{tr}(N_i N_j)}{a_i - a_j}, \qquad (2.22)$$

$$K_i := \frac{1}{4\pi i} \oint_{\lambda = a_i} (\lambda - a_i) \mathrm{tr}X^2(\lambda) \, d\lambda = \tfrac{1}{2}\mathrm{tr}(N_i^2), \qquad i = 1, \ldots n,$$

with the integrals $\oint_{\lambda = a_i}$ taken around a small circle enclosing only this pole. Then y, c and $\{K_i\}_{i=1,\ldots n}$ are all Casimir invariants and hence constant on each Ad_R^* orbit, and $\{G_i\}_{i=0,\ldots n}$ are independent elements of the Poisson commuting ring $\mathcal{I}_{a,\kappa}$. Denoting the time parameters for the Hamiltonian flows generated by $\{G_i\}_{i=0,\ldots n}$ as $\{t_i\}_{i=0,\ldots n}$, the AKS theorem implies that Hamilton's equations are of the form:

$$\begin{aligned}\frac{\partial X}{\partial t_0} &= \{X, G_0\} = [(dG_0)_+, X] \\ \frac{\partial X}{\partial t_i} &= \{X, G_i\} = -[(dG_i)_-, X], \qquad i = 1, \ldots n\end{aligned} \qquad (2.23)$$

where
$$(dG_0)_+ = Y, \qquad (dG_i)_- = -\frac{N_i}{\lambda - a_i}, \qquad i = 1, \ldots n. \qquad (2.24)$$

Evaluating residues at $\{\lambda = a_i\}_{i=1,\ldots n}$ and $\lambda = \infty$, we see that this is equivalent to:

$$\frac{\partial N_i}{\partial t_0} = [Y, N_i], \qquad \frac{\partial N_0}{\partial t_0} = [Y, N_0], \qquad \frac{\partial N_0}{\partial t_i} = [Y, N_i],$$

$$\frac{\partial N_i}{\partial t_j} = \frac{[N_i, N_j]}{a_i - a_j}, \quad i \neq j, \qquad \frac{\partial N_i}{\partial t_i} = \left[a_i Y + N_0 + \sum_{j \neq i} \frac{N_j}{a_i - a_j}, N_i\right], \qquad (2.25)$$

$$i, j = 1, \ldots n.$$

241

If we now reinterpret these as nonautonomous Hamiltonian systems, by identifying the constants $\{a_i\}$ with the flow parameters $\{t_i = a_i\}_{i=1,\ldots n}$, system (2.24) becomes (cf. [JMMS], [JMU]):

$$\frac{\partial N_i}{\partial a_0} = [Y, N_i], \quad \frac{\partial N_0}{\partial a_0} = [Y, N_0], \quad \frac{\partial N_0}{\partial a_i} = [Y, N_i],$$
$$\frac{\partial N_i}{\partial a_j} = \frac{[N_i, N_j]}{a_i - a_j}, \quad i \neq j, \quad \frac{\partial N_i}{\partial a_i} = [a_i Y + N_0 + \sum_{j \neq i} \frac{N_j}{a_i - a_j}, N_i], \quad (2.26)$$
$$i, j = 1, \ldots n.$$

(Here t_0 has similarly been renamed a_0, but it does not appear as a parameter in the system, and hence represents autonomous flow.) This system is equivalent to the commutativity of the following system of differential operators:

$$\mathcal{D}_\lambda := \frac{\partial}{\partial \lambda} - \lambda Y - N_0 - \sum_{i=1}^n \frac{N_i}{\lambda - a_i},$$
$$\mathcal{D}_i := \frac{\partial}{\partial a_i} + \frac{N_i}{\lambda - a_i}, \quad (2.27)$$
$$\mathcal{D}_0 := \frac{\partial}{\partial a_0} - Y.$$

It follows that the monodromy of the operator $\mathcal{D}_\lambda$ is preserved under the deformations parametrized by $\{a_0, a_1, \ldots, a_n\}$.

To obtain the system (1.26)–(1.29), we take the $\{a_j\}_{j=1}^n$ as real, $r = 2$, $n = 2m$, $Y = 0$ (so really, we are in the Poisson subspace with $k = 0$), and choose

$$C = \begin{pmatrix} 0 & \frac{\pi^2}{2} \\ -2 & 0 \end{pmatrix} \quad (2.28)$$

and the N_i's with Jordan normal form $\begin{pmatrix} 0 & 0 \\ 1 & 0 \end{pmatrix}$. The Ad^*_R orbit $\mathcal{O}_0 \subset \mathfrak{g}_{a,\kappa}$ then consists of elements of the form:

$$X(\lambda) = \begin{pmatrix} 0 & \frac{\pi^2}{2} \\ -2 & 0 \end{pmatrix} + \frac{1}{2} \sum_{i=1}^{2m} \frac{\begin{pmatrix} -x_i y_i & -y_i^2 \\ x_i^2 & x_i y_i \end{pmatrix}}{\lambda - a_i}. \quad (2.29)$$

Here, we have parametrized the orbit by elements of $M_0 := \mathbf{R}^{4m}/(\mathbf{Z}_2)^{4m}$, where $((x_1, \ldots, x_{2m}), (y_1, \ldots, y_{2m})) \in \mathbf{R}^{4m}$, and the $(\mathbf{Z}_2)^{4m}$ action is generated by reflections in the coordinate hyperplanes:

$$(\epsilon_1, \ldots, \epsilon_{2m}) : ((x_1, \ldots, x_{2m}), (y_1, \ldots, y_{2m}))$$
$$\longmapsto ((\epsilon_1 x_1, \ldots, \epsilon_{2m} x_{2m}), (\epsilon_1 y_1, \ldots, \epsilon_{2m} y_{2m})) \quad (2.30)$$

$$\{\epsilon_i = \pm 1\}_{i=1,\ldots 2m}.$$

The map $J_0 : M_0 \longrightarrow \tilde{\mathfrak{g}}_R^*$ defined by

$$J_0 : ((x_1, \ldots, x_{2m}), (y_1, \ldots, y_{2m})) \longmapsto \begin{pmatrix} 0 & \frac{\pi^2}{2} \\ -2 & 0 \end{pmatrix} + \frac{1}{2} \sum_{i=1}^{2m} \frac{\begin{pmatrix} -x_i y_i & -y_i^2 \\ x_i^2 & x_i y_i \end{pmatrix}}{\lambda - a_i} \quad (2.31)$$

is a symplectic embedding with respect to the standard symplectic structure:

$$\omega = \sum_{i=1}^{2m} dx_i \wedge dy_i \tag{2.32}$$

on M_0 and the orbital (Lie-Kostant-Kirillov) symplectic structure on $\mathcal{O}_0$. (This is a special case of the moment map embeddings into rational Poisson subspaces of the dual of loop algebras developed in [**AHP**], [**AHH3**].) Evaluating the Hamiltonians $\{G_i\}_{i=1,\ldots 2m}$ defined in (2.22) for this case gives those of eqs. (1.26) and the $\{a_j\}$, $j \neq 0$ components of the equations of motion (2.26) are equivalent to (1.29), up to quotienting by the $(\mathbf{Z}_2)^{4m}$ action (2.30).

To obtain the system (1.40)–(1.45), we again take $r = 2$, $n = 2m$, but choose

$$Y = \begin{pmatrix} 0 & -\frac{1}{2} \\ 0 & 0 \end{pmatrix}, \quad C = \begin{pmatrix} 0 & 0 \\ -2 & 0 \end{pmatrix}, \quad [\gamma_0, Y] = \begin{pmatrix} y_0 & \frac{x_0}{2} \\ 0 & -y_0 \end{pmatrix}, \tag{2.33}$$

with the N_i's again of Jordan normal form $\begin{pmatrix} 0 & 0 \\ 1 & 0 \end{pmatrix}$. In this case, the Ad_R^* orbit $\mathcal{O}_1 \subset \mathfrak{g}_{a,\kappa}$ consists of elements of the form:

$$X(\lambda) = \begin{pmatrix} y_0 & -\frac{\lambda}{2} + \frac{x_0}{2} \\ -2 & -y_0 \end{pmatrix} + \frac{1}{2}\sum_{i=1}^{2m} \frac{\begin{pmatrix} -x_i y_i & -y_i^2 \\ x_i^2 & x_i y_i \end{pmatrix}}{\lambda - a_i}. \tag{2.34}$$

Here, we have parametrized the orbit by elements of $M_1 := \mathbf{R}^2 \times \mathbf{R}^{4m}/(\mathbf{Z}_2)^{4m}$, where $(x_0, y_0) \in \mathbf{R}^2$, $((x_1,\ldots,x_{2m}),(y_1,\ldots,y_{2m})) \in \mathbf{R}^{4m}$, and the $(\mathbf{Z}_2)^{4m}$ action is again given by (2.30). The map $J_1 : M_1 \longrightarrow \tilde{\mathfrak{g}}_R^*$ defined by

$$J_1 : ((x_0, x_1,\ldots,x_{2m}),(y_0, y_1,\ldots,y_{2m}))$$

$$\longmapsto \begin{pmatrix} y_0 & -\frac{\lambda}{2} + \frac{x_0}{2} \\ -2 & -y_0 \end{pmatrix} + \frac{1}{2}\sum_{i=1}^{2m} \frac{\begin{pmatrix} -x_i y_i & -y_i^2 \\ x_i^2 & x_i y_i \end{pmatrix}}{\lambda - a_i} \tag{2.35}$$

is again a symplectic embedding, now with respect to the symplectic structure:

$$\omega = dx_0 \wedge dy_0 + \sum_{i=1}^{2m} dx_i \wedge dy_i \tag{2.36}$$

on M_1 and the orbital (Lie-Kostant-Kirillov) symplectic structure on $\mathcal{O}_1$. Evaluating the Hamiltonians $\{G_i\}_{i=0,\ldots 2m}$ defined in (2.22) for this case gives those of eqs. (1.41), (1.42) and the equations of motion (2.26) are equivalent to (1.45) – again, up to quotienting by the $(\mathbf{Z}_2)^{4m}$ action (2.30).

For this case, eqs. (2.21), (2.33), (2.35) give $y = 0$, $c = 1$ and $\mathrm{tr}(N_i^2) = 0$, so

$$-\det(X) = \tfrac{1}{2}\mathrm{tr}(X^2) = \lambda + G_0 + \sum_{i=1}^{2m} \frac{G_i}{\lambda - a_i} := \frac{P(\lambda)}{a(\lambda)}, \tag{2.37}$$

where $P(\lambda)$ is a monic polynomial of degree $2m+1$

$$P(\lambda) = \lambda^{2m+1} + P_{2m}\lambda^{2m} + \ldots + P_0 \tag{2.38}$$

with

$$P_{2m} = G_0 - \sum_{i=1}^{2m} a_i, \quad P_{2m-1} = \sum_{i=1}^{2m} G_i - G_0 \sum_{i=1}^{2m} a_i, \quad \text{etc.} \tag{2.39}$$

From eqs. (2.10)–(2.12), we have the following equation for the spectral curve Γ:

$$\mathcal{P}(\lambda, z) = z^2 - a(\lambda)P(\lambda) = 0, \tag{2.40}$$

which shows that it is hyperelliptic, of genus $g = 2m$, with $2m + 1$ of its branch points located at $\{a_1, \ldots a_{2m}, \infty\}$ and the others determined by the values of the roots of $P(\lambda)$. For the system (1.26)–(1.29) associated to the sine kernel, the leading term in (2.20) vanishes, so the polynomial $P(\lambda)$ is of degree $2m$ and the curve has genus $g = 2m - 1$, with no branch points at ∞. For the autonomous system (2.25), these curves are invariant under the flows. For the nonautonomous system (2.26), which in these cases reduces to (1.29) or (1.45), it would be of interest to determine the dependence of the spectral curves on the deformation parameters $\{a_i\}_{i=0,\ldots 2m}$ through a system of PDE's involving the spectral invariants $\{G_i\}_{i=0,\ldots 2m}$ alone. For the single finite interval case, this is given in [**JMMS**], [**TW1**] by the τ-function form of the Painlevé equation P_V, while for the single semi-infinite interval case, this is given in [**TW2**] by the corresponding τ-function form of P_{II}.

Acknowledgements

It is a pleasure to acknowledge E. L. Basor, F. J. Dyson, P. J. Forrester, A. Its, and M. L. Mehta for their helpful comments and encouragement and for sending us their preprints prior to publication. The second author wishes to thank the organizers of the program "Low Dimensional Topology and Quantum Field Theory" for their kind hospitality at the Isaac Newton Institute for Mathematical Sciences. Particular thanks to R. J. Baxter and H. Osborn for making the stay at Cambridge most pleasant.

REFERENCES

[AHH1] Adams, M. R., Harnad, J. and Hurtubise, J., "Isospectral Hamiltonian Flows in Finite and Infinite Dimensions II. Integration of Flows," *Commun. Math. Phys.* **134**, 555–585 (1990).

[AHH2] Adams, M. R., Harnad, J. and Hurtubise, J., "Darboux Coordinates and Liouville-Arnold Integration in Loop Algebras," preprint CRM (1992) (to appear in *Commun. Math. Phys.* (1993)) .

[AHH3] Adams, M. R., Harnad, J. and Hurtubise, J., "Dual Moment Maps to Loop Algebras," *Lett. Math. Phys.* **20**, 294–308 (1990).

[AHP] Adams, M. R., Harnad, J. and Previato, E., "Isospectral Hamiltonian Flows in Finite and Infinite Dimensions I. Generalised Moser Systems and Moment Maps into Loop Algebras," *Commun. Math. Phys.* **117**, 451–500 (1988).

[AvM] Adler, M. and van Moerbeke, P., "Completely Integrable Systems, Euclidean Lie Algebras, and Curves," *Adv. Math.* **38**, 267–317 (1980); "Linearization of Hamiltonian Systems, Jacobi Varieties and Representation Theory," *ibid.* **38**, 318–379 (1980).

[BTW] Basor, E. L., Tracy, C. A., and Widom, H., "Asymptotics of Level-Spacing Distributions for Random Matrices," *Phys. Rev. Letts.* **69**, 5–8 (1992).

[Dy] Dyson, F. J., "The Coulomb Fluid and the Fifth Painlevé Transcendent," IASSNSS-HEP-92/43 preprint, to appear in the proceedings of a conference in honor of C. N. Yang, ed. S.-T. Yau.

[Fo] Forrester, P. J., "The Spectrum Edge of Random Matrix Ensembles," preprint.

[H] Harnad, J., "Dual Isomonodromy Deformations and Moment Maps to Loop Algebras," preprint CRM-1844 (1993).

[IIKS] Its, A. R., Izergin, A. G., Korepin, V. E. and Slavnov, N. A., "Differential Equations for Quantum Correlation Functions," *Int. J. Mod. Phys.* **B4**, 1003–1037 (1990).

[IIKV] Its, A. R., Izergin, A. G., Korepin, V. E. and Varzugin, G. G., "Large Time and Distance Asymptotics of Field Correlation Function of Impenetrable Bosons at Finite Temperature," *Physica* **54D**, 351–395 (1992).

[JMMS] Jimbo, M., Miwa, T., Môri, Y. and Sato, M., "Density Matrix of an Impenetrable Bose Gas and the Fifth Painlevé Transcendent", *Physica* **1D**, 80–158 (1980).

[JMU] Jimbo, M., Miwa, T.,and Ueno, K., "Monodromy Preserving Deformation of Linear Ordinary Differential Equations with Rational Coeefficients I.," *Physica* **2D**, 306–352 (1981).

[JM] Jimbo, M., and Miwa, T., "Monodromy Preserving Deformation of Linear Ordinary Differential Equations with Rational Coeefficients II, III.," *Physica* **2D**, 407–448 (1981); *ibid.*, **4D**, 26–46 (1981).

[Mc] McCoy, B. M., "Spin Systems, Statistical Mechanics and Painlevé Functions," in *Painlevé Transcendents: Their Asymptotics and Physical Applications*, eds. D. Levi and P. Winternitz, Plenum Press, New York (1992), pgs. 377–391.

[Me1] Mehta, M. L., *Random Matrices*, 2nd ed., Academic Press, San Diego, (1991).

[Me2] Mehta, M. L., "A Non-linear Differential Equation and a Fredholm Determinant," *J. de Phys. I* **2**,1721–1729 (1992).

[MM] Mehta, M. L. and Mahoux, G., "Level Spacing Functions and Non-linear Differential Equations," SPh-T/92-107 preprint.

[Mo] Moore, G., "Matrix Models of 2D Gravity and Isomonodromic Deformation," *Prog. Theor. Phys. Suppl.* No. **102**, 255–285 (1990).

[MTW] McCoy, B. M., Tracy, C. A., and Wu, T. T., "Painlevé Functions of the Third Kind," *J. Math. Phys.* **18**, 1058–1092 (1977).

[Po] Porter, C. E., *Statistical Theory of Spectra: Fluctuations*, Academic Press, New York (1965).

[RS] Reiman, A. G., and Semenov-Tian-Shansky, M.A., "Reduction of Hamiltonian systems, Affine Lie algebras and Lax Equations I, II," *Invent. Math.* **54**, 81–100 (1979); *ibid.* **63**, 423–432 (1981).

[S] Semenov-Tian-Shansky, M. A., "What is a Classical R-Matrix," *Funct. Anal. Appl.* **17** 259–272 (1983); "Dressing Transformations and Poisson Group Actions," *Publ. RIMS Kyoto Univ.* **21** 1237–1260 (1985).

[SMJ] Sato, M., Miwa, T., and Jimbo, M., "Holonomic Quantum Fields I–V," *Publ. RIMS Kyoto Univ.* **14**, 223–267 (1978); *ibid.* **15**, 201–278 (1979); *ibid.* **15**, 577–629 (1979); *ibid.***15**, 871–972 (1979); *ibid.* **16**, 531–584 (1980).

[TW1] Tracy, C. A., and Widom, H., "Introduction to Random Matrices," UCD preprint ITD 92/93-10 (1992), to appear in *VIIIth Scheveningen Conf. Proc.*, Springer Lecture Notes in Physics.

[TW2] Tracy, C. A., and Widom, H., "Level Spacing Distributions and the Airy Kernel," UCD preprint ITD 92/93-9 (1992), to appear in *Commun. Math. Phys.*

[W] Widom, H., "The Asymptotics of a Continuous Analogue of Orthogonal Polynomials," to appear in J. Approx. Th.

[WMTB] Wu, T. T., McCoy, B. M., Tracy, C. A., and Barouch, E., "Spin-Spin Correlation Functions for the Two-Dimensional Ising Model: Exact Theory in the Scaling Region," *Phys. Rev.* **B13**, 316–374 (1976).

ON THE EXISTENCE OF POINTLIKE LOCALIZED FIELDS IN CONFORMALLY INVARIANT QUANTUM PHYSICS

Martin Jörß[*]

II. Institut für theoretische Physik
Universität Hamburg, 2 HH 50, Germany

ABSTRACT

In quantum field theory the existence of pointlike localizable objects called "fields" is a pre-assumption. Since charged fields are in general not observable this situation is unsatisfying from a quantum physics point of view. Indeed in any quantum theory the existence of fields should follow from deeper physical concepts and more natural first principles like stability, causality and symmetry.

In the framework of algebraic quantum field theory with Haag-Kastler nets of local observables this is presented for the case of conformal symmetry in 1+1 dimensions. Conformal fields are explicitly constructed as limits of observables localized in finite regions of space-time. These fields then allow to derive a geometric identification of modular operators, Haag duality in the vacuum sector, the PCT-theorem and an equivalence theorem for fields and algebras.

1 INTRODUCTION AND MOTIVATION

Quantum field theory tries to give a formulation of local quantum physics on space-time which is consistent with Einstein's theory of relativity. The general the-

---

[*]supported by the Deutsche Forschungsgemeinschaft (DFG)

ory of quantized fields is concerned with all model independent aspects of quantum field theory[14,18]. There are two main frameworks for a general theory of quantized fields: Haag-Kastler nets of C^*-algebras of local observables[19] and Wightman axioms for quantized fields[30,21].

The approach of Wightman starts with axioms for covariantly transforming distributions with values in unbounded operators on the physical Hilbert space. It has therefore an a priori notion of pointlike localizable fields. Most people working in the field of quantum physics rely on the validity of that strong formulation of localization in space-time and so does their work.

Our aim was to justify this ad hoc assumption in the case of conformally invariant quantum physics in 1+1 dimensions. We will explicitly construct conformal fields as limits of bounded operators representing observables localized in extended regions of space-time. The existence of pointlike localizable fields will thereby be derived from a more natural and physical set of first principles in the algebraic approach to quantum field theory of Haag and Kastler (see section 2).

Algebraic quantum field theory avoids the huge conceptual and technical difficulties caused by domain problems of unbounded operators of unobservable charged fields in the Wightman approach. It is solely based on the notion of measurement of local observables in finitely extended regions of space-time. The formulation of quantum field theory in terms of Haag-Kastler nets is therefore conceptually very satisfying. It is, too, very well suited for the analysis and discussion of structural problems. Doplicher, Haag and Roberts[10] and Buchholz and Fredenhagen[9] gave algebraic quantum field theory a certain degree of completion by fully analyzing the structure of superselection sectors and giving a intrinsic definition of particle statistics for 2, 3 and 4 space-time dimensions. They introduced the notion of localizable charges and gave a complete treatment of fractional statistics in massive theories (see ref. 13).

There are, however, some disadvantages attached to the Haag-Kastler approach. It is e.g. difficult to write down Haag-Kastler nets explicitly. Without a clear connection to classical field theory the algebraic approach turns out to have problems in model-building and lacks intuition to formulate a theory of elementary interaction forces. By now one is not able to check or reproduce the astonishingly successful predictions of standard Wightman quantum field theory gained by such mathematically unsatisfying methods as perturbation theory of Euclidean path integrals.

The partition of strength and weakness of both approaches to a general theory of quantized fields seems to be complementary. It is therefore tempting to work out the relationship of Wightman theory and Haag-Kastler theory and try to combine them to tackle open problems. There has been a lot of effort on that task[4,7,8,11,15,12,26,31,33]. You might use ref. 34 and the references therein as a review. Although usually one believes that both frameworks are physically equivalent there is no proof for that.

It is therefore promising that in our present work the proof of the coexistence and equivalence of pointlike localized fields and algebras of local observables can be used to solve another two previously open problems of algebraic quantum field theory:

We will prove Haag duality in the vacuum sector of a conformally covariant Haag-Kastler net, which is a maximality condition on local algebras and crucial for the analysis of Doplicher, Haag and Roberts mentioned above. Its general derivation in the framework of Haag and Kastler is unknown. Further on we will able to prove the PCT-theorem which is also not possible in the standard situation of algebraic quantum field theory. See, however, ref. 3.

Finally we can prove a theorem stating equivalence between the "field picture" and the "algebra picture" of 1+1-dimensional conformal quantum field theory.

The work presented here is based on a diploma thesis[20] and a intense collaboration with K. Fredenhagen. This conference report could only give a rough overview on our results and some spotlight on the proofs. A second article[16] is in preparation which will treat the subject in greater detail and will have a broader approach to related topics. One aim will be to give an algebraic derivation of the operator product expansions[24,25,29] in conformal quantum field theory.

2 THE ALGEBRAIC FRAMEWORK OF QUANTUM FIELD THEORY

The programme of Haag and Kastler for an algebraic formulation of quantum field theory shall be briefly presented. Starting point has been the idea to describe the structure of the set of all observables by a net of abstract C^*-algebras. Every local algebra indexed by a subset of space-time should be generated by all observables localized in the respective domain. All information on a physical system should be contained in the inclusion structure of the local algebras. The set of physical states on the algebra of all local observables then already completely determines the physical situation.

Therefore the basic structure is a net of C^*-algebras

$$\mathcal{A} : \mathcal{O} \to \mathcal{A}(\mathcal{O}), \mathcal{O} \in \mathcal{K},$$

where $\mathcal{A}(\mathcal{O})$ contains all observables localized in $\mathcal{O}$ and $\mathcal{K}$ denotes usually the set of all open double cones in space-time. The C^*-limit of the union of all local observables shall be called $\mathcal{A}$ und can be realized on a suitable Hilbert space

The net is assumed to satisfy the usual Haag-Kastler axioms:

1. Isotony
$$\mathcal{A}(\mathcal{O}_1) \subset \mathcal{A}(\mathcal{O}_2) \quad \text{for} \quad \mathcal{O}_1 \subset \mathcal{O}_2 \tag{1}$$

2. Locality
$$\mathcal{A}(\mathcal{O}_1) \subset \mathcal{A}(\mathcal{O}_2)' \quad \text{for} \quad \mathcal{O}_1 \subset \mathcal{O}_2' \tag{2}$$

3. Irreducibility
$$\bigcap_{\mathcal{O} \in \mathcal{K}} \mathcal{A}(\mathcal{O})' = \mathbf{C}\mathbf{1} \tag{3}$$

4. Covariance
$$\alpha_g(\mathcal{A}(\mathcal{O})) = \mathcal{A}(g(\mathcal{O})) \tag{4}$$

where $\mathcal{A}(\mathcal{O})'$ denotes the commutant of $\mathcal{A}(\mathcal{O})$, $\mathcal{O}'$ denotes the spacelike complement of $\mathcal{O}$, and $\alpha(G)$ is a automorphism representation of the group of space-time symmetries.

To guarantee stability one has to assume the existence of a vacuum representation $\pi_o(\mathcal{A})$ of the algebra of local observables with a unitary positive energy representation of translations and a unique translation invariant vacuum vector Ω.

Crucial for the analysis of the structure of superselection sectors, charges and particle statistics by Doplicher, Haag and Roberts[10] and Buchholz and Fredenhagen[9] will be a strengthened version of locality in the vacuum sector of the theory:

$$\pi_o(\mathcal{A}(\mathcal{O})) = \pi_o(\mathcal{A}(\mathcal{O}'))' \quad \text{for all } \mathcal{O} \in \mathcal{K} \qquad \text{"Haag duality"}. \tag{5}$$

This is a maximality condition for the local observables in the vacuum sector. It is likely to be fulfilled in absence of spontaneous breakdown of symmetry[28].

There is a recent book by Haag[18] which will answer everything you ever wanted to know about algebraic quantum field theory but never dared to ask.

3 CONFORMAL NETS OF LOCAL OBSERVABLES IN 1+1 DIMENSIONS

The classical conformal symmetry group in 1+1 dimensions is $\tilde{G}_2 = Diff(\mathbf{R}) \times Diff(\mathbf{R})$. After quantization only its subgroup $G_2 = SO(2,2)/\mathbf{Z}_2$, which leaves the vacuum invariant, remains a symmetry group.
In lightcone coordinates $(x-t, x+t)$ one easily sees the factorization

$$G_2 = G \times G = SL(2,\mathbf{R})/\mathbf{Z}_2 \times SL(2,\mathbf{R})/\mathbf{Z}_2. \tag{6}$$

G acts as follows on a single lightcone:

$$T(a,b,c,d): y \to \frac{ay+b}{cy+d} \quad \text{with} \quad a,b,c,d \in \mathbf{R} \quad ad-bc=1. \tag{7}$$

Let $\mathcal{K}$ denote the set of all open intervals on a single lightcone. A conformally covariant observable algebra generated by chiral conformal fields like the energy-momentum tensor and conserved current densities should then factorize in the following manner:

$$\mathcal{A}_0(I_l \times I_r) = \mathcal{A}_{0l}(I_l) \otimes \mathcal{A}_{0r}(I_r) \quad \text{for} \quad I_l, I_r \in \mathcal{K}. \tag{8}$$

Throughout this work we restrict the investigation to factorizing nets.

Let $\mathcal{A}: \mathcal{O} \to \mathcal{A}(\mathcal{O})$, $\mathcal{O} \in \mathcal{K}$, a Haag-Kastler net on $\mathbf{R}$, ω_o a vacuum state and H a separable Hilbert space. We will now define the vacuum sector of a conformally covariant net of algebras of local observables[5]:

$$\mathcal{B}(\mathcal{O}) := \mathcal{A}_{\omega_o}(\mathcal{O}) = \pi_{\omega_o}(\mathcal{A}(\mathcal{O}))'' \subset B(H), \quad \mathcal{O} \in \mathcal{K}. \tag{9}$$

The transformation law is

$$U(g)\mathcal{B}(\mathcal{O})U(g)^{-1} = \mathcal{B}(g(\mathcal{O})), \quad g \in G, \quad \mathcal{O}, g(\mathcal{O}) \in \mathcal{K}, \tag{10}$$

where $U(G)$ is a positive energy representation on H with finite multiplicities.

As we work in the vacuum sector one can rule out the possibility of nontrivial representations of the covering group $\tilde{G}$ and the spectrum of the conformal energy $H_C = P/2 + K/2$ is restricted to integer numbers. (P and K are the generators of the translations resp. of the special conformal transformations $y \to y/(1+cy)$.) Therefore the Hilbert space representation $(U(G), H)$ is completely reducible in irreducible positive energy modules $(U_n(G), H_n)$ which are well known from the "discrete series" of[1,23] G.

Since we are interested in fields on the light cone we now look for a realization of the discrete series on[22,17] $\mathbf{R}$:

Let $n \in \mathbf{N}$, $D_n := \{f \in C^\infty(\mathbf{R}) \,|\, x^{2n-2} f(1/x) \in C^\infty(\mathbf{R})\}$ and E_n the space of polynomials of degree $2n-2$. Then G can be represented irreducibly on $F_n := (D_n/E_n)^+ \subset C^\infty(\mathbf{R}) \cap L^2(\mathbf{R})$ by

$$\left(\tilde{U}(g)f\right)(x) := (a-cx)^{2n-2} f\left(\frac{dx-b}{a-cx}\right), \quad f \in F_n. \tag{11}$$

On $F_n \times F_n$ one can define the invariant scalar product

$$(f,g)_n := \lim_{\varepsilon \downarrow 0} \int_R \int_R dx\, dy\, \frac{\overline{f(x)}\,g(y)}{(x-y+i\varepsilon)^{2n}}, \quad f,g \in F_n. \tag{12}$$

250

Using $(\cdot,\cdot)$ we finally complete F_n canonically to get a unitary representation on $\overline{F_n}$. As a module of the discrete series of G with conformal energy n the representation $(\tilde{U}_n, \overline{F_n})$ is unique up to unitary equivalence. The unitary intertwiners V_n between $(\tilde{U}_n, \overline{F_n})$ and $(U_n(G), H_n)$ then introduce space-time structure in the physical Hilbert space H:

$$\Phi_n(f) := V_n f \qquad \text{`` } = \phi(f)\Omega \text{ ''} \qquad \text{for } f \in \overline{F_n}. \tag{13}$$

As these vectors resemble field operators acting on the vacuum one could call them "protofield-vectors".

4 NEW SCALING

A dense set of vectors in the physical representation space H can be rescaled. "New scaling" will turn out to be essential in the construction of pointlike localizable fields as limits of local observables.

Theorem: Let $n \in \mathbf{N}$, $g \in C^\infty(\mathbf{R}) \cap L^1(\mathbf{R}) \cap \overline{F_n}$. Then all vectors $\Phi_n(g)$ are scaleable, i.e. for $h \in C^\infty(\mathbf{R}) \cap L^1(\mathbf{R}) \cap \overline{F_n}$ we have

$$\lim_{\lambda \downarrow 0} \int_R dx\, h(x)\, U(x)\, D(\lambda)\, \lambda^{-n} \Phi_n(g) = \left(\int_R dz\, g(z) \right) \Phi_n(h). \tag{14}$$

Proof: To prove the theorem we will show weak convergence on the dense set $\{\Phi(f) \mid f \in C^\infty(\mathbf{R}) \cap L^1(\mathbf{R}) \cap \overline{F_n}\}$, uniform boundedness in λ and convergence of the square of the norm.

1. Some analysis leads to

$$\int_R dx\, h(x) (\Phi_n(f), U(x)\, D(\lambda)\, \lambda^{-n} \Phi_n(g))$$
$$= \int_R dx\, h(x) \lim_{\varepsilon \downarrow 0} \int_R \int_R dy\, dz\, \overline{f(y)}\, g(z)\, (y - x - \lambda z + i\varepsilon)^{-2n}$$
$$= \int_R dz\, g(z) \lim_{\varepsilon \downarrow 0} \int_R \int_R dx\, dy\, \overline{f(y)}\, h(x)\, (y - x - \lambda z + i\varepsilon)^{-2n}$$
$$= \int_R dz\, g(z)\, (\Phi_n(f), U(\lambda z) \Phi_n(h)).$$

2. As $U(\cdot)$ is strongly continuous, $(\Phi_n(f), U(\lambda z) \Phi_n(h))$ converges in the limit $\lambda \downarrow 0$ uniformly on compact domains in z to $(\Phi_n(f), \Phi_n(h))$. Therefore we arrive at

$$\lim_{\lambda \downarrow 0} \int_R dz\, g(z)\, (\Phi_n(f), U(\lambda z)\Phi_n(h)) = \left(\int_R dz\, g(z) \right) (\Phi_n(f), \Phi_n(h))$$

and weak convergence on a dense subspace is proven:

$$\lim_{\lambda \downarrow 0} \int_R dx\, h(x)\, (\Phi_n(f), U(x)\, D(\lambda)\, \lambda^{-n}\, \Phi_n(g)) = \left(\int_R dz\, g(z) \right) (\Phi_n(f), \Phi_n(h)).$$

3. Let now λ be a positive number. We get

$$\left\| \int_R dx\, h(x)\, U(x)\, D(\lambda)\, \lambda^{-n} \Phi_n(g) \right\|^2$$
$$= \int_R \int_R dx\, dx'\, \overline{h(x)}\, h(x')\, (U(x)\, D(\lambda)\, \Phi_n(g), U(x')\, D(\lambda)\, \Phi_n(g))\, \lambda^{-2n}$$
$$= \int_R dz\, g(z) \int_R dz'\, g(z')\, (U(\lambda z')\Phi_n(h), U(\lambda z)\Phi_n(h)).$$

This leads to

$$\lim_{\lambda \downarrow 0} \| \int_R dx\, h(x)\, U(x)\, D(\lambda)\, \lambda^{-n} \Phi_n(g) \|^2 = \left(\int_R dz\, g(z) \right)^2 \|\Phi_n(h)\|^2.$$

We showed that $\int_R dx\, h(x)\, U(x)\, D(\lambda)\, \lambda^{-n} \Phi_n(g)$ is uniformly in λ norm-bounded and does even converge for $\lambda \downarrow 0$ in the square of the norm. This proves the theorem.

5 LOCAL REGULARIZATION OF OBSERVABLES

We just proved a transformation law, called "new scaling", for vectors in the physical Hilbert space H. New scaling does work only on a dense set in each G-module. We want to use it for constructing pointlike localized fields as limits of local observables. In order to do so we got to find enough local observables that can be realized on the vacuum in terms of elements of the subset appropriate for new scaling, i.e. for all $n \in \mathbf{N}$ we would like to have

$$P_n A \Omega = \sum_{m=1}^{m_n} \Phi_{nm}(f_{A,n,m}) \quad \text{with} \quad f_{A,n,m} \in C^\infty(\mathbf{R}) \cap L^1(\mathbf{R}) \cap \overline{F}_n. \tag{15}$$

P_n is the projector on the direct sum of all modules $\Phi_{nm}(\cdot)$ of G with conformal energy $n \in \mathbf{N}$. Arbitrary local observables generally don't fulfill this condition. "Local regularization" will help to solve the problem.

Definition: Let B a local observable and $f \in C_o^\infty(G)$ a smooth function with compact support on G. Then $\mathcal{B}_{\text{reg}}$ is defined to be the set of all locally regularized operators of the form

$$\tilde{B} := \int_G dG\, f(g)\, U(g)\, B\, U(g)^*. \tag{16}$$

Remark: The vacuum vector is cyclic for $\mathcal{B}_{\text{reg}}$. In the next theorem we use instead of $\mathcal{B}_{\text{reg}}$ an appropriate subalgebra. Still cyclicity can be shown to hold.

Theorem: Smearing-out any local observable B in a vicinity of the unit $\mathbf{1} \in G$ with a suitable function gives a locally regularized operator $\tilde{B} \in \mathcal{B}_{\text{reg}}$ so that

$$f_{\tilde{B},n,m} \in C^\infty(\mathbf{R}) \cap L^1(\mathbf{R}) \cap \overline{F}_n. \tag{17}$$

Proof: We will give the proof of "local regularization" in three steps.
1. **Smoothing an arbitrary distribution**
Any distribution gets smoothed by local integration with the group of translations:
Let $n \in \mathbf{N}$, $f \in \overline{F}_n$, $g \in C_o^\infty(\mathbf{R})$. Then the rules of folding lead to

$$\int_R db\, g(b)\, \tilde{U}(b)\, f(\cdot) = \int_R db\, g(b)\, f(\cdot - b) = (g * f)(\cdot) \in C^\infty(\mathbf{R}) \cap \overline{F}_n.$$

2. **Damping a dense set of functions**
Any element of the dense set $F_n \subset \overline{F}_n$ gets damped by local integration with the group of special conformal Transformations, represented by the unitary operators $\tilde{W}(c)$:
Let $n \in \mathbf{N}$, $f \in F_n$ and $h \in C_o^\infty(\mathbf{R})$ with $0 \notin \operatorname{supp} h$. One can then verify

$$\hat{h}(x) := h(1/x)\, x^{2n-2} \in C_o^\infty(\mathbf{R}) \quad \text{and}$$

$$H(x) := \lim_{\varepsilon \downarrow 0} \int_R dc\, h(1/c)\, c^{2n-2} \frac{1}{(c+x+i\varepsilon)^{2n}}$$

$$= \frac{1}{(x-i0)^{2n}} * h(1/x)\, x^{2n-2} \in L^1 \cap C^\infty(\mathbf{R}).$$

Now we can prove the absolute integrability of $\int_R dc\, h(c)\, \tilde{W}(c)\, f$:
Choose h, f positive and additionally $\mathrm{supp} h \subset \mathbf{R}_+$. We then get

$$\int_R dx\, H(x)\, f(x) = \int_R dx \lim_{\varepsilon \downarrow 0} \int_R dc\, \frac{h(1/c)\, c^{2n-2}}{(c+x+i\varepsilon)^{2n}} f(x)$$

$$= \lim_{\varepsilon \downarrow 0} \int_R dc \int_R dx\, \frac{h(c)}{(1+cx+i\varepsilon)^{2n}} f(x)$$

$$= \lim_{\varepsilon \downarrow 0} \int_R dc \int_R dx\, h(c) \left(\frac{1-cx}{1+i\varepsilon}\right)^{2n-2} f\left(\frac{x}{1-cx}\right)$$

$$= \int_R dx \int_R dc\, h(c)\, (1-cx)^{2n-2}\, f\left(\frac{x}{1-cx}\right)$$

$$= \int_R dx \int_R dc\, h(c)\, \bigl(\tilde{W}(c) f\bigr)(x)\,.$$

For all $f \in F_n$, and all $h \in C_o^\infty(\mathbf{R})$ with $0 \notin \mathrm{supp} h$ we have therefore proven

$$\int_R dc\, h(c)\, \tilde{W}(c)\, f \;\in\; C^\infty(\mathbf{R}) \cap L^1(\mathbf{R}) \cap F_n \quad \text{and}$$

$$\int_R dx \int_R dc\, h(c)\, \bigl(\tilde{W}(c) f\bigr)(x) = (\hat{h}, f)_n$$

with the scalar product $(\cdot, \cdot)_n$ of $\overline{F}_n$.

3. Smoothing and damping an arbitrary distribution

Any distribution can get smoothed and damped by an appropriate local integration on G.

Let $n \in \mathbf{N}$, $f \in \overline{F}_n \cap C^\infty(\mathbf{R})$, $g, h \in C_o^\infty(\mathbf{R})$ with $0 \notin \mathrm{supp} h$. Any positive f can get pointwise approximated in F_n by a sequence $f_i \leq f$ so that

$$\int_R dx\, \left(g * \int_R dc\, h(c)\, \tilde{W}(c)\, f_i\right)(x) = \left(\int_R db\, g(b)\right) (\hat{h}, f_i)_n \leq \left(\int_R db\, g(b)\right) (\hat{h}, f)_n\,.$$

With monotone convergence in $L^1(\mathbf{R})$ we then get

$$\int_R db\, g(b)\, \tilde{U}(b) \int_R dc\, h(c)\, \tilde{W}(c)\, f \;\in\; C^\infty(\mathbf{R}) \cap L^1(\mathbf{R}) \cap \overline{F}_n\,.$$

Now we have finally proved local regularization:

For $n \in \mathbf{N}$, $f \in \overline{F}_n$, $g_1, g_2, h \in C_o^\infty(\mathbf{R})$ with $0 \notin \mathrm{supp}\, h$ we find

$$\int_R db_2\, g_2(b_2)\, \tilde{U}(b_2) \int_R dc\, h(c)\, \tilde{W}(c) \int_R db_1\, g_1(b_1)\, \tilde{U}(b_1)\, f$$

$$= g_2 * \int_R dc\, h(c)\, \tilde{W}(c)\, (g_1 * f) \;\in\; C^\infty(\mathbf{R}) \cap L^1(\mathbf{R}) \cap \overline{F}_n\,.$$

6 CONSTRUCTION OF UNBOUNDED FIELD OPERATORS

Let $n \in \mathbf{N}$, $B \in \mathcal{B}_{\mathrm{reg}}$, i.e. locally regularized. We proved the following representation:

$$P_n\, B\, \Omega = \sum_{m=1}^{m_n} \Phi_{nm}(f_{B,n,m}) \quad \text{with} \quad f_{B,n,m} \in C^\infty(\mathbf{R}) \cap L^1(\mathbf{R}) \cap \overline{F}_n\,,\; 1 \leq m \leq m_n\,. \quad (18)$$

We are now able to define pointlike localizable field operators on a dense domain.

Definition: Let $n \in \mathbf{N}$, $B \in \mathcal{B}_{\text{reg}}$, $\mathcal{O} \in \mathcal{K}$, $f \in \mathcal{S}(\mathcal{O}) := \{f \in \mathcal{S}(\mathbf{R}) \,|\, \text{supp}\, f \subset \mathcal{O}\}$.

$$\varphi_{B,n}(f)\,\Omega \;:=\; \lim_{\lambda \downarrow 0} \int_R dx\, f(x)\, U(x)\, D(\lambda)\, \lambda^{-n}\, P_n\, B\,\Omega$$

$$= \sum_{m=1}^{m_n} \left(\int_R dx\, f_{B,n,m}(x)\right) \Phi_{nm}(f) \tag{19}$$

$$\varphi_{B,n}(f)\, A'\, \Omega \;:=\; A'\, \varphi_{B,n}(f)\,\Omega, \quad A' \in \mathcal{B}(\mathcal{O})'. \tag{20}$$

We see that $D(\varphi_{B,n}(f)) = \mathcal{B}(\mathcal{O})'\Omega$ is a dense domain.

PROPERTIES OF THE FIELDS

Conformal covariance
The field operators transform covariantly under the conformal group G.
Theorem: Let $n \in \mathbf{N}$, $B \in \mathcal{B}_{\text{reg}}$, $g \in G$, $\mathcal{O}, g(\mathcal{O}) \in \mathcal{K}$, $f \in \mathcal{S}(\mathcal{O})$. One easily sees

$$U(g)\, \varphi_{B,n}(f)\, U(g)^{-1} = \varphi_{B,n}(\tilde{U}(g)\, f). \tag{21}$$

Proof: straightforward calculation.

Temperedness
The field operators on the vacuum are as regular as Wightman fields.
Theorem: Let $n \in \mathbf{N}$, $A \in \mathcal{B}_{\text{reg}}$. Then we have

$$\varphi_{A,n}(\cdot)\,\Omega \in B(\mathcal{S}(\mathbf{R}), H). \tag{22}$$

Proof: straightforward calculation.

Hermiticity and closeability
All constructed fields are hermitean. Every field operator is shown to be closeable. Closeability is the very test for a definition of unbounded field operators to be physically and mathematically sensible.
Theorem: Let $n \in \mathbf{N}$, $A \in \mathcal{B}_{\text{reg}}$, $\mathcal{O} \in \mathcal{K}$, $f \in \mathcal{S}(\mathcal{O})$, $B', C' \in \mathcal{B}(\mathcal{O})'$. Then we have

$$(B'\Omega, \varphi_{A,n}(f)\, C'\Omega) = (\varphi_{A^*,n}(\bar{f})\, B'\Omega, C'\Omega), \tag{23}$$

$$\varphi_{A,n}(f)^\dagger := \varphi_{A,n}(f)^* |_{\mathcal{B}(\mathcal{O})'\Omega} = \varphi_{A^*,n}(\bar{f}). \tag{24}$$

$\varphi_{A,n}(f)$ is closeable because $\varphi_{A,n}(f)^*$ has a dense domain.
Proof: The Casimir operator C_G belonging to the Lie-group $G = SL(2,\mathbf{R})/\mathbf{Z}_2$ and its representation $U(\cdot)$ is known to have the following spectral decomposition[23]:

$$C_G = \sum_{i=1}^{\infty} i(i-1)\, P_i.$$

Let $B_1, B_2 \in \mathcal{B}_{\text{reg}}$ local observables from causally disjunct domains. As C_G is a second order differential operator in G and B_1, B_2 commute we get

$$(B_1\,\Omega, C_G\, B_2\,\Omega) = (C_G\, B_2^*\,\Omega, B_1^*\,\Omega).$$

This is not known to be true for the single projectors P_n, which are not local.

Some algebra leads to

$(B'\Omega, \varphi_{A,n}(f) C'\Omega)$
$$= \lim_{\lambda \downarrow 0} \int_R dx\, f(x)\, (C'^* B'\Omega, U(x) D(\lambda) \lambda^{-n} P_n A\Omega)$$
$$= \lim_{\lambda \downarrow 0} \int_R dx\, f(x)\, (C'^* B'\Omega, U(x) D(\lambda) \lambda^{-n} \left(\prod_{i=1}^{n-1} \frac{C_G - i(i-1)}{n(n-1) - i(i-1)}\right) A\Omega)$$
$$= \lim_{\lambda \downarrow 0} \int_R dx\, f(x)\, (U(x) D(\lambda) \lambda^{-n} \left(\prod_{i=1}^{n-1} \frac{C_G - i(i-1)}{n(n-1) - i(i-1)}\right) A^*\Omega, B'^* C'\Omega)$$
$$= (\varphi_{A^*,n}(\bar{f}) B'^*\Omega, C'\Omega).$$

Affiliation to the local algebras
The (minimal) closures of the field operators are affiliated to the algebras of local observables.
   Theorem: Let $n \in \mathbb{N}$, $B \in \mathcal{B}_{\text{reg}}$, $\mathcal{O} \in \mathcal{K}$, and $f \in \mathcal{S}(\mathcal{O})$. Then $\overline{\varphi_{B,n}(f)} = \varphi_{B,n}(f)^{**}$ is affiliated to $\mathcal{B}(\mathcal{O})$, i.e. (see ref. 5)

$$\overline{\varphi_{B,n}(f)} \,\eta\, \mathcal{B}(\mathcal{O}) \tag{25}$$

Proof: straightforward calculation.

Cyclicity
There is a countably infinite set of orthogonal fields, which has the vacuum as a cyclic vector.
   Theorem: $\forall n \in \mathbb{N}_o\ \exists M_n \subset \mathcal{B}_{\text{reg}}$ so that

$$\overline{< \mathcal{F}\Omega >}_\mathbb{C} = H \quad \text{with} \quad \mathcal{F} := \bigcup_{n \in \mathbb{N}_o} \{\varphi_{B,n}(f)\,|\, B \in M_n \cup M_n^*,\, f \in \mathcal{S}(\mathbf{R})_{lok}\}. \tag{26}$$

Proof: Separability of H and straightforward calculation.

The Reeh-Schlieder theorem for $\mathcal{F}$
A field theory defined by $\mathcal{F}$ fulfills the Reeh-Schlieder theorem known from Wightman theory[30].
   Theorem: Let $\{\} \neq \mathcal{O} \subset \mathbf{R}$. Then we get

$$\overline{< \mathcal{F}(\mathcal{O})\Omega >}_\mathbb{C} = H \quad \text{with} \quad \mathcal{F}(\mathcal{O}) := \bigcup_{n \in \mathbb{N}_o} \{\varphi_{B,n}(f)\,|\, B \in M_n \cup M_n^*,\, f \in \mathcal{S}(\mathcal{O})\}. \tag{27}$$

Proof: similar to the proof in Wightman theory.

PCT-covariance on the vacuum
We can introduce an unique antiunitary and idempotent operator Θ, but we can't give him a geometric interpretation away from the vacuum as long as we don't yet know his action on the local algebras. This will be possible later.
   Theorem: Let $n \in \mathbb{N}$, $B \in M_n$, $f \in \mathcal{S}(\mathbf{R})$. Then

$$\Theta\,\Omega := \Omega \quad \text{and} \quad \Theta\,\varphi_{B,n}(f(\cdot))\,\Omega := (-1)^n\, \varphi_{B^*,n}(\bar{f}(-\cdot))\,\Omega \tag{28}$$

defines an antiunitary and idempotent operator which commutes PCT-covariantly with $U(G)$.
   Proof: straightforward calculation.

7 RECONSTRUCTION OF THE ALGEBRAS OF LOCAL OBSERVABLES

Starting from the field operators in $\mathcal{F}$ we will reconstruct a Haag-Kastler net of local algebras with conformal covariance. This new net is included in the one which originally defined our theory and it shares all its properties. To prove the identity between the new and the old net we still need some information to be gained in the next section.

Theorem: Let $\mathcal{O} \in \mathcal{K}$, $\overline{\mathcal{F}(\mathcal{O})} := \{\overline{\varphi} \mid \varphi \in \mathcal{F}(\mathcal{O})\}$. Then

$$\mathcal{C}(\mathcal{O}) := \overline{\mathcal{F}(\mathcal{O})}'' \subset \mathcal{B}(\mathcal{O}) \tag{29}$$

defines a net of von Neumann algebras $\mathcal{C} : \mathcal{O} \to \mathcal{C}(\mathcal{O})$, $\mathcal{O} \in \mathcal{K}$, which transforms covariantly under the positive energy representation $U(G)$, fulfills the Haag-Kastler axioms and all conditions that have been proven for $\mathcal{B} : \mathcal{O} \to \mathcal{B}(\mathcal{O})$, $\mathcal{O} \in \mathcal{K}$, e.g. cyclicity of the vacuum, the Reeh-Schlieder theorem and affiliation of $\mathcal{F}$.

Proof: Functional analysis and straightforward calculation.

Remark: Alternatively and equivalently $\mathcal{C}(\mathcal{O})$ can be defined as the von Neumann algebra generated by polar and spectral decomposition or by the bounded Borel functions of all elements of $\overline{\mathcal{F}(\mathcal{O})}$.

8 GEOMETRIC IDENTIFICATION OF MODULAR STRUCTURES

Tomita-Takesaki theory[32,5] assigns to every pair of a von Neumann algebra $\mathcal{A}$ and a cyclic and separating vector Ψ a closeable, antilinear operator:

$$S_o : A\Psi \mapsto A^*\Psi \quad \text{for all } A \in \mathcal{A}. \tag{30}$$

$S := \overline{S_o}$ has a polar decomposition $S = J\Delta^{1/2}$ and its components fulfill a couple of relations:

$$J = J^*, \quad J^2 = \mathbf{1}, \quad \Delta^{-1/2} = J\Delta^{1/2}J, \quad J\mathcal{A}J = \mathcal{A}'. \tag{31}$$

The set of operators Δ^{it}, $t \in \mathbf{R}$, generate the group of modular automorphisms:

$$\Delta^{it} \mathcal{A} \Delta^{-it} = \mathcal{A}. \tag{32}$$

If Ψ is chosen to be the vacuum vector Ω and $\mathcal{A}$ the algebra of the observables localized on the positive or negative half line $\mathbf{R}_+$ or $\mathbf{R}_-$ we are able to give a geometric identification of the abstractly defined modular structures. Without any a priori information on PCT-invariance or Wightman fields we will be able to regain the results of ref. 6, 2 for Wightman theories on $\mathbf{S}^1$.

Theorem: Let $S_+ = J_+ \Delta_+^{1/2}$ resp. $S_- = J_- \Delta_-^{1/2}$ the operators of Tomita-Takesaki theory assigned to the vacuum vector Ω and the half line algebras $\mathcal{A}_+ := \mathcal{B}(\mathbf{R}_+)$ (or $\mathcal{C}(\mathbf{R}_+)$) resp. $\mathcal{A}_- := \mathcal{B}(\mathbf{R}_-)$ (or $\mathcal{C}(\mathbf{R}_-)$). Let $V(\cdot)$ be the representation of the dilation group. For both Haag-Kastler nets we get

$$J_\pm = \Theta \quad \text{and} \quad \Delta_\pm^{1/2} = V(\pm i\pi). \tag{33}$$

The modular conjugation of the half lines is their reflection, the modular automorphism group is the dilation group.

Proof: We will give you a vague idea of the proof for the positive half line.

Let $n \in \mathbf{N}$, $B \in M_n$ and $\varphi(\cdot) := \varphi_{B,n}(\cdot)$. Let $A \in \mathcal{A}_+$ in the domain of $V(i\pi)$. Let $\mathbf{R}_+ \supset \mathcal{O} \in \mathcal{K}$, $f \in \mathcal{S}(\mathcal{O})$.

Positivity of the energy, analyticity properties, affiliation and dilation covariance lead in the present geometrically simple situation to the following relation:

$$(V(i\pi) A \Omega, \varphi(f) \Omega) = (-1)^n (A \Omega, \varphi(f(-\cdot)) \Omega) = (\Theta A^* \Omega, \varphi(f) \Omega).$$

This suffices to derive with some elementary arguments our result:

$$\Theta V(i\pi) = \overline{\Theta V(i\pi)|_{\mathcal{A}_+ \Omega}} = \overline{S_+|_{\mathcal{A}_+ \Omega}} = S_+ = J_+ \Delta_+^{1/2}.$$

$$J_+ = \Theta \quad \text{and} \quad \Delta_+^{1/2} = V(i\pi).$$

9 HAAG DUALITY AND PCT-INVARIANCE

The geometrical identification of the modular structures assigned to the vacuum vector and half line algebras gives us the lacking input to prove Haag duality, equivalence between the field approach and the algebra approach and the full PCT-theorem.

Theorem: The vacuum representation of any conformally covariant Haag-Kastler net on the real line fulfills Haag duality, i.e. $\mathcal{A}(\mathcal{O})' = \mathcal{A}(\mathcal{O}')$ for all $\mathcal{O} \in \mathcal{K}$.

Proof: Unlike ref. 2,6 we can't use the geometric action of Θ to prove Haag duality. Instead we use the identity of J_+ and J_-. We already knew

$$\mathcal{A}_+ \subset \mathcal{A}'_-, \quad \mathcal{A}_- \subset \mathcal{A}'_+, \quad J_+ \mathcal{A}_+ J_+ = \mathcal{A}'_+, \quad J_- \mathcal{A}_- J_- = \mathcal{A}'_- \quad \text{and} \quad J_+ J_+ = 1 = J_- J_-.$$

With $J_+ = J_-(= \Theta)$ we now get

$$\mathcal{A}'_- = J_- \mathcal{A}_- J_- \subset J_- \mathcal{A}'_+ J_- = J_- J_+ \mathcal{A}_+ J_+ J_- = \mathcal{A}_+ \subset \mathcal{A}'_-$$

which proves

$$\mathcal{A}_+ = \mathcal{A}'_- \quad \text{and} \quad \mathcal{A}_- = \mathcal{A}'_+. \tag{34}$$

Conformal covariance and some elementary geometry then imply Haag duality.

Remark: If observable algebras with unbounded localization domains are defined with translation covariance instead of conformal covariance only essential Haag duality can be proven for the conformal Haag-Kastler net. This corresponds[6,27] to the restriction of a conformal field theory from its space-time S^1 to $\mathbf{R} = S^1/\infty$.

With the result on Haag duality for half line algebras we succeed in proving the full PCT-theorem without any a priori knowledge on pointlike localization.

Theorem: In every Haag-Kastler theory on the real line with conformal covariance holds PCT-invariance, i.e. for all $\mathcal{O} \in \mathcal{K}$ we have

$$\Theta \mathcal{B}(\mathcal{O}) \Theta = \mathcal{B}(-\mathcal{O}). \tag{35}$$

Proof: Let $\mathcal{O} \in \mathcal{K}$, $g \in G$ with $g(\mathbf{R}_+) = \mathcal{O}$ and $\tilde{g} \in G$ with $\tilde{g}(\mathbf{R}_-) = -\mathcal{O}$. We already got $\Theta U(g) = U(\tilde{g}) \Theta$ and $J_\pm = \Theta$. This proves the theorem:

$$\Theta \mathcal{B}(\mathcal{O}) \Theta = U(\tilde{g}) \Theta \mathcal{B}(\mathbf{R}_+) \Theta U(\tilde{g})^{-1} = U(\tilde{g}) \mathcal{B}(\mathbf{R}_-) U(\tilde{g})^{-1} = \mathcal{B}(-\mathcal{O}).$$

Corollary: We then easily derive the PCT-theorem for the fields:
Let $n \in \mathbf{N}$, $B \in M_n$, $f \in \mathcal{S}(\mathcal{O})$. We get

$$\Theta \varphi_{B,n}(f) \Theta = (-1)^n \varphi_{B^*,n}(\bar{f}(-\cdot)). \tag{36}$$

The proof of Haag duality also helps to establish an equivalence of the "field picture" and the "algebra picture" in conformal field theory.

Theorem: Every conformally covariant Haag-Kastler net on the real line can equivalently and without any loss of information be described by a set of pointlike localizable fields $\mathcal{F}$. The net of local observables can be reconstructed, i.e. for all $\mathcal{O} \in \mathcal{K}$ we have

$$\mathcal{B}(\mathcal{O}) = \mathcal{C}(\mathcal{O}). \tag{37}$$

Proof: Inclusion and maximality imply identity. For all $\mathcal{O} \in \mathcal{K}$ we get

$$\mathcal{C}(\mathcal{O}) \subset \mathcal{B}(\mathcal{O}) = \mathcal{B}(\mathcal{O}')' \subset \mathcal{C}(\mathcal{O}')' = \mathcal{C}(\mathcal{O}).$$

10 CONCLUSION AND OUTLOOK

We explicitly constructed pointlike localizable fields as limits of local observables in a framework of conformally covariant Haag-Kastler nets in 1+1 dimensions. In order to do so we had to introduce the concepts of "new scaling" and "local regularization". The field operators turned out to be conformally covariant and closeable. Their minimal closures are affiliated to the local algebras. This sufficed to derive the result of Bisognano and Wichmann[2] on the geometric identification of modular structures. Finally Haag duality in the vacuum sector, the PCT-theorem and the equivalence between "field picture" and "algebra picture" could be proven.

All arguments and proofs did go through without any knowledge or use of products of unbounded field operators. One would like to control these products to understand the operator product expansions of conformal field theory or to simply check the Wightman axioms for our fields. This should be possible by evaluating the representation theoretical information on the conformal three-point functions. One should remember that is was mainly our complete knowledge of the conformal two-point functions which let us control the limits in the definition of the field operators.

A second direction of generalization could be the definition of charged fields by studying the representation theory of the universal covering group $\widetilde{G}$ of $G = SL(2, \mathbf{R})/\mathbf{Z}_2$.

At last we would like to mention our hope that a better understanding of the relation between "field aspects" and "algebra aspects" of the theory might help in the classification programme of conformal quantum field theory.

Acknowledgement

This work is based on a collaboration with K. Fredenhagen and even the parts that have been contributed by the author heavily depended on fruitful common discussions. The author would like to express his gratitude for that.

REFERENCES

[1] A. O. Barut, R. Rączka: "Theory of group representations and applications", (1986).

[2] J. J. Bisognano, E. H. Wichmann, *J. Math. Phys.* 16 (1975) 985-1007; *J. Math. Phys.* 17 (1976) 303-321.

[3] H. J. Borchers, *Commun. Math. Phys.* 134 (1992) 315.

[4] H. J. Borchers, J. Yngvason, *Commun. Math. Phys.* 127 (1990) 607.

[5] O. Bratteli, D. W. Robinson, "Operator algebras and quantum statistical mechanics. I", (1979).

[6] D. Buchholz, H. Schulz-Mirbach, *Rev. Math. Phys.* 2 (1990) 105.

[7] D. Buchholz, *J. Math. Phys.* 31 (1990) 1839-1846.

[8] D. Buchholz, K. Fredenhagen, *J. Math. Phys.* 18 (1977) 1107-1111.

[9] D. Buchholz, K. Fredenhagen, *Commun. Math. Phys.* 84 (1982) 1-64

[10] S. Doplicher, R. Haag, J. E. Roberts, *Commun. Math. Phys.* 13 (1969) 1-23; *Commun. Math. Phys.* 15 (1969) 173-200; *Commun. Math. Phys.* 23 (1971) 199-230; *Commun. Math. Phys.* 35 (1974) 49-85.

[11] W. Driessler, J. Fröhlich, *Ann. Inst. Henri. Poincaré* 27 (1977) 221-236

[12] W. Driessler, S. J. Summers, E. H. Wichmann, *Commun. Math. Phys.* 105 (1986) 49-84.

[13] K. Fredenhagen, *in:* Proc. Schladming 1990.

[14] K. Fredenhagen, *in:* Proc. Leipzig 1991.

[15] K. Fredenhagen, J. Hertel, *Commun. Math. Phys.* 80 (1981) 555-561.

[16] K. Fredenhagen, M. Jörß: in preparation.

[17] I. M. Gel'fand, M. I. Graev, N. Y. Vilenkin, "Generalized Functions V, Integral geometry and representation theory", (1966).

[18] R. Haag, "Local Quantum Physics", (1992).

[19] R. Haag, D. Kastler, *J. Math. Phys.* 5 (1964) 848-861.

[20] M. Jörß, FU Berlin, diploma thesis (1991).

[21] R. Jost, "The general theory of quantized fields", (1963).

[22] J. Kupsch, W. Rühl, B. C. Yunn, *Ann. of Phys.* 89 (1975) 115-148.

[23] S. Lang, "$SL_2(\mathbf{R})$", (1975).

[24] M. Lüscher, *Commun. Math. Phys.* 50 (1976) 23-52.

[25] G. Mack, *Commun. Math. Phys.* 53 (1976) 155.

[26] J. Rehberg, M. Wollenberg, *Math. Nachr.* 125 (1985) 259-274.

[27] C. Rigotti, Université Marseille, preprint (1977).

[28] J. E. Roberts, *in:* Proc. Camerino 1974.

[29] B. Schroer, J. A. Swieca, A. H. Völkel, *Phys. Rev.* D11 (1974) 1509.

[30] R. Streater, A. S. Wightman, "PCT, Spin & Statistics and all that", (1964).

[31] S. J. Summers, *Helv. Phys. Act.* 60 (1987) 1004.

[32] M. Takesaki, "Tomita's theory of modular Hilbert algebras and its applications", (1970).

[33] M. Wollenberg, *Math. Nachr.* 128 (1986) 287-298; Akad. d. Wiss. Berlin, preprint P-Math-12/87; *J. Math. Phys.* 29 (1988) 2106-2111.

[34] J. Yngvason, *in:* Proc. Swansea 1988.

THE PHASE SPACE OF THE WESS-ZUMINO-WITTEN MODEL

G. Papadopoulos[1] and B. Spence[2]

[1]Department of Physics
Queen Mary and Westfield College
London E1 4NS

[2]Blackett Laboratory
Imperial College
London SW7 2BZ

ABSTRACT

We prove that the covariant and Hamiltonian phase spaces of the Wess-Zumino-Witten model on the cylinder are diffeomorphic and we derive the Poisson brackets of the theory.

INTRODUCTION

The Wess-Zumino-Witten (WZW) model [1] is a two-dimensional non-linear sigma model with Wess-Zumino term whose target manifold is a compact, connected Lie group G, and it is a fundamental conformal field theory. Recently there have been a number of proposals for deriving a quantum group structure in this model [2], [3], [4], [5], [6], [7], [8], [9]. As a consequence of this work, there has been interest in the phase space structure of the WZW model [8], [9], [10]. For example, the authors of Ref. [9] have argued that the quantisation of the Poisson bracket algebra of a set of variables leads to an exchange algebra. However, there has been little study of the *global* structure of the phase space of the WZW model. This global structure is crucial

---

[1] Address from Oct. 1st 1992: Dept. Mathematics, Kings College, London WC2R 2LS.
[2] Address from Oct. 1st 1992: Dept. Physics, University of Melbourne, Victoria 3052 Australia.

to the definition of the symplectic form and the derivation of the Poisson brackets of the theory. We will discuss this in the following. Some of this work has also been reported in Ref. [11]

There are two approaches to the definition of the phase space of a field theory. The first is the Hamiltonian definition. In this approach, the space-time is a topological product $M = \Sigma \times \mathbf{R}$ and the momentum of the theory is defined as follows:

$$P_i = \frac{\partial L(X)}{\partial \partial_t X^i}. \tag{1}$$

X are the fields and L is the Lagrangian of the theory quadratic in the time t derivatives of the fields. If there are no constraints in the theory, the Hamiltonian phase space P_H is the co-tangent bundle T^*Q of the configuration space Q of the model. The Poisson brackets of the theory are

$$\{X^i(x), P_j(y)\} = \delta^i_j(x,y), \tag{2}$$

where $x, y \in \Sigma$. These brackets correspond to the standard symplectic structure of T^*Q. To describe the other definition of the phase space of a field theory, we start from the Lagrangian L and define the symplectic current

$$S^\mu = \delta X^i \, \delta \frac{\partial L}{\partial \partial_\mu X^i}. \tag{3}$$

This current is conserved subject to the Lagrangian equations of motion of the theory. The symplectic form is

$$\Omega = \int_\Sigma S_\mu \, d\Sigma^\mu. \tag{4}$$

Ω is closed and it does not depend on the choice of the Cauchy surface Σ (see Ref. [12] for discussions of the Lagrangian approach to the phase space). If the theory does not have constraints, it is possible to relate the Lagrangian and Hamiltonian definition of the symplectic form by a Legendre transform. One way to parameterise the symplectic form Ω is in terms of the initial data $f(x) = X(x,0)$ and $w(x) = (\partial_t X)(x,0)$. However, if the solutions of the Lagrangian equations of motion are known, then it is possible to parameterise Ω in terms of the parameters of the solutions of the theory. The space of solutions of the Lagrangian equations of motion of a theory equipped with the symplectic form of eqn (4) is called the covariant phase space P_C (see Ref. [13] for recent discussions). The covariant phase space is particularly suited to the study of field theories where we know the space of classical solutions, such as in the case of the WZW model.

In the following, we will give a new parameterisation of the space of solutions of the WZW model, and use this to prove that the Hamiltonian and covariant phase spaces are diffeomorphic. The Poisson brackets of the theory will also be derived. Finally, we will compare our approach with other formulations that have appeared in the literature.

THE WZW MODEL

In the Hamiltonian approach, one may consider the WZW model as a two-dimensional non-linear sigma model with Wess-Zumino term, whose target space is a group manifold G. Applying the usual Hamiltonian analysis to this sigma model action, one finds directly that the phase space P_H of the WZW model is the co-tangent bundle of its configuration space LG, i.e. $P_H = T^*LG$ (see Ref. [14]).

Now we consider the covariant approach to the phase space. The symplectic form of the WZW model is (see Ref. [9], for example)

$$\Omega = -\frac{\kappa}{8\pi} \int_0^1 dx \, \mathrm{tr} \left((g^{-1}\delta g)\partial_+(g^{-1}\delta g) - (\delta g \, g^{-1})\partial_-(\delta g \, g^{-1}) \right), \tag{5}$$

where κ is the coupling constant of the model, (x,t), $0 \leq x < 1$, $-\infty < t < +\infty$ are the co-ordinates of the cylinder $S^1 \times \mathbf{R}$ and $x^\pm = x \pm t$. This symplectic form is closed and time independent (we take $t = 0$ in the following).

One way to parameterise Ω is in terms of the initial conditions $f(x) = g(x,0)$ and $w(x) = (g^{-1}\partial_t g)(x,0)$ on the Cauchy surface $t = 0$. In terms of the functions f and w, the symplectic form Ω becomes

$$\Omega = -\frac{\kappa}{8\pi} \int_0^1 dx \, \mathrm{tr} \left(\frac{1}{2} f^{-1}\delta f \partial_x(f^{-1}\delta f) - \frac{1}{2}\delta f f^{-1} \partial_x(\delta f f^{-1}) \right. \\ \left. + (f^{-1}\delta f)^2 w + f^{-1}\delta f \delta w \right). \tag{6}$$

To construct the covariant phase space P_C, we should introduce a parameterisation of the space of solutions of the WZW model. The equations of motion are

$$\partial_-(\partial_+ g \, g^{-1}) = 0. \tag{7}$$

The equations of motion are invariant under the semi-local transformations $g \to l(x^+) \, g \, r^{-1}(x^-)$ and the corresponding currents are

$$J_+ = -\frac{\kappa}{4\pi} \partial_+ g g^{-1}, \qquad J_- = \frac{\kappa}{4\pi} g^{-1} \partial_- g. \tag{8}$$

There are several suggestions in the literature as to how to parameterise the space of solutions of the WZW model. In the following section we will discuss the parameterisation of Ref. [11], and in the last section we will compare this parameterisation with others in the literature.

THE POISSON BRACKETS

In Ref. [11], we have parameterised the space of solutions to the field equations of the WZW model as follows:

$$g(x,t) = U(x^+) \mathcal{W}(A; x^+, x^-) V(x^-),$$
$$\mathcal{W}(A; x^+, x^-) = P \exp \int_{x^-}^{x^+} A(s) ds, \tag{9}$$

where U and V are periodic maps from the real line $\mathbf{R}$ to the group G, and the field A in the path-ordered exponential is a $(\text{Lie}G)^*$-valued periodic one-form on the real line. The expression for $g(x,t)$ in Eqn. (9) is then periodic in x and solves the field equations (7). To prove the periodicity of g, it is enough to show that $\mathcal{W}(A; x^+ + 1, x^- + 1) = \mathcal{W}(A; x^+, x^-)$. One way to verify this is to use the power series expansion of $\mathcal{W}$ and change variables. This gives $\mathcal{W}(A; x^+ + 1, x^- + 1) = \mathcal{W}(\hat{A}; x^+, x^-)$ where $\hat{A}(x) = A(x+1)$. Using the periodicity of A we can then prove that g is periodic. An alternative way is to use the formula $\mathcal{W}(A; x^+ + 1, x^- + 1) = m(x^+)\mathcal{W}(A; x^+, x^-)m^{-1}(x^-) = \mathcal{W}(A^m; x^+, x^-)$, where $m(x) = \mathcal{W}(A; x+1, x)$ and $A^m = \partial_x m(x) m^{-1}(x) + m(x)A(x)m^{-1}(x)$. However $\partial_x m(x) = A(x+1)m(x) - m(x)A(x)$, and from the periodicity of A we get $A^m = A$. This again proves the periodicity of g.

Next, using the parallel transport equation

$$\partial_s \mathcal{W}(A; s, x^-) = A(s)\,\mathcal{W}(A; s, x^-), \qquad (10)$$

we can prove that g in eqn. (9) satisfies the equations of motion of the WZW model. Choosing a point x_0 on the real line, we can write the solution given in eqn. (9) in a chirally factorised form $g(x,t) = u(x^+, x_0)v(x^-, x_0)$, by using the identity $\mathcal{W}(A; x^+, x^-) = \mathcal{W}(A; x^+, x_0)\mathcal{W}(A; x_0, x^-)$. However, this factorisation depends on the choice of the point x_0.

Inserting the solution (9) into the symplectic form (5) gives

$$\Omega = -\frac{\kappa}{8\pi}\int_0^1 dx\, \text{tr}\left((U^{-1}\delta U)\partial_x(U^{-1}\delta U) + 2(U^{-1}\delta U)^2 A + 2(U^{-1}\delta U)\delta A \right.$$
$$\left. - (\delta VV^{-1})\partial_x(\delta VV^{-1}) - 2(\delta VV^{-1})^2 A + 2(\delta VV^{-1})\delta A\right). \qquad (11)$$

The solution g of the WZW equations of motion given in the parameterisation (9) is invariant under the transformations

$$U(x) \to U(x)h(x), \quad V(x) \to h^{-1}(x)V(x),$$
$$A(x) \to -h^{-1}(x)\partial_x h(x) + h^{-1}(x)A(x)h(x), \qquad (12)$$

where $h \in LG$. To prove this, we observe that under these transformations $\mathcal{W}(A; x^+, x^-) \to h^{-1}(x^+)\mathcal{W}(A; x^+, x^-)h(x^-)$. The phase space P_C of the WZW model is then the space of fields $\{U, V, A\}$, modulo the transformations (12). This is $\frac{LG \times LG \times \mathcal{A}}{LG}$ where $\mathcal{A}$ is the space of G-connections over the circle. [3] This is diffeomorphic to T^*LG, i.e. it is the same as the phase space P_H derived from the Hamiltonian treatment of the theory.

The symplectic form (11) is degenerate along the directions of the action (12) of the loop group LG. We may deal with this by gauge-fixing or by enhancing the phase space of the theory and then imposing constraints [11]. In the following, we will use the gauge-fixing method. We may choose as a gauge fixing condition $U = e$ where

---

[3] There two ways to think about the field A. The first is as a periodic one-form on the real line, valued in $(\text{Lie}G)^*$. The second is as a connection over the circle S^1. We have identified $\text{Lie}G$ with $(\text{Lie}G)^*$ using the invariant metric.

e is the identity element of the loop group LG. This is a good gauge choice, as LG acts freely and transitively on the space of U's, $\{U\}$. The symplectic form (11) then becomes

$$\Omega = -\frac{\kappa}{8\pi} \int_0^1 dx \, \text{tr} \left(-(\delta VV^{-1})\partial_x(\delta VV^{-1}) - 2(\delta VV^{-1})^2 A + 2(\delta VV^{-1})\delta A \right). \quad (13)$$

This symplectic form is not degenerate and is invertible.[4]

The simplest way to invert the form (13) is to first rewrite it in terms of a local parameterisation $X^i(x)$ for the maps V ($V = V(X)$). This gives

$$\Omega = -\frac{\kappa}{8\pi} \int_0^1 dx \left(-(R_i^a \delta X^i)\partial_x(R_j^a \delta X^j) - f_{ab}{}^c R_i^a R_j^b A_c \, \delta X^i \, \delta X^j + 2R_i^a \, \delta X^i \delta A_a \right), \quad (14)$$

where $\delta VV^{-1} = R^a t_a$. The remarkable feature of this expression for the form Ω is that one does not need to invert any differential operator in order to invert the form (c.f. Refs. [9],[10]), where in order to invert the symplectic form it was necessary to find the inverse of the operator ∂_x on the circle). The gauge $U = e$ parameterises the symplectic form on T^*LG in terms of the right trivialisation of T^*LG and the gauge $V = e$ parameterises the same symplectic form in terms of the left trivialisation. The inversion of the form (14) is straightforwardly carried out, and leads to the Poisson brackets ($\beta = -\frac{4\pi}{\kappa}$)

$$\{X^i(x), X^j(y)\} = 0,$$
$$\{X^i(x), A_a(y)\} = \beta R_a^i X(x)\delta(x,y), \quad (15)$$
$$\{A_a(x), A_b(y)\} = \beta \left(\delta_{ab}\partial_x + f_{ab}{}^c A_c(x) \right) \delta(x,y),$$

where $\delta(x,y)$ is the delta function on S^1.

Using Eqn. (15), we can calculate Poisson brackets involving V and A – for example $\{V(x) \overset{\otimes}{,} V(y)\} = 0$, $\{V(x), A_a(y)\} = \beta V(x)t_a \delta(x,y)$. In this gauge, the WZW currents (8) become $J_+ = -\frac{\kappa}{4\pi}A$, $J_- = \frac{\kappa}{4\pi}(V^{-1}\partial_x V - V^{-1}AV)$, and it can be verified by a straightforward calculation that their Poisson bracket algebra is isomorphic to two commuting copies of a Kac-Moody algebra with a central extension.

DISCUSSION

We would now like to discuss how our results relate to other work in the literature. In Ref. [1], Witten proposed that the general solution of the WZW equations of motion is $g(x,t) = U(x^+)V(x^-)$ where U and V are maps from the circle into the group G. In this parameterisation the covariant phase space is $LG \times LG$ where LG is the loop group of the group G. Recently, another parameterisation of the space of solutions of the WZW model was proposed by Chu et al in Ref. [9]. In this parameterisation the solution of the WZW model factorises as $g(x,t) = u(x^+)v(x^-)$.

---

[4] Notice that if we set $A = \frac{1}{2}(\partial_x f f^{-1} + fwf^{-1})$ and $V = f$ in the symplectic form (6) then we get the symplectic form (13).

However, the functions u and v are not periodic and satisfy the conditions $u(x+1) = u(x)M$, $v(x+1) = M^{-1}v(x)$, where $M \in G$ is the monodromy. In addition, the authors of Ref. [9] observed that the solution $g = uv$ of the WZW model remains invariant under the action $u(x) \to u(x)k$, $v(x) \to k^{-1}v(x)$, $M \to k^{-1}Mk$ of the group G, with $k \in G$. An equivalent way to write the solution g of the WZW model in this parameterisation is $g(x,t) = U(x^+)M^{2t}V(x^-)$, where U, V are periodic. The covariant phase space of the WZW model in the parameterisation of Ref. [9] is $\frac{LG \times LG \times G}{G}$. Neither this covariant phase space, nor that of Ref. [1], are diffeomorphic to the Hamiltonian phase space T^*LG of the WZW model – for example, $\pi_2(LG \times LG) = \pi_2(\frac{LG \times LG \times G}{G}) = \mathbf{Z} \oplus \mathbf{Z}$, whereas $\pi_2(T^*LG) = \mathbf{Z}$ (for G simple and simply connected).

To relate the parameterisation of the space of solutions of the WZW model of Ref. [9] to the parameterisation of Section Three, the symmetries of the space of solutions of the WZW model (Eqn. (12)) can be treated by choosing the gauge-fixing conditions to be different from those considered in Section Three above. For example in the parameterisation (9) one can gauge-fix the connection A so that it is a *constant* connection over the circle. The residual transformations for this gauge-fixing are the constant gauge transformations. The constant gauge transformations are parameterised by the elements of the group G and they act on the parameters of the solutions as $U \to Uk$, $V \to k^{-1}V$ and $A \to k^{-1}Ak$ where $k \in G$ and A is a constant connection. This parameterisation is the one given in Ref. [9]. The resulting phase space of the theory is $\frac{LG \times LG \times G}{G}$, if we parameterise the solutions in terms of the monodromy M of the connection A. The reason that this phase space is different from T^*LG is that there is a Gribov-Singer ambiguity associated with this gauge fixing; note that this is the one-dimensional analogue of the four-dimensional Yang-Mills Gribov-Singer ambiguity [15]. The k-symmetry just mentioned can be further gauge fixed by choosing A to be in the Cartan subalgebra $\mathbf{h}$ of Lie$G$, and this parameterisation was used to calculate the Poisson brackets of this theory in Refs. [9], [10]. Finally, the parameterisation of Ref. [1] corresponds to the gauge-fixing choice $A = 0$. However, not all connections can be brought into this form by a gauge transformation.

An alternative way to compare the different parameterisations of the space of solutions of the WZW model is to study the initial values $f(x) = g(x,0)$ and $w(x) = (g^{-1}\partial_t g)(x,0)$ of the field g that correspond to these parameterisations. In general f and w are independent functions. It is easy to see that the parameterisation of eqn. (9) corresponds to the most general Cauchy data of the theory. However, the parameterisations of Refs. [9] and [1] are associated to a *subset* of the available Cauchy data of this theory. The restriction that the parameterisation of Ref. [9] imposes on the Cauchy data is that the holonomy of the connection $\hat{w} = \frac{1}{2}(f^{-1}\partial_x f - w)$ does not depend on the point chosen to evaluate it, i.e. $\partial_x \mathcal{W}(\hat{w}; x+1, x) = 0$. The constraint that the parameterisation of Ref. [1] imposes on the Cauchy data is that the holonomy of the connection $\hat{w}$ on the circle S^1 must be the identity group.

In conclusion, the parameterisation of the solutions of the WZW model given in Section Three (Eqn. (9)) is general in the sense that it is invariant under a larger symmetry than other parameterisations considered in the literature, and the latter can be thought of as locally-valid gauge-fixed versions of it. In our parameterisation, the covariant canonical phase space of the WZW model is diffeomorphic to the

Hamiltonian phase space of the theory, and the calculation of the Poisson brackets is straightforward.

Acknowledgements

This work was supported by the SERC.

REFERENCES

[1] E. Witten, Commun. Math. Phys. **92** (1984) 455.
[2] B. Blok, Phys. Lett. **233B** (1989) 359.
[3] A.Yu. Alekseev and S. Shatashvili, Commun. Math. Phys. **128** (1990) 197; **133** (1990) 353.
[4] L.D. Faddeev, Commun. Math. Phys. **132** (1990) 131.
[5] J. Balog, L. Dąbrowski and L. Fehér, Phys. Lett. **244B** (1990) 227.
[6] G. Felder, K. Gawędski and A. Kupiainen, Nucl. Phys. **B299** (1988) 355; Commun. Math. Phys. **117** (1988) 127.
[7] K. Gawędski, Commun. Math. Phys. **139** (1991) 201.
[8] G. Bimonte, P. Salomonson, A. Simoni and A. Stern, Int. J. Mod. Phys. A **7** (1992) 6159.
[9] M.F. Chu, P. Goddard, I. Halliday, D. Olive and A. Schwimmer, Phys. Lett. **B266** (1991) 71.
[10] G. Papadopoulos and B. Spence, Phys. Lett. **B292** (1992) 321.
[11] G. Papadopoulos and B. Spence, Phys. Lett. **B295** (1992) 44.
[12] R. Abraham and J.E. Marsden, *Foundations of Mechanics*, Benjamin/Cummings Publishing Company, 1978.
[13] G. Zuckerman, in *Mathematical Aspects of String Theory*, ed. S.-T. Yau, World Scientific, Singapore 1987;
Č. Crnković and E. Witten, in *Three Hundred Years of Gravitation*, eds. S.W. Hawking and W. Israel, C.U.P. Cambridge, 1987.
[14] I. Bakas and D. McMullan, Phys. Lett. **B189** (1987) 141.
[15] V.N. Gribov, Nucl. Phys. **B139** (1978) 1;
I. Singer, Commun. Math. Phys. **60** (1978) 7.

REGULARIZATION AND RENORMALIZATION OF CHERN-SIMONS THEORY

G. Giavarini[1], C. P. Martin[2] and F. Ruiz Ruiz[3]

[1]L.P.T.H.E., Universités Paris VI-VII,
Place Jussieu, 75251 Paris Cedex 05, France.
[2]Department of Mathematics and Statistics,
University of Guelph,
Guelph, Ontario N1G 2W1, Canada.
[3]Niels Bohr Institute, University of Copenhagen,
Blegdamsvej 17, DK-2100 Copenhagen Ø, Denmark.

The topological field theory most familiar both to physicists and to mathematicians is surely Chern-Simons theory. The classical Chern-Simons euclidean action, for a principal G-bundle P over an oriented three manifold M, is given by

$$S_{\text{CS}}[A] = -\frac{ik}{4\pi} \int_M \text{Tr}\left(A \wedge dA + \frac{2}{3} A \wedge A \wedge A\right), \tag{1}$$

where A is the gauge connection on $P \to M$ and Tr represents a G-invariant bilinear form on $\mathfrak{g}$, the Lie algebra of G. In the following we shall assume $G = SU(N)$ and A taking values in the fundamental representation of $\mathfrak{su}(N)$, with the (antihermitian) generators T^a normalized so that $\text{Tr}(T^a T^b) = \frac{1}{2}\delta^{ab}$. The field theory with action (1) is "topological" in the sense that it does not depend on the metric chosen on M and general covariance is thus manifest.

The action (1) is invariant under the infinitesimal gauge transformation $A \to A + D\eta$, where η is a $\mathfrak{su}(N)$ valued function on M and $D = d + A$ is the covariant derivative with respect to A. The field theory described by $S_{\text{CS}}[A]$ is then particularly appealing since it provides an alternative to the usual Yang-Mills action for describing gauge fields in three dimensions. It is by now a well established fact that matter fields coupled to a Chern-Simons gauge field give rise to particles with fractional statistics (the so called anyons) that might be relevant for the description of phenomena such as the fractional quantum Hall effect or high T_c superconductivity[1]. In the following however, our main concern will be the pure gauge theory with action given by (1).

As usual in field theory, one is interested in the computation of the vacuum expectation value of observables. It is clear that gauge invariant local operators are not necessarily generally covariant and consequently they are not appropriate for the case at hand. One must then look for non-local, metric independent and gauge invariant objects. Physicists working on more standard gauge theories (like the old good QCD in 4 dimensions) have been for a long time dealing with the computation of observables having the above properties. They called these observables "Wilson loops", but mathematicians seem to prefer the name of "holonomies" of the connection A. Given a closed curve C in M, a Wilson loop is defined as

$$W(C) = \text{Tr}\, P \exp \oint_C A \ .$$

The expectation value of a collection $\{C_i\}$ of Wilson loops is given by the Feynman path integral

$$Z(M; C_i) = \int \mathcal{D}A \prod_i W(C_i) e^{-S_{\text{CS}}[A]} \ . \tag{2}$$

In the simple case in which no Wilson loop appears in (2) one gets the partition function $Z(M)$ of the theory.

Owing to the intrinsically topological character of $S_{\text{CS}}[A]$ one expects the functors $Z(M; C_i)$ to give back topological invariants of M and $\{C_i\}$. To make sure this is indeed the case we might content ourselves with the study of the formal properties of the path integral (2). However, if some explicit result has to be obtained, one cannot avoid to give concrete meaning to eq.(2). This in practice signifies that we must quantize Chern-Simons theory (CST). Taking $M = S^3$, it was shown in ref. 2, by using input results coming from conformal quantum field theory, that the expectation values of Wilson loops satisfy the same skein relation as the Jones polynomials[3] and can be identified with the latter. More precisely, $Z(S^3; C)/Z(S^3)$ with the standard framing is the Jones polynomial for C in the variable $q = \exp\{2\pi i/(k + \text{sign}(k)c_V)\}$, where k is the classical parameter appearing in (1) and c_V is the second Casimir operator in the adjoint representation of $SU(N)$. With our normalization choice for the group generators, c_V is simply N. In this way CST provides a three dimensional framework for computing knot invariants. The partition function itself is a topological invariant of M, as explicitly checked in ref. 4.

The fact that the observables of CST are functions of the combination $k + c_V$, rather than k, has a clear quantum mechanical interpretation. It relies on the precise correspondence between CST on M and the Wess-Zumino-Witten model on ∂M, as first displayed in ref. 2. Indeed, k can be identified with the level of the affine $SU(N)$ current algebra on ∂M. Then the Sugawara form of the energy-momentum tensor implies the shift

$$k \to k + \text{sign}(k)c_V \ . \tag{3}$$

We shall discuss in a moment the reasons leading to quantization conditions for the parameter k in CST which, just like the central charge in the representation theory of affine algebras, must take integer values.

One may think of getting a clue to the understanding of the shift (3) without resorting to the aid of conformal field theory. In a semiclassical quantization of CST, eq.(3) can be recovered as the radiative correction to the parameter k at order $\hbar$ (see ref. 2). From this simple remark, it should appear clear the relevance of perturbation theory to achieve a complete understanding of the quantum theory defined by (1). Perturbation theory, moreover, provides a very direct tool for obtaining intrinsically three dimensional integral representations of knot and link invariants that generalize Gauss' formula for the linking number of two closed curves, as illustrated in ref. 5.

Quite a number of papers dealing with the issue of perturbative quantization of CST have already appeared [6-10]. This has led to some controversy about the exact meaning of the shift (3) in the perturbative framework. Our main concern here will be to provide a sound perturbative setting for the quantization of CST. A careful analysis of some of the features of the perturbative approach will result in a proper understanding of eq. (3). Our study passes through the determination of the effective action $\Gamma[A]$, the quantum analogue of the classical $S_{CS}[A]$. The effective action is the generating functional of the 1PI Green functions and therefore the starting point for the computation of any other quantum observable.

We have already mentioned the invariance of $S_{CS}[A]$ under infinitesimal gauge transformations. Acting with a *finite* gauge transformation h, $S_{CS}[A]$ is transformed (apart from surface terms) into[11] $S_{CS}[{}^hA] = S_{CS}[A] + 2\pi k\, w(h)$. The quantity $w(h)$ is an integer, being the winding number of h. In order to have a single valued partition function, k must obey the quantization condition $k \in \mathbb{Z}$. Notice that this is a non-perturbative requirement, however, since finite gauge transformations lie beyond the perturbative regime. Furthermore, at the quantum level, the same requirement of monodromy should not be imposed on the classical k but rather on its quantum or renormalized counterpart. We shall come back to this point later on.

To quantize perturbatively CST one must first fix a gauge. We shall work in the Landau gauge, which is known to be free of infrared divergences. With the standard Faddeev-Popov construction this amounts to adding to $S_{CS}[A]$ the term

$$S_{GF} = 2 \int_M d^3x\, \sqrt{g}\, \mathrm{Tr}\left[(J^\mu - \partial^\mu \bar{c})D_\mu c - b\partial_\mu A^\mu - \frac{1}{2}H[c,c]\right].$$

As customary, b denotes the Lagrange multiplier for the gauge condition $\partial_\mu(\sqrt{g}A^\mu) = 0$, c and $\bar{c}$ are Faddeev-Popov ghosts and J and H are auxiliary fields introduced for later convenience. The relevant aspect of S_{GF} is that it necessarily picks a metric g on M. Thus the resulting gauge-fixed action $S = S_{CS} + S_{GF}$ is no longer gauge invariant nor metric independent. The action is, however, BRS invariant. As S_{GF} is a pure BRS variation, it is not observable. Therefore, at least at the classical level, not only gauge invariance but also the topological character of the theory is recovered. One might then wonder if the same result holds true for the quantized theory.

To carry out our quantization program we choose to work on $M = \mathbb{R}^3$ endowed with the flat metric $g_{\mu\nu} = \delta_{\mu\nu}$. Although by naive power counting the theory appears renormalizable, it is in fact ultraviolet (UV) finite[12,13], that is, the beta function and the anomalous dimensions of the fields vanish to all orders in perturbation theory. This result should not be very surprising since CST on S^3, being topological, has no physical local excitations. It was subsequently questioned if UV finiteness of CST implied the vanishing of radiative corrections to the parameter k, something which would be in disagreement with the well-established non-perturbative result (3). We shall see that this is not going to be the case.

UV perturbative finiteness of Green functions does not imply that the corresponding Feynman diagrams should also be finite. On the contrary, since power counting predicts divergences for individual diagrams, the expected scenario is a cancellation of the divergences order by order in perturbation theory when summing over all diagrams contributing to a given Green function. Although divergences cancel out in the final answer, for practical computational reasons the divergent integrals must be made mathematically manageable by means of a suitable regulator. In this respect CST does not differ from ordinary renormalizable theories.

Let us denote with Λ the regulator (or set of regulators) needed to regularize CST at any perturbative order. Since the theory is UV finite, the limit $\Lambda \to \infty$ in which the regulator is removed defines a "minimal" renormalization scheme where renormalized quantities equal bare ones. In this scheme, that we call renormalized=bare, the renormalized effective action Γ and Green functions are defined as

$$\Gamma = \lim_{\Lambda \to \infty} \Gamma_\Lambda \, , \qquad G(p_1, \ldots, p_E) = \lim_{\Lambda \to \infty} G_\Lambda(p_1, \ldots, p_E) \, , \qquad (4)$$

where Γ_Λ and G_Λ are the corresponding regularized quantities. The value of the Green functions so obtained depends in principle on the particular regulatization employed and so does the value of the observables of the theory. However, we shall see that all the known regularization methods satisfying certain invariance requirements lead to observable one-loop radiative corrections that reproduce the non-perturbative results. Thus, these invariance requirements, along with the renormalization scheme above, provide a "natural" parametrization of the perturbative theory.

The BRS symmetry of the gauge fixed action is what is left of the original symmetries of the theory. Classically, the cohomology of the BRS operator guarantees that, when computing observables, the original symmetries still hold true. It is then clear that if we set some hope on obtaining the same picture at the quantum level, we cannot but rely on the BRS symmetry, which is known to be non anomalous in this case[12,13]. We then say that BRS is *fundamental*, meaning with this that it must be enforced at the quantum level in order to make sense out of the quantum theory. It will appear that for CST in the Landau gauge, once BRS symmetry has been implemented on the renormalized theory, the parameter k occurs to be the only actual free parameter. The "natural" parametrization choice is then in terms of the bare (or classical) k. This does not specify completely a renormalization scheme because finite renormalizations of the fields are still allowed. However, since the value of the observables is unchanged under finite wave function rescalings, the quantum theory is unambiguously defined. Actually, we shall make recourse to this freedom of rescaling the fields when comparing Green functions obtained with different regulators. The simplest way to make sure that we get a BRS invariant quantum theory is to start with a BRS invariant regulator so that the effective action we get from (4) in the scheme renormalized=bare satisfies automatically the BRS identity. It will turn out that the observables as functions of the bare k are the same functions of $k + c_V$ for all BRS invariant regularization schemes. Obviously, the agreement between perturbative and non-perturbative results in this case originates from the thorough gauge invariance of both quantization methods. It is worth mentioning that with explicitly BRS breaking regularizations, such as the one in ref. 6, the shift (3) is not observed as consequence of the loss of BRS invariance at the regularized level, despite the latter is restored when the regulator is removed.

We then move on to the computation of the most general BRS invariant effective action. It is convenient to introduce the functional

$$\bar{\Gamma} = \Gamma + 2 \int d^3x \; \text{Tr}(b \, \partial_\mu A^\mu)$$

which, owing to the Landau gauge condition and the antighost equation, is independent of b and depends on J_μ and $\bar{c}$ only through the combination $G_\mu = J_\mu - \partial_\mu \bar{c}$ (see ref. 8.) The BRS identity for $\bar{\Gamma}$ takes then the form

$$\int d^3x \, \text{Tr} \left(\frac{\delta \bar{\Gamma}}{\delta A^\mu} \frac{\delta \bar{\Gamma}}{\delta G_\mu} + \frac{\delta \bar{\Gamma}}{\delta c} \frac{\delta \bar{\Gamma}}{\delta H} \right) = 0 \, . \tag{5}$$

Inserting in eq.(5) the loop expansion $\bar{\Gamma} = \sum_{n=0}^{\infty} \hbar^n \bar{\Gamma}_n$, we get a tower of equations that must be satisfied order by order in $\hbar$. At first order we have

$$\Delta \bar{\Gamma}_1 = 0 \, , \tag{6}$$

where Δ is the Slavnov-Taylor operator

$$\Delta = \int d^3x \, \text{Tr} \left[\frac{\delta \bar{\Gamma}_0}{\delta A^\mu} \frac{\delta}{\delta G_\mu} + \frac{\delta \bar{\Gamma}_0}{\delta G_\mu} \frac{\delta}{\delta A^\mu} + \frac{\delta \bar{\Gamma}_0}{\delta c} \frac{\delta}{\delta H} + \frac{\delta \bar{\Gamma}_0}{\delta H} \frac{\delta}{\delta c} \right] \, .$$

The operator Δ is nilpotent and is the quantum generalization of the classical BRS operator.

Eq. (6) is formally identical to the usual stability equation for BRS. However, the cohomological problem to be solved here is by far more difficult, since $\bar{\Gamma}_1$ might contain not only local but also non-local contributions. Under the additional hypothesis that contributions involving fewer than four fields be purely local, the local and non-local sectors can be proven to decouple in eq.(6) and we can easily solve for the local part[8]. This locality requirement is justified a posteriori by the explicit computation of the Green functions involving up to three external legs. Observe that the latter are the only Green functions that by power counting need to be regularized. Consequently, the local part of $\bar{\Gamma}_1$ encodes all the arbitrariness of the regularization scheme used. Of course, terms involving four or more fields are necessarily non-local, if not zero, for dimensional reasons and, being finite, do not depend on the regularization. For the local part of $\bar{\Gamma}_1$ we then get[8]

$$W(\alpha_1, \beta_1, \gamma_1) = -\frac{ik}{4\pi} \int d^3x \, \text{Tr} \left[(\alpha_1 + 2\beta_1) \, A \wedge dA + \frac{2}{3} (\alpha_1 + 3\beta_1) \, A \wedge A \wedge A \right]$$
$$- 2 \int d^3x \, \text{Tr} \left[\beta_1 \, G_\mu \partial^\mu c - \gamma_1 \, G_\mu (D^\mu c) + \frac{1}{2} \gamma_1 \, H[c, c] \right] \, ,$$

where α_1, β_1 and γ_1 are regularization dependent (finite) coefficients. They correspond to the freedom of renormalizing k and the fields. The (local part of the) one loop effective action can then be recast in the form

$$\Gamma^{\text{loc}} = (1 + \alpha_1) S_{CS}[A] - 2 \int d^3x \, \text{Tr}(b \partial_\mu A^\mu) + \Delta X(\beta_1, \gamma_1) \, . \tag{7}$$

Here $X(\beta_1, \gamma_1) = 4 \int d^3x \, \text{Tr} \left[\beta_1 G^\mu A_\mu - (1 + \gamma_1) H c \right]$. The effective action receives two different kinds of contributions. One (corresponding to α_1) is gauge invariant, metric independent and provides a monodromy parameter $(1 + \alpha_1) k$. The other is the term ΔX which, being cohomologically trivial, does not contribute to to the expectation values of Wilson loops[10]. These properties can be made manifest by the wave function renormalization $\Phi_R = Z_\Phi \Phi$ ($\Phi = A, b, c, H, G$), with $Z_A = Z_G^{-1} = Z_b^{-1} = 1 + \beta_1$, $Z_c = Z_H^{-1} = 1 + \gamma_1$, so that Γ^{loc} takes the form

$$\Gamma^{loc} = (1 + \alpha_1) S_{CS}[A_R] + S_{GF}[\Phi_R] \ . \tag{8}$$

Since α_1, the only observable one-loop correction, could still depend on the regularization, one would conclude that the emerging quantum theory is not unique. As anticipated, however, all BRS invariant regulators used so far for CST yield the same value for α_1. In the following table we collect the values (in units of c_V/k) of the one-loop parameters for the following BRS invariant regularization schemes: large mass limit of dimensionally regularized topologically massive Yang-Mills (TMYM) theory[8], η-function regularization[2], higher covariant derivatives (HCD) and Pauli-Villars fields[7,10] and geometric regularization[9].*

Table 1. One loop parameters for BRS invariant regularization schemes

Regularization Method	α_1	β_1	γ_1
Large m of TMYM + dimensional reg.	1	2/3	0
η-function regularization	1	0	0
HCD + Pauli-Villars	1	2/9	0
Geometric regularization	1	$4/(3\pi) I_n$	*

Geometric quantization makes use of ghost generations different from the standard Faddeev-Popov ones, so only the pure gauge sector of Γ^{loc} can be compared; the quantity I_n is defined as $I_n = \int_0^\infty dp \, \frac{(1+p^2)^n}{1+p^2(1+p^2)^{2n}}$, with $n > 1$ an arbitrary integer. As it appears evident, different regularizations yield different Γ's but all of them predict the same observable shift $k \to k + c_V$ and therefore the same physical theory. This is in agreement with the non-perturbative result (3) and leads to a self-consistent, single valued quantum theory, since the monodromy parameter $k + c_V$ keeps being an integer.

At this point the obvious question arises: what does it happen at higher perturbative orders? Now that we obtained the solution (7) for the one loop effective action, we can proceed and solve the BRS identity at order $\hbar^2$. If $\bar{\Gamma}_2$ has the same locality properties as $\bar{\Gamma}_1$, its local part is found to be[8]

$$\bar{\Gamma}_2^{loc} = W(\alpha_2, \beta_2, \gamma_2) + \frac{ik}{4\pi} \int d^3x \, \beta_1(\alpha_1 + 3\beta_1 - \gamma_1) \, \mathrm{Tr}(A \wedge dA)$$

so that the (local part of) the two-loop effective action reads

$$\Gamma^{loc} = \Gamma_0 + W(\alpha_1 + \alpha_2, \beta_1 + \beta_2, \gamma_1 + \gamma_2) + \frac{ik}{4\pi} \int d^3x \, \beta_1(\alpha_1 + 3\beta_1 - \gamma_1) \, \mathrm{Tr}(A \wedge dA) \tag{9}$$

The three new parameters α_2, β_2 and γ_2 are second order and regularization scheme dependent. The apparently awkward expression (9) takes a simple and familiar form if

---

* The values given here for HCD+Pauli-Villars are those computed in ref. 10, rather than those in ref. 7, where strictly speaking only Pauli-Villars fields and no HCD are used. In ref. 7 one gets $\alpha_1 = 1$, $\beta_1 = \gamma_1 = 0$.

we subtract the one loop contributions corresponding to $\Delta X(\beta_1, \gamma_1)$ entering the two-loop diagrams through one loop subdiagrams. This is accomplished by means of the same wave function renormalization leading to eq. (8). In terms of the renormalized fields we have

$$\Gamma^{\text{loc}} = (1 + \alpha_1 + \tilde{\alpha}_2) S_{CS}[A_R] - 2 \int d^3x \; \text{Tr}(b_R \, \partial_\mu A_R^\mu) + \Delta_R X_R(\tilde{\beta}_2, \gamma_2) \; ,$$

where the subscript R indicates renormalized quantities and $\tilde{\alpha}_2 = \alpha_2 - 6(\beta_1)^2 - 3\alpha_1\beta_1 + 3\beta_1\gamma_1$, $\tilde{\beta}_2 = \beta_2 + \beta_1^2 - \beta_1\gamma_1$. Just like at one loop, BRS invariance implies that out of three parameters, only one, $\tilde{\alpha}_2$, is gauge invariant hence observable as a shift of k. Thus, from a perturbative viewpoint a two-loop correction to k would be allowed. Note that now $\tilde{\alpha}_2$ is order $(c_V/k)^2$ and cannot possibly lead to an integer shift for any integer choice of k. Finding a non-zero $\tilde{\alpha}_2$ would imply an incompatibility between the non-perturbative request of single valuedness and perturbative quantization. The computation of $\tilde{\alpha}_2$ in a BRS invariant regularization scheme is then particularly intriguing. In ref. 8 we showed how to compute the two-loop effective action for the first BRS invariant regularization scheme listed in the table above. In the remaining part of this note we shall briefly summarize the ideas and the results of the method, warmly inviting the interested reader to refer back to the original paper where more details can be found.

Dimensional regularization ensures BRS invariance and algebraic consistency if we start with a D dimensional extended action $S_{CS}[A]$ with the following prescription for the $\epsilon_{\mu\nu\rho}$ symbol (showing up once $S_{CS}[A]$ is written in coordinates). The D-dimensional analogue of $\epsilon_{\mu\nu\rho}$ is defined as a completely antisymmetric object in its indices satisfying the following identities[14]:

$$\epsilon_{\mu_1\mu_2\mu_3}\epsilon_{\nu_1\nu_2\nu_3} = \sum_{\pi \in S_3} \text{sign}(\pi) \prod_{i=1}^3 \tilde{g}_{\mu_i\nu_{\pi(i)}} \; , \quad \epsilon_{\mu_1\mu_2\mu_3}\hat{g}^{\mu_3\mu_4} = 0 \; . \tag{10}$$

Here $g_{\mu\nu} = \tilde{g}_{\mu\nu} \oplus \hat{g}_{\mu\nu}$ is the euclidean metric in D dimensions, with $\tilde{g}_{\mu\nu}$ and $\hat{g}_{\mu\nu}$ its 3- and $(D-3)$-dimensional projections respectively. Furthermore, given a D-dimensional vector u^μ we define $\hat{u}^\mu = \hat{g}^{\mu\nu} u_\nu$ and $\tilde{u}^\mu = \tilde{g}^{\mu\nu} u_\nu$. Notice that objects with a hat vanish when $D \to 3$; they are called evanescent.

Unfortunately, pure CST theory with the above prescription for the $\epsilon_{\mu\rho\nu}$ symbol, has a non-invertible kinetic term, even in a general α-gauge, because of the zero modes $z(p) = f(p)(\hat{p}^2 \hat{g}_{\mu\nu} - \hat{p}_\mu \hat{p}_\nu)$. This hinders a perturbative analysis. To get out of this impasse we can add to $S_{CS}[A]$ a Yang-Mills term

$$S_{\text{YM}}[A] = \frac{k}{8\pi m} \int d^3x \; \text{Tr}(F_{\mu\nu}F^{\mu\nu}) \; ,$$

where $F_{\mu\nu}$ is the field strength of the gauge field A_μ. The theory defined by the action $S_m[A] = S_{CS}[A] + S_{\text{YM}}[A]$ corresponds to so the called Topologically Massive Yang Mills theory (TMYM), proposed in ref. 11 to provide a way (alternative to the usual Higg mechanism) for making the gauge field massive without losing gauge invariance. The parameter m is the bare topological mass of the gauge field and CST is recovered in the limit of infinite mass gap. The TMYM theory, differently from CST, is superrenormalizable and only one- and two-loop diagrams are divergent. Therefore, from our viewpoint, m can be envisioned as an UV regulator. The regularized CST is then defined in the limits $m \to \infty$ and $D \to 3$. These two limits do not commute and, to have a consistent BRS invariant regularization prescription, one must take $D \to 3$ first. Only in the case that the limit $D \to 3$ does not give rise to divergences, is one allowed to let $m \to \infty$. Therefore, if our regularization prescription has to make sense, TMYM theory must be UV finite. That this is indeed the case has been shown in ref. 8.

The addition of a Yang-Mills term to the Chern-Simons action entails several side effects. From a non-perturbative, functional point of view, the wild oscillatory behaviour of the path integral (2) is tamed by the presence of the Yang-Mills contribution. Thus the TMYM path integral is in fact a regularized version of Chern-Simons one. The price paied for this regulating effect is a total change in the structure of the Hilbert space. Indeed TMYM theory is not topological and the gauge fields excitations are propagating. In the limit of infinite mass gap the propagating modes decouple from the non-propagating ones, and CST is reobtained. What we are left with is thus the (degenerate) ground state of TMYT which precisely corresponds to the zero-energy Hilbert space of CST[15]. This picture is reminiscent of Klauder's regularization prescription of the path integral for quantum mechanical systems, which is known to be completely equivalent to geometric quantization[16].

The definition (10) for the D-dimensional $\epsilon_{\mu\nu\rho}$ makes the formal theory invariant under $SO(3)\otimes SO(D-3)$, rather than $SO(D)$. As a result, the Feynman rules involve hatted and twiddled objects in a non-trivial way. For the gauge propagator we have (dropping colour indices): $D_{\mu\nu}(\tilde{p},\hat{p}) = \Delta_{\mu\nu}(p) + R_{\mu\nu}(\tilde{p},\hat{p})$, where $\Delta_{\mu\nu}(p)$ and $R_{\mu\nu}(\tilde{p},\hat{p})$ are given by

$$\Delta_{\mu\nu}(p) = \frac{4\pi}{k} \frac{m}{p^2(p^2+m^2)} \left(m\epsilon_{\mu\rho\nu} p^\rho + p^2 g_{\mu\nu} - p_\mu p_\nu \right),$$

$$R_{\mu\nu}(\tilde{p},\hat{p}) = \frac{4\pi}{k} \frac{m^3}{p^2[(p^2)^2 + m^2 \tilde{p}^2]} \left[\frac{\hat{p}^2}{p^2+m^2} \left(\epsilon_{\mu\rho\nu} p^\rho + p^2 g_{\mu\nu} + \frac{m^2}{p^2} p_\mu p_\nu \right) \right.$$
$$\left. + \tilde{p}^2 \hat{g}_{\mu\nu} + \hat{p}_\mu \hat{p}_\nu - p_\mu \hat{p}_\nu - \hat{p}_\mu p_\nu \right].$$

Notice that $R_{\mu\nu}(\tilde{p},\hat{p})$ is vanishing in the limit $D \to 3$. It is obvious that vanishing objects do not contribute at the tree level. At higher perturbative orders however, they could combine with divergent terms and yield finite contributions. A careful study[8] of the convergence properties of 1PI Green functions having at least one $R_{\mu\nu}(\tilde{p},\hat{p})$ insertion leads to the conclusion that they are finite and hence vanishing at $D = 3$. Therefore we can use $\Delta_{\mu\nu}(p)$ as the "effective" free gauge field propagator.

The determination of $\bar{\Gamma}_2^{\text{loc}}$ necessitates of the knowledge of three independent Green functions. It is wise to choose the simplest ones, namely the vacuum polarization, the ghost self-energy and the Hcc vertex. The explicit computation of these Green functions presents, from the point of view of the algebra involved, the same degree of complexity as QCD. The situation is much worse from the point of view of integration, since now we are faced with massive denominators. The key observation that makes the computation feasible is that we are only interested in the asymptotic behaviour of the integrals for large values of m. We will then use two theorems that enable us to tell if an integral is vanishing in the limit $m \to \infty$ or not.

A general integral from a L-loop Feynman diagram is of the form

$$I(p,m) = m^\beta \int dk \, \frac{M(k)}{\prod_i (l_i^2)^{n_i} \prod_j (l_j^2 + m^2)^{n_j}}, \tag{11}$$

where the integration measure is $dk = d^3k_1 \cdots d^3k_L$, β denotes an arbitrary real number and n_i are positive integers. The momenta l_i are linear combinations of internal and external momenta. The numerator $M(k)$ is a monomial of degree n_k in the components of the vectors $k_1, \ldots, k_L$. We call d to the mass dimension of $I(p,m)$ and denote by $\underline{\omega}_{\min}$ the minimum between zero and the lowest infrared degree at zero external momenta of all the subintegrals of $I(p,m)$, including $I(p,m)$ itself. Then the following theorem holds:

Theorem 1. If the integral $I(p,m)$ is both UV and IR convergent by power counting at non-exceptional external momenta, and the mass dimension d and $\underline{\omega}_{\min}$ defined above satisfy $d - \underline{\omega}_{\min} < 0$, then $I(p,m)$ vanishes when m goes to ∞.

To formulate the second vanishing theorem we introduce the notation $[n] = 0$ for n even and $[n] = 1$ for n odd. The theorem states then:

Theorem 2. If the integral $I(p,m)$ in (11) is absolutely convergent at zero external momenta and its mass dimension d satisfies $[n_k] > d$, then $I(p,m) \to 0$ as $m \to \infty$.

Concerning the hypothesis of UV convergence in these two theorems, we must mention that it can in fact be relaxed[8] for the case of TMYM, at least up to two loops, to the much weaker condition that the integral (11) be finite in dimensional regularization.

In the limits $D \to 3$ and $m \to \infty$, the result we obtained for the second order coefficients appearing in the two-loop effective action (9) are the following

$$\alpha_2 = \frac{14}{3} \left(\frac{c_V}{k}\right)^2, \quad \beta_2 = \frac{169 + L}{72} \left(\frac{c_V}{k}\right)^2, \quad \gamma_2 = 0,$$

with $L = 528 \ln 2 - 567 \ln 3$. Plugging these values into the expression of $\tilde{\alpha}_2$ and taking into account the values of the one-loop parameters in the table, we conclude that $\tilde{\alpha}_2$ is zero and therefore the two-loop corrections do not have observable consequences. Stated differently, there is no two-loop shift of the parameter k owing to the cohomologically trivial character of the second order corrections. We conjecture that this picture holds true also at higher perturbative corrections, despite an explicit calculation is still lacking.

It would be also nice to have at our disposal the corresponding two-loop results for the other BRS invariant regularizations listed in the previous table, so to get a thorough check of the uniqueness of the quantum theory parameterized in terms of the bare k. Unfortunately, what we have presented here is, at present, the only instance where a two-loop computation has been carried out for CST with a manifestly BRS invariant regularization scheme.

REFERENCES

1. For a recent review see for example:
   R. Iengo and K. Lechner, *Phys. Rep.* **213** (1992) 179.
2. E. Witten, *Commun. Math. Phys.* **121** (1989) 351.
3. V.F.R. Jones, *Ann. Math.* **126** (1987) 335.
4. D.S. Freed and R.E. Gompf, *Phys. Rev. Lett.* **66** (1991) 1255.
5. E. Guadagnini, M. Martellini and M. Mintchev, *Nucl. Phys.* **B330** (1990) 575.
6. E. Guadagnini, M. Martellini and M. Mintchev, *Phys. Lett.* **224** (1989) 489.
7. L. Alvarez-Gaumé, J.M.F. Labastida and A.V. Ramallo, *Nucl. Phys.* **B334** (1990) 103.
8. G. Giavarini, C.P. Martin and F. Ruiz Ruiz, *Nucl. Phys.* **B381** (1992) 222.
9. M. Asorey and F. Falceto, *Phys. Lett.* **241** (1990) 31; *Int. J. Mod. Phys.* **A7** (1992) 235.
10. G. Giavarini, F. Ruiz Ruiz, C.P. Martin, *Perturbative quantization of Chern-Simons theory*, preprint.
11. S. Deser, R. Jackiw and S. Templeton, *Ann. Phys.* **140** (1982) 372.
12. A. Blasi and R. Collina, *Nucl. Phys.* **B345** (1990) 472.
13. F. Delduc, C, Lucchesi, O. Piguet and S.P. Sorella, *Nucl. Phys.* **B346** (1990) 513.
14. P. Breitenlohner and D. Maison, *Commun. Math. Phys.* **52** (1977) 11.
15. A.P. Polychronakos, *Ann. of Phys.* **203** (1990) 231.
16. J. Klauder and E. Onofri, *Int. J. Mod. Phys.* **A4** (1989) 15.

RAY-SINGER TORSION, TOPOLOGICAL FIELD THEORIES AND THE RIEMANN ZETA FUNCTION AT $s = 3$

Charles Nash and D. J. O' Connor

Department of Mathematical Physics
St. Patrick's College
Maynooth, Ireland
and
School of Theoretical Physics
Dublin Institute for Advanced Studies
10 Burlington Road, Dublin 4, Ireland

ABSTRACT

Starting with topological field theories we investigate the Ray-Singer analytic torsion in three dimensions. For the lens Spaces $L(p;q)$ an explicit analytic continuation of the appropriate zeta functions is constructed and implemented. Among the results obtained are closed formulae for the individual determinants involved, the large p behaviour of the determinants and the torsion, as well as an infinite set of distinct formulae for $\zeta(3)$: the ordinary Riemann zeta function evaluated at $s = 3$. The torsion turns out to be trivial for the cases $L(6,1)$, $L((10,3)$ and $L(12,5)$ and is, in general, greater than unity for large p and less than unity for a finite number of p and q.

1. INTRODUCTION

The torsion studied in this paper has its origins in the 1930's, cf. Franz [1], where it was combinatorially defined and used to distinguish various lens spaces from one another. Given a manifold M and a representation of its fundamental group $\pi_1(M)$ in a flat bundle E, this Reidemeister-Franz torsion is a real number which is defined as a particular product of ratio's of volume elements V^i constructed from the cohomology groups $H^i(M;E)$.

Since volume elements are essentially determinants then, for any alternative definition of a determinant, an alternative definition of the torsion can be given. Now if one uses de Rham cohomology to compute $H^i(M;E)$ then these determinants become determinants of Laplacians Δ_p^E on p-forms with coefficients in E. But zeta functions for elliptic operators can be used to give finite values to such infinite dimensional determinants and so an analytic definition of the torsion results and this is the analytic torsion of Ray and Singer [2,3,4] given in the 1970's; furthermore this torsion was proved by them to be independent of the Riemannian metric used to define the Laplacian's Δ_p^E.

This analytic torsion coincided, for the case of lens spaces, with the combinatorially defined Reidemeister-Franz torsion. Finally Cheeger and Müller [5,6] independently proved that the analytic Ray-Singer torsion coincides with the combinatorial Reidemeister-Franz torsion in all cases.

Infinite dimensional determinants also occur naturally in quantum field theories when computing correlation functions and partition functions. In 1978 Schwarz [7] showed how to construct a quantum field theory on a manifold M whose partition function is a power of the Ray-Singer torsion on M.

Schwarz's construction uses an Abelian gauge theory but in three dimensions a non-Abelian gauge theory—the $SU(2)$ Chern-Simons theory—can be constructed and has deep and important properties established by Witten in 1988: Its partition function is the Witten invariant for the three manifold M and the correlation functions of Wilson loops give the Jones polynomial invariant for the link determined by the Wilson loops—cf. [8,9]. Finally the weak coupling limit of the partition function is a power of the Ray-Singer torsion.

We shall be concerned here with the special situation of three dimensions and with the case where the three manifold M is a lens space. In the next two sections we describe the precise setting and the analytic continuation while the final section contains our concluding remarks.

2. TOPOLOGICAL FIELD THEORIES, ANALYTIC TORSION AND LENS SPACES

Quantum field theories of the type alluded to in the previous section are usually referred to as topological quantum field theories or simply topological field theories.

It turns out that more than one topological field theory can be used to give the torsion, for an excellent review of this question cf. Birmingham et al. [10]. For example one can take the action

$$S[\omega] = \int_M \omega_n d\omega_n, \qquad \dim M = 2n+1, \qquad (2.1)$$

where ω_n is an n-form. The partition function is then

$$Z[M] = \int \mathcal{D}\omega\, \mu[\omega] \exp[-S[\omega]]. \qquad (2.2)$$

$S[\omega]$ has a gauge invariance whereby $S[\omega] = S[\omega + d\lambda]$ and therefore to define the partition function it is necessary to integrate over only inequivalent field configurations. The measure $\mathcal{D}\omega\mu[\omega]$ thus contains functional delta functions which constrain the integration and play the role of gauge fixing, together with their associated determinants. This measure can be constructed using, for example, the Batalin Vilkovisky BRST construction [11,12]. We wish to devote more space here to lens spaces and the computation of their torsion and so we turn to that now.

To define the Ray-Singer torsion, or simply torsion, we take a closed compact Riemannian manifold M over which we have a flat bundle E. Let M have a non-trivial fundamental group $\pi_1(M)$ which is represented on E—this latter property arises very naturally in the physical gauge theory context where it corresponds simply to the space of flat connections all of whose content resides in their holonomy—In any case the torsion is then the real number $T(M, E)$ where

$$\ln T(M, E) = \sum_0^n (-1)^q q \ln \det \Delta_q^E, \quad n = \dim M. \tag{2.3}$$

The metric independence of the torsion requires that we assume, in the above definition, that the cohomology ring $H^*(M; E)$ is trivial; this means that the Laplacians Δ_q^E have empty kernels and so are strictly positive definite. Given this fact one may use zeta functions to define $\det \Delta_q^E$ in the standard way. Recall that if P is a positive elliptic differential or pseudo-differential operator with spectrum $\{\mu_n\}$ and degeneracies Γ_n then its associated zeta function $\zeta_P(s)$ is a meromorphic function of s, regular at $s = 0$, which is given by

$$\zeta_P(s) = \sum_{\mu_n} \frac{\Gamma_n}{\mu_n^s} \tag{2.4}$$

and its determinant $\det P$ is defined by

$$\ln \det P = -\left.\frac{d\zeta_P(s)}{ds}\right|_{s=0}. \tag{2.5}$$

Using this we have

$$\ln T(M, E) = -\sum_0^n (-1)^q q \left.\frac{d\zeta_{\Delta_q^E}(s)}{ds}\right|_{s=0}. \tag{2.6}$$

Next we turn to lens spaces. For general background on lens spaces cf. [13,14] and references therein—briefly, a lens space can be constructed as follows: Take an odd dimensional sphere S^{2n-1}, considered as a subset of $\mathbf{C}^n$, on which a finite cyclic group of rotations G, say, acts. The quotient S^{2n-1}/G of the sphere under this action is a lens space. More precisely, suppose that G is of order p, $(z_1, \ldots, z_n) \in \mathbf{C}^n$ and the group action takes the form

$$(z_1, \ldots, z_n) \longmapsto (\exp(2\pi i q_1/p)z_1, \ldots, \exp(2\pi i q_n/p)z_n) \tag{2.7}$$
with $q_1, \ldots, q_n$ integers relatively prime to p

then the quotient S^{2n-1}/G is a lens space often denoted by $L(p; q_1, \ldots, q_n)$. A formula for the torsion of these spaces was first worked out by Ray [2]. We wish to focus on the situation that obtains when $n = 2$ and G is the group $\mathbf{Z}_p \equiv \mathbf{Z}/p\mathbf{Z}$. For the most part we shall deal with the lens space $L(p; 1, 1)$ which, for simplicity, we shall denote by $L(p)$; we shall also use the notation $L(p; q)$ to denote the Lens space $L(p; 1, q)$. In passing we note that when $p = 2$ we have $L(2) = \mathbf{R}P^3 \simeq SO(3)$.

The group action above defines a representation V, say, of $\pi_1(L(p))$ and also determines a flat bundle $F = (V \times S^3)/\mathbf{Z}_p$, over $L(p)$. It is the torsion of this F over $L(p)$ with which we are concerned here. Using zeta functions the torsion of these lens spaces is therefore given by

$$\ln T(L(p), F) = -\sum_0^3 (-1)^q q \left.\frac{d\zeta_{\Delta_q^F}(s)}{ds}\right|_{s=0}. \tag{2.8}$$

As an aid to the calculation of $\ln T(L(p), F)$ it is useful to introduce the notation

$$\tau(p,s) = -\sum_{0}^{3}(-1)^q q\zeta_{\Delta_q^F}(s), \qquad T(p) = T(L(p), F). \qquad (2.9)$$

For $\tau(p, s)$ itself we now have, using Poincaré duality,

$$\tau(p,s) = \zeta_{\Delta_1^F}(s) - 2\zeta_{\Delta_2^F}(s) + 3\zeta_{\Delta_3^F}(s) = 3\zeta_{\Delta_0^F}(s) - \zeta_{\Delta_1^F}(s), \qquad (2.10)$$

Making use of the triviality of the kernels of Δ_q^F we further obtain [2] the formula

$$\tau(p,s) = 2\zeta_{d^*d_0}(s) - \zeta_{d^*d_1}(s), \qquad (2.11)$$

For the individual zeta functions we denote the eigenvalues and their degeneracies by $\lambda_n(q,p)$ and $\Gamma_n(q,p)$ respectively giving the expressions

$$\zeta_{d^*d_0}(s) = \sum_n \frac{\Gamma_n(0,p)}{\lambda_n^s(0,p)}, \qquad \zeta_{d^*d_1}(s) = \sum_n \frac{\Gamma_n(1,p)}{\lambda_n^s(1,p)}. \qquad (2.12)$$

It remains to compute these eigenvalues and degeneracies cf. [15]. The eigenvalues are

$$\lambda_n(0,p) = n(n+2), \qquad \lambda_n(1,p) = (n+1)^2, \; n = 1, 2, \ldots. \qquad (2.13)$$

To calculate the degeneracies is more difficult; we make use of the fact that S^3 is a group manifold and proceed as follows: Consider the Laplacians d^*d_q on S^3, and $d^*d_q^F$ on $L(p)$ also, if λ is an eigenvalue, denote the corresponding eigenspaces by $\Lambda_q(\lambda)$ and $\Lambda_q^F(\lambda)$ respectively. Let

$$v(z) \in \Lambda_q(\lambda), \text{ with } z \in S^3 \subset \mathbf{C}^2, \text{ and } g \in \mathbf{Z}_p, \qquad (2.14)$$

where $g \equiv \exp[2\pi i j/p]$, $0 \leq j \leq (p-1)$. The element g acts on $v(z)$ to give $g \cdot v(z)$ where

$$g \cdot v(z) = v(gz) \quad \text{where } gz = (\exp[2\pi i j/p]z_1, \exp[2\pi i j/p]z_2). \qquad (2.15)$$

The above definitions allow us to define the projection $P(\lambda)$ on $\Lambda_q(\lambda)$ by

$$P(\lambda)v = \frac{1}{p} \sum_{g \in \mathbf{Z}_p} \exp[-2\pi i j/p] g \cdot v. \qquad (2.16)$$

Evidently $[P(\lambda), d^*d_q] = 0$ and so $P(\lambda)$ projects the space $\Lambda_q(\lambda)$ onto the space $\Lambda_q^F(\lambda)$. Finally this means that we obtain a formula for the degeneracy $\Gamma_n(q,p)$, namely

$$\Gamma_n(q,p) = \operatorname{tr}\left(P|_{\Lambda_q^F(\lambda)}\right) = \frac{1}{p}\sum_{j=0}^{(p-1)} \exp[-2\pi i j/p] \operatorname{tr}\left(g|_{\Lambda_q^F(\lambda)}\right). \qquad (2.17)$$

To actually apply this formula we now add in the fact that S^3 is the group manifold for $SU(2)$. The Peter–Weyl theorem tells us, in this case where all representations are self-conjugate, that

$$L^2(S^3) = L^2(SU(2)) = \bigoplus_\mu c_\mu D_\mu = \bigoplus_\mu D_\mu \otimes D_\mu, \qquad (2.18)$$

where c_μ measures the multiplicity of the representation μ which must therefore be $\dim D_\mu$. But Hodge theory gives us the alternative decomposition

$$L^2(S^3) = \bigoplus_\lambda \Lambda_0(\lambda) . \tag{2.19}$$

In addition the Casimir operator for $SU(2)$ is a multiple of the Laplacian and, if the representation label μ is taken to be the usual half-integer j, then we know that this Casimir has eigenvalues $j(j+1)$, and also that $\dim D_j = 2j+1$. These facts identify the Laplacian $\Delta_0 = d^*d_0$ as four times the Casimir and identify $\Lambda_0(\lambda)$ as $\dim D_j$ copes of D_j. Thus if we set $n = 2j$, so that n is always integral, then we have the degeneracy formula

$$\Gamma_n(0,p) = \frac{(n+1)}{p} \sum_{j=0}^{(p-1)} \exp[-2\pi i j/p] \, \chi^{n/2}(2\pi j/p) \tag{2.20}$$

where $\chi^j(\theta)$ denotes the $SU(2)$ character, on D_j, for rotation through the angle θ; i.e.

$$\chi^j(\theta) = \frac{\sin((2j+1)\theta)}{\sin(\theta)} . \tag{2.21}$$

Hence our explicit degeneracy formula for 0-forms on $L(p)$ is

$$\Gamma_n(0,p) = \frac{(n+1)}{p} \sum_{j=0}^{(p-1)} \exp[-2\pi i j/p] \, \frac{\sin(2\pi(n+1)j/p)}{\sin(2\pi j/p)} . \tag{2.22}$$

We now have to find the analogous formula for the 1-forms. The formula that results is

$$\Gamma_n(1,p) = \frac{1}{p} \sum_{j=0}^{(p-1)} \exp[-2\pi i j/p] \left\{ n\chi^{(n+1)/2}(2\pi j/p) + (n+2)\chi^{(n-1)/2}(2\pi j/p) \right\} . \tag{2.23}$$

To simplify the notation we introduce the 'p-averaged character' $\langle \chi^j \rangle_p$ which we define by

$$\langle \chi^j \rangle_p = \frac{1}{p} \sum_{j=0}^{(p-1)} \exp[-2\pi i j/p] \, \chi^j(2\pi j/p) . \tag{2.24}$$

Finally this gives us a concrete expression for $\tau(p,s)$, i.e.

$$\tau(p,s) = \sum_n \left\{ \frac{n \langle \chi^{(n+1)/2} \rangle_p + (n+2) \langle \chi^{(n-1)/2} \rangle_p}{(n+1)^{2s}} - \frac{2(n+1) \langle \chi^{n/2} \rangle_p}{\{n(n+2)\}^{2s}} \right\} \tag{2.25}$$

$$= \tau_+(p,s) - \tau_-(p,s),$$

with the obvious definition for $\tau_+(p,s)$ and $\tau_-(p,s)$.

To actually compute the torsion we need to be able to evaluate these p-averaged characters. This is a somewhat non-trivial combinatorial task and it is necessary to divide n up into its conjugacy classes mod p by writing $n = pk - j$, $k \in \mathbf{Z}$, $j = 0, 1, \ldots, (p-1)$ We eventually discover that

$$\left\langle \chi^{(pk-j)/2} \right\rangle_p = \begin{cases} \begin{cases} k & \text{for } j = 0, 2, \ldots, (p-1) \\ k & \text{for } j = 1 \\ (k-1) & \text{for } j = 3, 5, \ldots, (p-2) \end{cases} & \text{if } p \text{ is odd,} \\ \begin{cases} 0 & \text{for } j = 0, 2, \ldots, (p-2) \\ 2k & \text{for } j = 1 \\ (2k-1) & \text{for } j = 3, 5, \ldots, (p-1) \end{cases} & \text{if } p \text{ is even.} \end{cases} \quad (2.26)$$

We must now construct the analytic continuation of the series for $\tau(p, s)$. For the details we refer the reader to [15]. We shall just describe here the $p = 2$ case.

3. THE ANALYTIC CONTINUATION

The series to be continued is

$$\tau(p, s) = \sum_n \left\{ \frac{n \left\langle \chi^{(n+1)/2} \right\rangle_p + (n+2) \left\langle \chi^{(n-1)/2} \right\rangle_p}{(n+1)^{2s}} - \frac{2(n+1) \left\langle \chi^{n/2} \right\rangle_p}{\{n(n+2)\}^{2s}} \right\} \quad (3.1)$$

and it already converges for $\operatorname{Re} s > 3/2$; however a calculation of the torsion requires us to work at $s = 0$, hence we see the need for, and the extent of, the continuation.
With $p = 2$ we have

$$\tau(2, s) = \sum_n \left\{ \frac{2(n+1) \left\langle \chi^{n/2} \right\rangle_2}{\{n(n+2)\}^s} - \frac{n \left\langle \chi^{(n+1)/2} \right\rangle_2 + (n+2) \left\langle \chi^{(n-1)/2} \right\rangle_2}{(n+1)^{2s}} \right\}, \quad (3.2)$$

but using (2.26) we find that

$$\left\langle \chi^{(n+1)/2} \right\rangle_2 = \left\langle \chi^{(2k-j+1)/2} \right\rangle_2, \quad (n = 2k - j)$$

$$= \begin{cases} 0, & j = 1 \\ 2k+2, & j = 0 \end{cases} \equiv \begin{cases} 0, & n \text{ odd,} \\ 2k+2, & n \text{ even.} \end{cases} \quad (3.3)$$

Similarly

$$\left\langle \chi^{(n-1)/2} \right\rangle_2 = \left\langle \chi^{(2k-j)/2} \right\rangle_2, \quad (n = 2k - j)$$

$$= \begin{cases} 2k, & j = 1 \\ 0, & j = 0 \end{cases} \equiv \begin{cases} (n+1), & n \text{ odd,} \\ 0, & n \text{ even.} \end{cases} \quad (3.4)$$

Thus $\tau(2, s)$ becomes

$$\tau(2, s) = \tau_+(2, s) - \tau_-(2, s) = \sum_{n \text{ odd}} \frac{2(n+1)^2}{\{n(n+2)\}^s} - \sum_{n \text{ even}} \frac{2n(n+2)}{(n+1)^{2s}}. \quad (3.5)$$

Setting $n = (2m-1)$ in $\tau_+(2,s)$ and $n = 2m$ in $\tau_-(2,s)$ we have

$$\tau(2,s) = \sum_{m=1}^{\infty} \frac{8m^2}{(4m^2-1)^s} - \sum_{m=0}^{\infty} \frac{2m(2m+2)}{(2m+1)^{2s}}$$

$$= \sum_{m=1}^{\infty} \frac{8m^2}{(4m^2-1)^s} - \sum_{m=0}^{\infty} \frac{1}{(2m+1)^{(2s-2)}} + \sum_{m=0}^{\infty} \frac{1}{(2m+1)^{2s}}$$
(3.6)

Now if $\zeta(s)$ is the usual Riemann zeta function we can use the fact that

$$\sum_{n=1,3,5,\ldots} \frac{1}{n^s} = (1 - 2^{-s})\zeta(s) \tag{3.7}$$

then we get

$$\tau(2,s) = \sum_{m=1}^{\infty} \frac{8m^2}{(4m^2-1)^s} - 2(1 - 2^{-(2s-2)})\zeta(2s-2) + 2(1 - 2^{-2s})\zeta(2s) . \tag{3.8}$$

The only term in (3.8) without a well defined continuation is the first term. To this end we define the quantity $A(m,s)$ by

$$A(m,s) = \frac{4m^2}{(4m^2-1)^s} = \frac{4m^2}{(4m^2)^s}\left(1 - \frac{1}{4m^2}\right)^{-s} = \frac{1}{(4m^2)^{(s-1)}}\left\{1 + \frac{s}{4m^2} + \cdots\right\}$$

$$= \frac{1}{(4m^2)^{(s-1)}} + \frac{s}{(4m^2)^s} + R(m,s), \quad \text{(def. of } R(m,s)) .$$
(3.9)

Hence the remainder term $R(m,s)$ is given by

$$R(m,s) = A(m,s) - \frac{1}{(4m^2)^{(s-1)}} - \frac{s}{(4m^2)^s}$$

$$= \frac{4m^2}{(4m^2-1)^s} - \frac{1}{(4m^2)^{(s-1)}} - \frac{s}{(4m^2)^s}$$
(3.10)

The definition of the remainder term is chosen to ensure that

$$|R(m,s)| \le \frac{(\ln m)^\alpha}{m^2} \tag{3.11}$$

and this has the vital consequence that the operations d/ds (at $s = 0$) and $\sum_m$ *commute* when applied to $R(m,s)$.

Returning to $\tau(s,2)$ itself we have

$$\tau(2,s) = 2\sum_{m=1}^{\infty} A(m,s) - 2(1 - 2^{-(2s-2)})\zeta(2s-2) + 2(1 - 2^{-2s})\zeta(2s)$$

$$\Rightarrow \tau(2,s) = 2\sum_{m=1}^{\infty} \frac{1}{(4m^2)^{(s-1)}} + 2s\sum_{m=1}^{\infty} \frac{1}{(4m^2)^s} + 2\sum_{m=1}^{\infty} R(m,s)$$

$$- 2(1 - 2^{-(2s-2)})\zeta(2s-2) + 2(1 - 2^{-2s})\zeta(2s)$$
(3.12)

Defining

$$R(s) = \sum_{m=0}^{\infty} R(m,s) \tag{3.13}$$

285

gives a series for $R(s)$ which is guaranteed to be convergent and the analytic continuation is now complete; thus we can now take the final step which is to differentiate and obtain the torsion $T(2)$. The result that we get is that

$$\ln T(2) = \frac{d\tau(2,0)}{ds} = 28\zeta'(-2) + 2(1+\ln 4)\zeta(0) + 2R'(0) , \qquad (3.14)$$

but it is easy to check that $\zeta(0) = -1/2$ and $\zeta'(-2) = -\zeta(3)/4\pi^2$ and by our remark above concerning the motive for our choice of definition for $R(m,s)$ we have

$$R'(0) = \frac{d}{ds} \sum_m R(m,s)\big|_{s=0}$$

$$\Rightarrow R'(0) = \sum_m \frac{dR(m,s)}{ds}\bigg|_{s=0} \qquad (3.15)$$

$$= \sum_m \left[4m^2 \{\ln(4m^2) - \ln(4m^2 - 1)\} - 1\right]$$

$$= -\sum_m \left[4m^2 \ln(1 - 1/4m^2) + 1\right] .$$

Hence

$$\ln T(2) = -\frac{7}{\pi^2}\zeta(3) - 1 - 2\ln(2) - 2\sum_m \left[4m^2 \ln(1 - 1/4m^2) + 1\right] \qquad (3.16)$$

However the series for $R'(0)$ can be expressed as a trigonometric integral cf. [15]. In fact we have

$$\sum_{m=1}^{\infty} \left[4m^2 \ln(1 - 1/4m^2) + 1\right] = -\frac{1}{2} + \frac{4}{\pi^2} \int_0^{\pi/2} dz\, z^2 \cot(z) \qquad (3.17)$$

which means that

$$\ln T(2) = -\frac{7}{\pi^2}\zeta(3) - 2\ln(2) - \frac{8}{\pi^2} \int_0^{\pi/2} dz\, z^2 \cot(z) . \qquad (3.18)$$

This formula (3.18) for $T(2)$ can be pushed even further; by using Ray's expression [2] for the torsion we can deduce that

$$\ln T(p) = -\frac{4}{p} \sum_{j=1}^{(p-1)} \sum_{k=1}^{p} \cos(\frac{2jk\pi}{p}) \ln(2\sin(\frac{2k\pi}{p})) \exp[\frac{2k\pi i}{p}] = -4\ln(2\sin(\frac{\pi}{p})) \qquad (3.19)$$

which, for $p = 2$, becomes simply

$$\ln T(2) = -4\ln(2) . \qquad (3.20)$$

Consequently we straightaway have that

$$-4\ln(2) = -\frac{7}{\pi^2}\zeta(3) - 2\ln(2) - \frac{8}{\pi^2} \int_0^{\pi/2} dz\, z^2 \cot(z) , \qquad (3.21)$$

or
$$\zeta(3) = \frac{2\pi^2}{7} \ln(2) - \frac{8}{7} \int_0^{\pi/2} dz\, z^2 \cot(z) \tag{3.22}$$

in other words our computation of the torsion has given us a formula for $\zeta(3)$.

In [15] we construct the continuation for arbitrary p but here we limit ourselves to quoting the torsion formula for p odd which (recall that $\ln T(p) = -4\ln(2\sin(\pi/p)))$ is

$$\ln T(p) = -\frac{(p^3-1)\,\zeta(3)}{p} - \frac{2}{\pi}(p-2)\int_0^{\frac{\pi}{p}} dz\, z \cot(z) + \frac{2}{\pi}(p-2)\int_0^{\frac{(p-1)\pi}{p}} dz\, z \cot(z)$$
$$- \frac{2p}{\pi^2}\int_0^{\frac{\pi}{p}} dz\, z^2 \cot(z) - \frac{2p}{\pi^2}\int_0^{\frac{(p-1)\pi}{p}} dz\, z^2 \cot(z) - \frac{4\ln(2\sin(\frac{\pi}{p}))}{p}$$
$$+ \frac{16}{\pi}\sum_{l=1}^{(p-3)/2} l \int_0^{\frac{2l\pi}{p}} dz\, z \cot(z) - \frac{4p}{\pi^2}\sum_{l=1}^{(p-3)/2} \int_0^{\frac{2l\pi}{p}} dz\, z^2 \cot(z)$$
$$- \frac{4}{p}\sum_{l=1}^{(p-3)/2} 4l^2 \ln(2\sin(\frac{2l\pi}{p})), \qquad \text{for } p \text{ odd}. \tag{3.23}$$

4. CONCLUDING REMARKS

These formulae have yet to be elucidated further.

A thought provoking fact is that $\zeta(3)$ occurs in a recent paper of Witten [16] where, after multiplication by a known constant, it gives the volume of the symplectic space of flat connections over a *non-orientable* Riemann surface. The corresponding calculation for *orientable* surfaces (where the volume element is a rational cohomology class) allows a cohomological rederivation of the irrationality of $\zeta(2), \zeta(4), \ldots$. This paper also involves the torsion but in two dimensions rather than three. The proof that $\zeta(3)$ is irrational was only obtained in 1978 cf. [17] and the rationality of $\zeta(5), \zeta(7), \ldots$ is at present open.

Our technique, applied in five dimensions instead of three would yield formulae for $\zeta(5)$ but their nature is as yet unclear.

Further interesting results are that the $T(p)$ is trivial (i.e unity) when $p = 6$; and that if we work with $L(p,q)$ rather than $L(p)$ then the only other three dimensional lens spaces for which the torsion is trivial are $L(10,3)$ and $L(12,5)$. The large p behaviour of the determinants is also computable: $T(p)$ grows as $p^4/(2\pi)^4 + p^2/6(2\pi)^2$ for large p, while the determinants grow much faster than quartically.

REFERENCES

[1] Franz W., Über die Torsion einer Überdeckung, J. R. Angew. Math., **173**, 245–254, (1935).
[2] Ray D. B., Reidemeister torsion and the Laplacian on lens spaces, Adv. in Math., **4**, 109–126, (1970).
[3] Ray D. B. and Singer I. M., R-torsion and the Laplacian on Riemannian manifolds, Adv. in Math., **7**, 145–201, (1971).
[4] Ray D. B. and Singer I. M., Analytic Torsion for complex manifolds, Ann. Math., **98**, 154–177, (1973).
[5] Cheeger J., Analytic torsion and the heat equation, Ann. Math., **109**, 259–322, (1979).

[6] Müller W., Analytic torsion and the R-torsion of Riemannian manifolds, Adv. Math., **28**, 233, (1978).

[7] Witten E., *Quantum field theory and the Jones polynomial*, I. A. M. P. Congress, Swansea, 1988, edited by: Davies I., Simon B. and Truman A., Institute of Physics, (1989).

[8] Witten E., Quantum field theory and the Jones polynomial, Commun. Math. Phys., **121**, 351–400, (1989).

[9] Nash C., *Differential Topology and Quantum Field Theory*, Academic Press, (1991).

[10] Birmingham D., Blau M., Rakowski M. and Thompson G., Topological field theory, Phys. Rep., **209**, 129–340, (1991).

[11] Batalin I. A. and Vilkovisky G. A., Quantisation of Gauge Theories with linearly independent generators, Phys. Rev., **D28**, 2567–2582, (1983).

[12] Batalin I. A. and Vilkovisky G. A., Existence theorem for gauge algebras, Jour. Math. Phys., **26**, 172–184, (1985).

[13] Rolfsen D., *Knots and Links*, Publish or Perish, (1976).

[14] Bott R. and Tu L. W., *Differential Forms in Algebraic Topology*, Springer-Verlag, New York, (1982).

[15] Nash C. and O' Connor D. J., in preparation.

[16] Witten E., On quantum gauge theories in two dimensions, Commun. Math. Phys., **141**, 153–209, (1991).

[17] Poorten Alfred van der, A proof that Euler missed ... Apéry's proof of the irrationality of $\zeta(3)$. An informal report., The Mathematical Intelligencer, **1**, 195–203, (1979).

MONSTROUS MOONSHINE AND THE UNIQUENESS OF THE MOONSHINE MODULE

Michael P. Tuite

Department of Mathematical Physics
University College, Galway, Ireland
and
Dublin Institute for Advanced Studies
10 Burlington Road, Dublin 4, Ireland

ABSTRACT

In this talk we consider the relationship between the conjectured uniqueness of the Moonshine module $\mathcal{V}^\natural$ of Frenkel, Lepowsky and Meurman and Monstrous Moonshine, the genus zero property for Thompson series discovered by Conway and Norton. We discuss some evidence to support the uniqueness of $\mathcal{V}^\natural$ by considering possible alternative orbifold constructions of $\mathcal{V}^\natural$ from a Leech lattice compactified string. Within these constructions we find a new relationship between the centralisers of the Monster group and the Conway group generalising an observation made by Conway and Norton. We also relate the uniqueness of $\mathcal{V}^\natural$ to Monstrous Moonshine and argue that given this uniqueness, then the genus zero properties hold if and only if orbifolding $\mathcal{V}^\natural$ with respect to a monster element reproduces $\mathcal{V}^\natural$ or the Leech theory.

THE MOONSHINE MODULE

The Moonshine module [1] of Frenkel, Lepowsky and Meurman (FLM) is the first example of an orbifold CFT [2] and is constructed from a string compactified to R^{24}/Λ where Λ is the Leech lattice, the unique 24 dimensional even self-dual lattice without roots i.e. $\lambda^2 \neq 2$ cf. [3]. The orbifolding is then based on the Z_2 reflection automorphism of Λ.

Let $\mathcal{V}^\Lambda$ denote the set of vertex operators $\{\phi(z)\}$ for the Leech lattice CFT which forms a closed meromorphic operator product algebra (OPA) with central charge 24 [1,4]

$$\phi_i(z)\phi_j(w) \sim \sum_k C_{ijk}(z-w)^{h_k - h_i - h_j} \phi_k(w) + \ldots . \qquad (1)$$

Low-Dimensional Topology and Quantum Field Theory,
Edited by H. Osborn, Plenum Press, New York, 1993

We will represent such an OPA schematically by $\phi\phi \sim \phi$. The 1-loop partition function $Z(\tau) = \text{Tr}_{\mathcal{V}^\Lambda}(q^{L_0})$ is a modular invariant and meromorphic function of τ with a unique simple pole at $q = e^{2\pi i \tau} = 0$ and is given by the unique (up to an additive constant) modular invariant function $J(\tau)$

$$Z(\tau) = J(\tau) + 24 ,$$
$$J(\tau) = \frac{E_2^3}{\eta^{24}} - 744 = \frac{1}{q} + 0 + 196884 q + \ldots . \quad (2)$$

The constant 24 reflects the existence of 24 massless (conformal weight 1) operators in this theory. $\eta(\tau) = q^{1/24} \prod_n (1 - q^n)$ and $E_2(\tau)$ is the Eisenstein modular form of weight 4 [5].

The FLM Moonshine module [1] is an orbifold CFT based on the Z_2 lattice reflection automorphism $\bar{r} : \lambda \to -\lambda$ for $\lambda \in \Lambda$. $\bar{r}$ lifts to a family of Z_2 automorphisms of $\mathcal{V}^\Lambda$ preserving (1) from which family one automorphism r is chosen. Defining the projection $\mathcal{P}_r = (1 + r)/2$, the set of operators $\mathcal{P}_r \mathcal{V}^\Lambda$ then also form a closed meromorphic OPA. However, the corresponding partition function $\text{Tr}_{\mathcal{P}_r \mathcal{V}^\Lambda}(q^{L_0}) = \frac{1}{2}(\begin{smallmatrix} 1 \\ 1 \end{smallmatrix}\square + {}^r\!\!\begin{smallmatrix} \\ 1 \end{smallmatrix}\square)$ is not modular invariant, employing standard notation for the worldsheet torus boundary conditions e.g. [6]. Thus, under a modular transformation $\tau \to -1/\tau$, ${}^r\!\!\begin{smallmatrix} \\ 1 \end{smallmatrix}\square = 1/\eta_{\bar{r}}(\tau) \to \begin{smallmatrix} 1 \\ r \end{smallmatrix}\square = 2^{12} \eta_{\bar{r}}(\tau/2) = 2^{12} q^{1/2} + \ldots$ where $\eta_{\bar{r}}(\tau) = [\eta(2\tau)/\eta(\tau)]^{24}$. Therefore the introduction of a 'twisted' sector with vacuum energy $1/2$ and degeneracy 2^{12} is necessary to form a modular invariant theory [1,2]. The states of this sector are constructed from twisted vertex operators $\mathcal{V}_r = \{\psi(z)\}$ acting on the untwisted vacuum. Thus $\mathcal{V}^\Lambda$ is enlarged by $\mathcal{V}_r$ to $\mathcal{V}' = \mathcal{V}^\Lambda \oplus \mathcal{V}_r$ which forms a closed non-meromorphic OPA [1,7,8,9] where (schematically)

$$\phi\phi \sim \phi, \qquad \phi\psi \sim \psi, \qquad \psi\psi \sim \phi, \quad (3)$$

$\bar{r}$ can also be lifted to an automorphism r of (3) where the operators of $\mathcal{P}_r \mathcal{V}_r$ have integral conformal weight. Then $\mathcal{V}^\natural = \mathcal{P}_r \mathcal{V}'$ forms a closed meromorphic OPA, the FLM Moonshine module [1]. The r projection ensures the absence of untwisted massless operators whereas the twisted sector operators are all massive since the twisted vacuum energy is $1/2$. Thus the orbifold partition function is

$$\text{Tr}_{\mathcal{V}^\natural}(q^{L_0}) = \mathcal{P}_r \begin{smallmatrix} \\ 1 \end{smallmatrix}\square + \mathcal{P}_r \begin{smallmatrix} \\ r \end{smallmatrix}\square = J(\tau) . \quad (4)$$

The absence of massless operators in $\mathcal{V}^\natural$ sets the Moonshine module apart from other CFTs. Usually such operators are present and form a closed Kac-Moody algebra. However, the 196884 conformal weight 2 operators in $\mathcal{V}^\natural$, including the energy-momentum tensor $T(z)$ can be used to define a closed non-associative commutative algebra. FLM demonstrated [1] that this algebra is an affine version of the 196883 dimensional Griess algebra [10] together with $T(z)$. The automorphism group of the Griess algebra is the Monster M. FLM showed that M is the automorphism group for the OPA of $\mathcal{V}^\natural$ where the operators of $\mathcal{V}^\natural$ of a given conformal weight form a (reducible) representation of M. This demonstrates an observation of McKay and Thompson [11] that the coefficients of $J(\tau)$ are positive sums of dimensions of irreducible representations of M e.g. the coefficient of q is $196884 = 1 + 196883$, the sum of the trivial and adjoint representation.

We may identify an involution $i \in M$, defined like a 'fermion number', under which all untwisted (twisted) operators have eigenvalue $+1(-1)$ where i also respects

(3). The centraliser of i can be found [1] to give $C(i|M) = \{g \in M | ig = gi\} = 2_+^{1+24}.Co_1$ where Co_1 is the Conway simple group (the automorphism group Co_0 of Λ modulo the reflection automorphism $\bar{r}$), 2_+^{1+24} is an extra-special group and $A.B$ denotes a group with normal subgroup A with $B = A.B/A$. This result is an essential part of the FLM construction since M is generated by $2_+^{1+24}.Co_1$ and a second involution σ [10]. FLM constructed σ, which mixes the untwisted and twisted sectors, from a hidden triality symmetry [1,12] and hence showed that the automorphism group of $\mathcal{V}^\natural$ is M.

The automorphisms i and r can be said to be 'dual' to each other in the sense that they are both automorphisms of $\mathcal{V}'$ and that the subsets invariant under i and r, $\mathcal{V}^\Lambda$ and $\mathcal{V}^\natural$ respectively, form meromorphic OPAs. In addition, we may 'reorbifold' $\mathcal{V}^\natural$ with respect to i to reproduce $\mathcal{V}^\Lambda$. Thus

$$\begin{array}{ccc} & \mathcal{V}' & \\ {}^{\mathcal{P}_i}\swarrow & & \searrow^{\mathcal{P}_r} \\ \mathcal{V}^\Lambda & \underset{i}{\overset{r}{\rightleftarrows}} & \mathcal{V}^\natural \end{array} \quad (5)$$

where the horizontal arrows denote an orbifolding and the diagonal arrows a projection [13].

MONSTROUS MOONSHINE

The operators of $\mathcal{V}^\natural$ of a given conformal weight form reducible representations of M. The Thompson series $T_g(\tau)$ for $g \in M$ is defined by the trace

$$T_g(\tau) = \text{Tr}_{\mathcal{V}^\natural}(gq^{L_0}) = \frac{1}{q} + 0 + [1 + \chi(g)]q + \ldots , \quad (6)$$

which depends only on the conjugacy class of g where $\chi(g)$ is the character in the 196883 dimensional irreducible representation. Thus for i defined above, it is easy to show $T_i(\tau) = [\eta_{\bar{r}}(\tau)]^{-1} + 24$.

The Thompson series for the identity element is $J(\tau)$ which is unique (up to a constant) for the following reasons. Let $\mathcal{F} = H/\Gamma$ be the fundamental region where $\Gamma = SL(2, Z)$ is the full modular group and H is the upper half complex plane. Adding the point at infinity, the compactification $\overline{\mathcal{F}}$ is isomorphic to the Riemann sphere of genus zero where the function $J(\tau)$ realises this isomorphism. Such a function is called a *hauptmodul for the genus zero modular group* Γ. A modular invariant meromorphic function is a hauptmodul if and only if it possesses a unique simple pole. Once the location of this pole is specified, this function is itself unique up to a constant cf. [5,14].

Based on 'experimental' evidence, Conway and Norton [15] conjectured that each $T_g(\tau)$ is a hauptmodul for a genus zero modular group Γ_g. This has recently been rigorously demonstrated by Borcherds although the origin of the genus zero property remains obscure [16]. In general, for g of order n, $T_g(\tau)$ is found to be invariant up to phases of order (at most) h under $\Gamma_0(n) = \{\begin{pmatrix} a & b \\ nc & d \end{pmatrix} | \det = 1\}$ where $h|n$ and $h|24$. $T_g(\tau)$ is fixed by Γ_g with $\Gamma_0(N) \subseteq \Gamma_g \subseteq \mathcal{N}(N) = \{\rho \in SL(2, R) | \rho\Gamma_0(N) = \Gamma_0(N)\rho\}$, the normaliser of $\Gamma_0(N)$ in $SL(2, R)$ where $N = nh$. Furthermore, Γ_g is a genus zero modular group and $T_g(\tau)$ is the corresponding hauptmodul with a simple pole at $q = 0$. Consider the elements of prime order $n = p$. Apart from one class of order 3 with $h = 3$, we have $h = 1$ in each case. Thus either $\Gamma_g = \Gamma_0(p)$ or $\Gamma_0(p)+$, generated by $\Gamma_0(p)$ and the Fricke involution $W_p : \tau \to -1/p\tau$ with $W_p^2 = 1$, the only non-trivial element of $\mathcal{N}(p)$. $\Gamma_0(p)$ is of genus zero when $(p-1)|24$ ($p = 2, 3, 5, 7, 13$) where the hauptmodul is $[\eta(\tau)/\eta(p\tau)]^{2d} + 2d$ with $2d = 24/(p-1)$. There is a class of M denoted by $p-$ for each such prime with this Thompson series e.g. the involution i belongs to the class $2-$. $\Gamma_0(p)+$ is of genus zero for $2 \leq p \leq 31$ or $p = 41, 47, 59, 71$, which constitute all the prime divisors of the order of M [17]. Similarly, there is a class of M, denoted by $p+$, for each such prime with Thompson series fixed by $\Gamma_0(p)+$.

It is natural to interpret the Thompson series $T_g(\tau)$ as a contribution to the partition function for a further orbifolding of $\mathcal{V}^\natural$ with respect to g [18,14]. In particular, we expect that under $\tau \to -1/\tau$, $T_g(\tau)$ transforms to the partition function for a g twisted sector $\mathcal{V}_g$ as follows:

$$T_g(\tau) = g \,\square^\natural_{\,1} \to 1 \,\square^\natural_{\,g} = N_g q^{E_0^g} + \ldots . \tag{7}$$

where $\natural$ denotes a trace contribution to the orbifolding of $\mathcal{V}^\natural$ and $\mathcal{V}_g$ has vacuum energy E_0^g and degeneracy N_g. For many classes of M, the method of construction of $\mathcal{V}_g$ is not known. However, for certain elements discussed below and some others, a construction can be given [14,13].

Consider now this orbifold picture of $T_g(\tau)$ for the prime classes $p+$ and $p-$, although the analysis given can be generalised to all classes [14,19,13]. Under a modular transformation $\gamma : \tau \to \frac{a\tau+b}{c\tau+d}$ we find $g \,\square^\natural_{\,1} \to g^{-d} \,\square^\natural_{\,g^c}$ assuming that no extra global phase occurs [20] (such a phase corresponds to $h \neq 1$ in the original Moonshine conjectures [14,13]). For $\gamma \in \Gamma_0(p)$ with $c = 0 \mod p$ we find $\gamma : T_g(\tau) \to T_{g^{-d}}(\tau) = T_g(\tau)$ since d and p are relatively prime and $T_g(\tau)$ is $\Gamma_0(p)$ invariant.

The genus zero property can be also understood as follows. $T_g(\tau)$ always has a simple pole at $q = 0$ ($\tau = \infty$). The only other possible pole for $T_g(\tau)$ is at $\tau = 0$ since the fundamental region $\mathcal{F}_p = H/\Gamma_0(p)$ for $\Gamma_0(p)$ has only these two cusp points [21]. From (7), $T_g(\tau)$ has a pole at $\tau = 0$ if and only if $E_g^0 < 0$. Thus $T_g(\tau)$ is a hauptmodul for $\Gamma_0(p)$ if and only if $E_g^0 \geq 0$. Also from (7), $T_g(W_p(\tau)) = 1 \,\square^\natural_{\,g}(p\tau)$, so that $T_g(\tau)$ is a hauptmodul for $\Gamma_0(p)+$ if and only if $E_g^0 = -1/p$ and $N_g = 1$.

For classes of type $p+$, $T_g(\tau) = 1 \,\square^\natural_{\,g}(p\tau)$ is a series in q with non-negative coefficients since the RHS of (7) is the $\mathcal{V}_g$ partition function. For classes of type $p-$, $T_g(\tau)$ has coefficients of mixed sign. In general, all classes of M can be divided into two such types i.e. classes with Thompson series invariant (or not invariant) under a Fricke involution $W_N : \tau \to -1/N\tau$ which are called Fricke (or non-Fricke) classes. There are a total of 121 Fricke classes all of which have non-negative coefficient Thompson series and 51 non-Fricke classes with mixed sign coefficients for similar reasons to the prime ordered classes described. This division of the classes of M will be important below.

THE FLM UNIQUENESS CONJECTURE

FLM have conjectured that $\mathcal{V}^\natural$ is characterised (up to isomorphism) as follows [1]: $\mathcal{V}^\natural$ *is the unique meromorphic conformal field theory with modular invariant partition function $J(\tau)$ and central charge 24*. This is analogous to the uniqueness property of the Leech lattice as being the only even self-dual lattice in 24 dimensions without roots.

Let us now consider orbifold models based on other automorphisms a of the untwisted Leech lattice theory $\mathcal{V}^\Lambda$ lifted from automorphisms $\bar{a} \in Co_0$ [19,22]. $\bar{a}$ will be chosen so that each model contains no massless operators, has a meromorphic OPA and is modular invariant with partition function $J(\tau)$ and hence, should reproduce $\mathcal{V}^\natural$. Each $\bar{a} \in Co_0$ can be parameterised as follows

$$\det(x - \bar{a}) = \prod_{k|n}(x^k - 1)^{a_k}, \tag{8a}$$

$$\sum_{k|n} a_k = 0, \tag{8b}$$

with $\sum_{k|n} k a_k = 24$ where $k|n$ denotes that k divides n, the order of $\bar{a}$ and $\{a_k\}$ are integers. (8b) is imposed to ensure the absence of fixed points for $\bar{a}$ so that no massless operators in $\mathcal{V}^\Lambda$ survive the $\mathcal{P}_a$ projection. For $n = p$ prime, we have $a_p = -a_1 = 2d$ where $(p - 1)2d = 24$.

Since a is an OPA automorphism for $\mathcal{V}^\Lambda$, the a invariant subspace $\mathcal{P}_a \mathcal{V}^\Lambda$ also forms a closed meromorphic OPA. The partition function $\text{Tr}_{\mathcal{P}_a \mathcal{V}^\Lambda}(q^{L_0})$ is not modular invariant, as before, necessitating the introduction of sectors $\mathcal{V}_a$ twisted by a. Thus under $\tau \to -1/\tau$

$$a\,\square_1 = \frac{1}{\eta_{\bar{a}}(\tau)} \to {}_1\square_a = D_a^{1/2} \prod_{k|n} \eta(\tau/k)^{-a_k} = D_a^{1/2} q^{E_0^a}(1 + O(q^{1/n})) \tag{9}$$

with $\eta_{\bar{a}}(\tau) = \prod_k \eta(k\tau)^{a_k}$ and $D_a = \det(1 - \bar{a})$ where $D_a^{1/2}$ and $E_0^a = -\frac{1}{24}\sum_k \frac{a_k}{k}$ are the degeneracy and energy of the a twisted vacuum. Under $\tau \to \tau + n$, the a twisted partition function is invariant up to a phase $\exp(2\pi i n E_0^a)$. For modular consistency of the orbifold partition function we must have $nE_0^a = 0$ mod 1 i.e. there is no global phase anomaly [20]. In addition, if $E_0^a > 0$, then the a twisted sector does not reintroduce massless states. We therefore restrict ourselves to $\bar{a} \in Co_0$ obeying [19]

$$\sum_{k|n} a_k = 0, \tag{10a}$$

$$E_0^a > 0, \tag{10b}$$

$$nE_0^a = 0 \text{ mod } 1. \tag{10c}$$

There are a total of 38 classes of Co_0 [23] that obey these constraints [19]. If we relax condition (10c) then a further 13 classes of Co_0 obey only (10a-b) [24,13]. Each of these 13 classes is characterised by some $h \neq 1$ where $h|24$ with $h|k$ for all $a_k \neq 0$. In all 51 cases the parameters $\{a_k\}$ obey $a_k = -a_{nh/k}$ and so $E_0^a = 1/nh$ which violates (10c) for $h \neq 1$. ${}^a\square_1$ is invariant up to phases of order h under $\Gamma_0(n)$ and is a hauptmodul for Γ_a with $\Gamma_0(N) \subseteq \Gamma_a \subset \mathcal{N}(N)$, $N = nh$, where Γ_a is one of the genus zero modular groups considered by Conway and Norton. Furthermore, since $E_0^a > 0$, ${}^a\square_1$ cannot be Fricke invariant and hence these 51 hauptmoduls are the 51 non-Fricke Monster group hauptmoduls. Thus there is a correspondence between 51 classes $\{\bar{a}\}$ of Co_0 and the 51 non-Fricke classes of M. We will explicitly identify an element $g_n \in M$ of each such class below.

$\mathcal{V}_a$ with the partition function ${}_1\square_a$ of (9) has a standard construction [25]. Likewise, $\mathcal{V}_{a^k}$ twisted sectors must be introduced for modular invariance and OPA closure. Then the following intertwining non-meromorphic OPA should hold (schematically)

$$\psi_{a^j} \psi_{a^k} \sim \psi_{a^{j+k}} \tag{11}$$

with $\psi_{a^k}(z) \in \mathcal{V}_{a^k}$. Apart from the original Z_2 case, this OPA has only been rigorously constructed in the prime ordered cases [22]. We will assume that it is true in general. We therefore enlarge $\mathcal{V}^\Lambda$ by the introduction of $\mathcal{V}_{a^k}$ to $\mathcal{V}' = \mathcal{V}^\Lambda \oplus \mathcal{V}_a \oplus ... \oplus \mathcal{V}_{a^{n-1}}$ which forms a closed non-meromorphic OPA. The projection $\mathcal{V}^a_{\mathrm{orb}} = \mathcal{P}_a \mathcal{V}'$ then forms a meromorphic OPA. (10c) is sufficient to guarantee the modular invariance of the corresponding partition function. (10b) can be also shown to be sufficient to ensure no massless operators appear in $\mathcal{P}_a \mathcal{V}_{a^k}$ [19,13]. Thus, for the 38 automorphisms obeying (10a-c), the partition function is modular invariant and is given by $Z_{\mathrm{orb}}(\tau) = J(\tau)$. Therefore $\mathcal{V}^a_{\mathrm{orb}} \equiv \mathcal{V}^\natural$ according to the FLM uniqueness conjecture. Let us now consider some evidence to support this.

Let M^a_{orb} be the automorphism group of the OPA for $\mathcal{V}^a_{\mathrm{orb}}$ where we expect $M = M^a_{\mathrm{orb}}$. We define $i_a \in M^a_{\mathrm{orb}}$ of order n (which generalises the involution i in the original FLM construction) under which all the operators of $\mathcal{V}_{a^k}$ are eigenstates with eigenvalue $e^{2\pi i k/n}$. i_a is also an automorphism of $\mathcal{V}'$ and is 'dual' to the automorphism a where $\mathcal{P}_a \mathcal{V}' = \mathcal{V}^a_{\mathrm{orb}}$ and $\mathcal{P}_{i_a} \mathcal{V}' = \mathcal{V}^\Lambda$. Furthermore, we may reorbifold $\mathcal{V}^a_{\mathrm{orb}}$ with respect to i_a to reproduce $\mathcal{V}^\Lambda$ as before [13]

$$\begin{array}{ccc} & \mathcal{V}' & \\ {}^{\mathcal{P}_{i_a}}\swarrow & & \searrow^{\mathcal{P}_a} \\ \mathcal{V}^\Lambda & \underset{i_a}{\overset{a}{\rightleftarrows}} & \mathcal{V}^a_{\mathrm{orb}} \end{array} \qquad (12)$$

Thus if $\mathcal{V}^a_{\mathrm{orb}} \equiv \mathcal{V}^\natural$, we can explicitly construct the twisted sectors $\mathcal{V}_{i_a^k}$ assumed earlier for $i_a \in M$. We may also compute the Thompson series for $i_a \in M^a_{\mathrm{orb}}$ by taking the trace over $\mathcal{V}^a_{\mathrm{orb}}$ to obtain

$$T^{\mathrm{orb}}_{i_a}(\tau) = \mathrm{Tr}_{\mathcal{V}^a_{\mathrm{orb}}}(i_a q^{L_0}) = \frac{1}{\eta_{\bar{a}}(\tau)} - a_1 \qquad (13)$$

which is the hauptmodul for the genus zero modular group Γ_a introduced earlier [19]. Thus each $i_a \in M^a_{\mathrm{orb}}$ dual to a has the same Thompson series as a corresponding non-Fricke element of M e.g. for $\bar{a}$ of prime order p, $T^{\mathrm{orb}}_{i_a}(\tau) = [\eta(\tau)/\eta(p\tau)]^{2d} + 2d = T_{p-}(\tau)$. Note also, from (7), that $\mathcal{V}_{i_a}$ has vacuum energy $E^{i_a}_0 = 0$ and degeneracy $-a_1 > 0$. (13) may be generalised to the other 13 classes violating (10c) where $\bar{a}^h$, of order $n' = n/h$, can be employed to construct an orbifold with partition function $J(\tau)$. Let g_n denote the lifting of $\bar{a}$ where $g_n^h = i_{a^h}$ is dual to a^h, a lifting of $\bar{a}^h$ (for $h = 1$, $g_n = i_a$). We may then compute the Thompson series for g_n as a trace over $\mathcal{V}^{a^h}_{\mathrm{orb}}$ to show that (13) again holds so that g_n has the same Thompson series as a non-Fricke element of M [13].

We may also compute the centraliser $C(g_n|M^{a^h}_{\mathrm{orb}}) = \{g \in M^{a^h}_{\mathrm{orb}} | g_n^{-1} g g_n = g\}$. For the 38 classes with $h = 1$ this consists of automorphisms that do not mix the sectors $\mathcal{P}_a \mathcal{V}_{a^k}$. For the other 13 automorphisms g_n, $C(g_n|M^{a^h}_{\mathrm{orb}}) \subset C(i_{a^h}|M^{a^h}_{\mathrm{orb}})$. In general, $c \in C(i_a|M^a_{\mathrm{orb}})$ must commute with a and therefore c is lifted from the automorphism $\bar{c} \in G_n = C(\bar{a}|\mathrm{Co}_0)/\langle\bar{a}\rangle$. One can then show that [24,13]

$$C(g_n|M^{a^h}_{\mathrm{orb}}) = \hat{L}_{\bar{a}} . G_n \qquad (14)$$

where $\hat{L}_{\bar{a}} = n.L_{\bar{a}}$, an extension of $L_{\bar{a}} = \Lambda/(1-\bar{a})\Lambda$ by a cyclic group of order n. $\hat{L}_{\bar{a}}$ arises from the vacuum structure of $\mathcal{V}_a$ where $D_a = |L_{\bar{a}}|$. With $M^a_{\mathrm{orb}} = M$, (14) generalises a a well-known observation of Conway and Norton concerning the 5 prime classes where $C(p-|M) = p^{1+2d}_+.G_p$ and $i_a = p-$ [15]. For the other 46 classes, there are 11 cases for which (14) can be checked using the available information about these centralisers [15,26]. In general, the order of these groups agrees with (14) in each case supporting the very likely validity of the result.

Both (13) and (14) support the conjecture that $\mathcal{V}^a_{\text{orb}} \equiv \mathcal{V}^\natural$. This can only be proved by finding a generalised version of σ in the FLM construction which mixes the untwisted and twisted sectors [1,12] i.e. there should exist some permutation group Σ_n which mixes the sectors of $\mathcal{V}^a_{\text{orb}}$ where $C(g_n|M)$ and Σ_n generate M. In the prime cases $p \neq 2$, Σ_p has been recently constructed and it has been rigorously shown that $M^a_{\text{orb}} = M$ for $p = 3$ and almost so for $p = 5, 7, 13$ [22].

MONSTROUS MOONSHINE FROM THE UNIQUENESS OF $\mathcal{V}^\natural$.

Let us now assume that the FLM Uniqueness conjecture is correct. We can then argue that *Thompson series are hauptmoduls if and only if orbifolding $\mathcal{V}^\natural$ with respect to elements of M reproduces $\mathcal{V}^\natural$ or $\mathcal{V}^\Lambda$*. Thus Monstrous Moonshine is intimately linked to the uniqueness of $\mathcal{V}^\natural$.

From (12), orbifolding $\mathcal{V}^\natural$ with respect to the 38 non-Fricke elements i_a dual to a reproduces $\mathcal{V}^\Lambda$. We may similarly consider the orbifolding of $\mathcal{V}^\natural$ with respect to the Fricke elements $\{f\}$ with $h = 1$ which lead to a modular invariant theory $\mathcal{V}^f_{\text{orb}}$ [14,13], given that the operators $\mathcal{V}_{f^k}$ can be constructed. Assuming that the Thompson series are hauptmoduls we find that $\mathcal{V}^f_{\text{orb}} \equiv \mathcal{V}^\natural$ i.e. orbifolding $\mathcal{V}^\natural$ with respect to a Fricke automorphism reproduces $\mathcal{V}^\natural$ again. Thus we have [13]

$$\mathcal{V}^\Lambda \; \underset{i_a}{\overset{a}{\rightleftarrows}} \; \mathcal{V}^\natural \; \overset{f}{\longleftrightarrow} \; \mathcal{V}^\natural \;. \tag{15}$$

For example, consider f an element of a prime class $p+$. Fricke invariance implies $1\square^\natural_{f^k} = T_f(\tau/p) = q^{-1/p} + 0 + O(q^{1/p})$ so that there is a 'gap' in the spectrum of $\mathcal{V}_{f^k}$ and no massless operators are reintroduced in orbifolding $\mathcal{V}^\natural$. Thus the modular invariant partition function for $\mathcal{V}^f_{\text{orb}}$ is $J(\tau)$ and hence $\mathcal{V}^f_{\text{orb}} \equiv \mathcal{V}^\natural$. A similar argument can be made in the general case [13].

The converse to the above also holds i.e. assuming that (15) is true for all automorphisms of M that define a modular consistent theory, then the Thompson series are hauptmoduls. To see this, firstly consider an orbifolding with respect to $i_a \in M$ which reproduces $\mathcal{V}^\Lambda$. i_a must be dual to one of the 38 automorphisms obeying (10a-c) and has non-Fricke invariant Thompson series (13) which is the hauptmodul for a genus zero group. Similarly, as discussed above, the other non-Fricke automorphisms can also be found with a corresponding genus zero Thompson series. For the remaining Fricke classes of M we provide an argument for f an element of prime order. We wish to show that $\mathcal{V}_f$ has the correct vacuum structure so that $T_f(\tau)$ is a hauptmodul for $\Gamma_0(p)+$. In the orbifolding of $\mathcal{V}^\natural$ with respect to f which reproduces $\mathcal{V}^\natural$, let $i_f \in M$ be dual to f with eigenvectors $\mathcal{V}_{f^k}$ for eigenvalue $e^{2\pi i k/n}$. Then it can be shown that $T_{i_f}(\tau) = T_f(\tau)$ so that i_f is in the same class as f. Furthermore, the centralisers obey $C(f|M) \subseteq C(i_f|M)$ with the necessary equality only when the $\mathcal{V}_f$ vacuum is unique i.e. $N_f = 1$. Since the twisted sector $\mathcal{V}_f$ does not reintroduce massless operators, the vacuum energy obeys either (a) $E^0_f = -1/p$ or (b) $E^0_f > 0$. (a) is possible because the absence of massless operators in $\mathcal{V}^\natural$ allows for a similar 'gap' in the spectrum of $\mathcal{V}_g$. If (b) holds, then $T_f(\tau)$ has a unique simple pole at $q = 0$ and must be a hauptmodul for $\Gamma_0(p)$ with $(p-1)|24$ and $T_f(\tau) = [\eta(\tau)/\eta(p\tau)]^{2d} + 2d$. However, this is impossible since then $E^0_f = 0$ with $N_f = 2d$ from (7). Thus (15) implies that $\mathcal{V}_f$ has vacuum structure $E^0_f = -1/f$ with $N_f = 1$ and hence, as described before, $T_f(\tau)$

is a hauptmodul for the genus zero group $\Gamma_0(p)+$ and f is of class $p+$. A similar argument can be given for the other Fricke classes [13].

REFERENCES

[1] Frenkel, I., Lepowsky, J. and Meurman, A., Proc. Natl. Acad. Sci. USA **81** (1984) 3256; J. Lepowsky *et al.* (eds.), Vertex operators in mathematics and physics, (Springer Verlag, New York, 1985); Vertex operator algebras and the monster, (Academic Press, New York, 1988).

[2] Dixon, L., Harvey, J.A., Vafa, C., and Witten, E., Nucl. Phys. **B261** (1985) 678; Nucl. Phys. **B274** (1986) 285.

[3] Conway, J.H. and Sloane, N.J.A., Sphere packings, lattices and groups, (Springer Verlag, New York, 1988).

[4] Goddard, P., Proceedings of the CIRM Luminy conference, 1988, (World Scientific, Singapore, 1989).

[5] Serre, J-P., A course in arithmetic, (Springer Verlag, New York, 1970).

[6] Ginsparg, P., Les Houches, Session XLIX, 1988, "Fields, strings and critical phenomena", ed. E. Brezin and J. Zinn-Justin, Elsevier Science Publishers (1989).

[7] Dixon, L., Friedan, D., Martinec, E. and Shenker, S., Nucl. Phys. **B282** (1987) 13.

[8] Corrigan, E. and Hollowood, T.J., Nucl. Phys. **B304** (1988) 77.

[9] Dolan, L., Goddard, P. and Montague, P., Nucl. Phys. **B338** (1990) 529.

[10] Griess, R., Inv. Math. **68** (1982) 1.

[11] Thompson, J.G., Bull. London Math. Soc.**11** (1979) 347.

[12] Dolan, L., Goddard, P. and Montague, P., Phys. Lett. **B236** (1990) 165.

[13] Tuite, M.P., DIAS preprint 1992, in preparation.

[14] Tuite, M.P., Comm. Math. Phys. **146** (1992) 277.

[15] Conway, J.H. and Norton, S.P., Bull. London. Math. Soc. **11** (1979) 308.

[16] Borcherds, R., Univ. Cambridge DPMMS preprint 1989.

[17] Ogg, A., Bull. Soc. Math. France **102** (1974) 449.

[18] Dixon, L. Ginsparg, P. and Harvey, J.A., Comm. Math. Phys. **119** (1988) 285.

[19] Tuite, M.P., DIAS-STP-90-30, to appear in Comm. Math. Phys..

[20] Vafa, C., Nucl. Phys. **B273** (1986) 592.

[21] Gunning, R.C., Lectures on modular forms, (Princeton University Press, 1962).

[22] Dong, C. and Mason, G., U.C. Santa Cruz preprint 1992.

[23] Kondo, T., J. Math. Soc. Japan **37** (1985) 337.

[24] Tuite, M.P., DIAS-STP-91-25 Aug 1991.

[25] Lepowsky, J., Proc. Natl. Acad. Sci. USA **82** (1985) 8295; Kac, V. and Peterson, D., Proceedings of the Argonne symposium on anomalies, geometry, topology, 1985 (World Scientific, Singapore, 1985).; Corrigan, E. and Hollowood, T.J., Nucl. Phys. **B303** (1988) 135.

[26] Wilson, R., J. Alg. **85** (1983) 144.

LIE ALGEBRAS AND POLYNOMIAL SOLUTIONS OF DIFFERENTIAL EQUATIONS

A. Turbiner

Theoretische Physik,
ETH-Honggerberg, Zurich

ABSTRACT

A general method of obtaining linear differential equations having polynomial solutions is proposed. The method is based on an equivalence of the spectral problem for an element of the universal enveloping algebra of some Lie algebra in the "projectivized" representation possessing an invariant subspace and the spectral problem for a certain linear differential operator. As examples, some polynomials connected to $sl_2(\mathbf{R})$, $sl_2(\mathbf{R})_q$, $osp(2,2)$ and so_3 are briefly discussed.

INTRODUCTION

In this talk I am going to speak about new connections between spectral theory, finite-dimensional representations of Lie algebras in differential operators and the theory of Riemann surfaces.

Let us consider the eigenvalue problem for a family of the Schrödinger operators depending holomorphically on some parameter a:

$$\mathcal{H}(a)\Psi(y) = E\Psi(y) , \quad \mathcal{H}(a) = -\Delta + V(y,a) , \tag{1}$$

where $y \in \mathbf{R}^n, \Psi(y) \in \mathcal{L}_2(\mathbf{R}^n)$. Assume, that for almost any real a, the problem (1) has an infinite discrete spectra. In general, the original problem (1) has a transcendental nature being equivalent to the problem of diagonalization of an infinite-dimensional Hamiltonian matrix.

Several years ago[1] a new class of the Schrödinger equations – *quasi-exactly-solvable Schrödinger equations* – was discovered. These problems are characterised by the property that the first several eigenvalues are roots of some explicit algebraic equation for any a and can be found algebraically, while the remaining eigenvalues satisfy a transcendental equation. This property means that these first several eigenvalues form a finite-sheeted Riemann surface as a function of the parameter a in the Schrödinger operator and, in fact, they are branches of some multi-valued analytic function of the parameter a. Also it may serve as an indication that the corresponding Schrödinger operator possesses an invariant sub-space with a basis formed by the eigenfunctions of those eigenvalues. In other words, one can say that the original transcendental problem (1) has an algebraic sub-structure. One of the simplest quasi-exactly-solvable problems is given by the one-dimensional Schrödinger operator with the potential

$$V_0(y) = a^2 y^6 + 2aby^4 + [b^2 - (4n+3)a]y^2, \quad (2)$$

where a, b are real constants, for which the first $(n+1)$ eigenfunctions of positive parity (even in x) can be found algebraically. For instance, if $n = 0$, then the ground state is given by

$$\Psi_0(y) = \exp(-\tfrac{1}{4}ay^4 - \tfrac{1}{2}by^2), \quad E_0 = b, \quad (3)$$

and if $n = 1$, then the two lowest eigenstates of positive parity are known

$$\Psi_\pm(y) = (2ay^2 + b \pm \sqrt{b^2 + 2a})\exp(-\tfrac{1}{4}ay^4 - \tfrac{1}{2}by^2), \quad E_\pm = 3b \mp 2\sqrt{b^2 + 2a}. \quad (4)$$

In general, the eigenvalues of $\mathcal{H}(b)$ with the potential (2) form an infinitely-sheeted Riemann surfaces as a function of parameter the b. However, for the case (3), the branch corresponding to the lowest eigenvalue splits off from the whole Riemann surface corresponding to the eigenvalues of positive parity. For the case (4), the double-sheeted Riemann surface of the first two eigenvalues of positive parity splits off. For an arbitrary, non-negative n in (2), the first $(n+1)$ eigenvalues of positive parity form a $(n+1)$-sheet Riemann surface in b. It is worth emphasising that the first $(n+1)$ eigenfunctions in (1),(2) can also be obtained in a purely algebraic manner. The original problem (1) with the potential (2) is non-trivial, it may serve as a paradigm both for a certain check of non-perturbative methods and for a study of non-perturbative phenomena.[*]

Now let us describe the reason which ensures the algebraic nature of the first $(n+1)$ eigenstates of the Schrödinger operator with the potential (2). The explanation is very striking: the Schrödinger operator $\mathcal{H}_0 = -\Delta + V_0(y)$ after a gauge transformation and change of variable $x = y^2$ becomes a second-order differential operator, which can be represented as the second order polynomial

$$g^{-1}\mathcal{H}g = -4J^0 J^- + 4aJ^+ + 4bJ^0 - 2(n+1)J^-, \quad (5)$$

where $g = \exp(-\tfrac{1}{4}ay^4 - \tfrac{1}{2}by^2)$, in the generators

$$\begin{aligned} J^+ &= x^2\partial_x - nx, \\ J^0 &= x\partial_x - \tfrac{1}{2}n, \\ J^- &= \partial_x, \end{aligned} \quad (6)$$

obeying $sl_2(\mathbf{R})$ commutation relations. If the parameter n in (6) is a non-negative integer, a finite-dimensional representation of dimension $(n+1)$ appears with basis

$$\mathcal{P}_n = \langle 1, x, x^2, \ldots, x^n \rangle. \quad (7)$$

---

[*]This model can be considered as an interacting field theory in 0+1 dimensions. The existence of explicit analytic expressions for some eigenfunctions has already led to quite surprising and sudden phenomenon[2]: discontinuities in the coupling constant a. It turns out there are two analytically disconnected spectral problems corresponding to the potential (2) defined at positive and negative a, respectively.

This sheds light on the appearance of a certain algebraic sub-structure in the transcendental operator $\mathcal{H}$. In fact, it is nothing but the existence of a finite-dimensional invariant sub-space of $\mathcal{H}$ with the polynomial basis (7) multiplied by the factor $\Psi_0(y)$ (see (3)). Evidently, the existence of $\mathcal{H}$ with an invariant sub-space with an explicit basis reduces the problem of finding solutions of the Schrödinger equation belonging to this sub-space to a problem of linear algebra.

Since the existence of second-order differential operators possessing an invariant sub-space (sub-spaces) with an explicitly determined basis is rather important for applications,[†] Let us ask the general question: how can we describe (classify) second-order differential operators having such a property?

1 ORDINARY DIFFERENTIAL OPERATORS

Consider the space of all polynomials of order n with a basis

$$\mathcal{P}_n = \langle 1, x, x^2, \ldots, x^n \rangle , \qquad (8)$$

where n is a non-negative integer and $x \in \mathbf{R}$.

Definition. Let us call the second order differential operator, $T_2(x)$ **quasi-exactly-solvable**, if it preserves the space $\mathcal{P}_n$. Correspondingly, the operator $E_2(x)$, which preserves the infinite flag $\mathcal{P}_0 \subset \mathcal{P}_1 \subset \mathcal{P}_2 \subset \ldots \subset \mathcal{P}_n \subset \ldots$ of spaces of all polynomials, is called **exactly-solvable**.

Lemma 1.1 (i) Suppose $n > 1$. Any quasi-exactly-solvable operator $T_2(x)$ can be represented by a 2-nd degree polynomial of the generators (6) of the algebra $sl_2(\mathbf{R})$. If $n \leq 1$, the part of the quasi-exactly-solvable operator $T_2(x)$ containing derivatives up to the order n can be represented by a n-th degree polynomial in the generators (6).

(ii) Conversely, any polynomial in (6) is quasi-exactly solvable.

(iii) Among quasi-exactly-solvable operators there exist exactly-solvable operators $E_2(x) \subset T_2(x)$.

The proof is based on irreducibility of the space (7) with respect to the action of the operators (6) and the Burnside theorem.

Comment 1. If we define the universal enveloping algebra U_g of a Lie algebra g as the algebra of all polynomials in the generators with a certain ordering, then T_2 for $n > 1$ is simply an element of the universal enveloping algebra $U_{sl_2(\mathbf{R})}$ of the algebra $sl_2(\mathbf{R})$ taken in representation (6). If $n \leq 1$, then T_2 is represented as an element of $U_{sl_2(\mathbf{R})}$ plus $B \frac{d^{n+1}}{dx^{n+1}}$, where B is any linear differential operator of order not higher than $(1-n), n = 0, 1$.

It is quite important to describe under what conditions a quasi-exactly-solvable operator becomes exactly-solvable. In order to do this, let us introduce a grading of the generators (6):

$$\deg(J^+) = +1 , \ \deg(J^0) = 0 , \ \deg(J^-) = -1 , \qquad (9)$$

hence

$$\deg[(J^+)^{n_+}(J^0)^{n_0}(J^-)^{n_-}] = n_+ - n_- . \qquad (10)$$

This grading allows the classification of the operators T_k in a Lie-algebraic sense.

[†]Higher-order differential operators were considered in refs. 3.

Lemma 1.2 A quasi-exactly-solvable operator $T_2 \subset U_{sl_2(\mathbf{R})}$ has no terms of positive grading, iff it is an exactly-solvable operator.

From Lemma 1.1 it follows that the quasi-exactly-solvable operator T_2 has the form

$$T_2 = c_{++}J^+J^+ + c_{+0}J^+J^0 + c_{+-}J^+J^- + c_{0-}J^0J^- + c_{--}J^-J^-$$
$$+ c_+J^+ + c_0J^0 + c_-J^- + c, \qquad (11)$$

where the c's are constants.‡ The number of free parameters is $par(T_2(x)) = 9$. Under the condition $c_{++} = c_{+0} = c_+ = 0$, the operator $T_2(x)$ becomes exactly-solvable (see Lemma 1.2) and the number of free parameters is $par(E_2(x)) = 6$. After substitution of the explicit expression for the generators (6) to (11), we obtain the general form of a quasi-exactly-solvable operator of second-order

$$T_2(x) = \sum_{j=0}^{2} P_{k+j}(x) \partial_x^j \qquad (12)$$

and an exactly-solvable operator of second-order

$$E_2(x) = \sum_{j=0}^{2} P_j(x) \partial_x^j \qquad (13)$$

where the coefficient functions P_m are polynomials of the order m with coefficients connected to the c's.

Among second-order quasi-exactly-solvable operators there exists a quite special particular case, when $T_2(x)$ has two invariant sub-spaces.

Lemma 1.3 If the operator (11) is such that

$$c_{++} = 0 \quad \text{and} \quad c_+ = (\tfrac{1}{2}n - m)c_{+0}, \text{ for } m = 0, 1, 2, \ldots \qquad (14)$$

then the operator $T_2(x)$ preserves both $\mathcal{P}_n$ and $\mathcal{P}_m$. In this case the number of free parameters is $par(T_2(x)) = 7$.

Now we can formulate the main statement of this section

Theorem 1.1 Take the eigenvalue problem for a linear differential operator of second order in one variable

$$T_2\varphi = \varepsilon\varphi, \qquad (15)$$

where T_2 is symmetric. The problem (15) has $(n+1)$ linear independent eigenfunctions in the form of a polynomial in the variable x of order not higher than n, iff T_2 is quasi-exactly-solvable. The problem (11) has an infinite sequence of polynomial eigenfunctions, iff the operator is exactly-solvable.

Comment 2. The "if" part of the first statement is obvious. The "only if" part is a direct corollary of Lemma 1.1

It turns out that following this Theorem, the differential equations possessing classical orthogonal polynomials as eigenfunctions are reproduced as some particular cases of $E_2(x)$. Also all one-dimensional quasi-exactly-solvable problems correspond to $T_2(x)$.

---
‡Due to the irreducibility of the representation (7), the quadratic Casimir operator is a number. Therefore we eliminated the term $c_{00}J^0J^0$ in (11).

The above considerations show the non-existence of other exactly- and quasi-exactly-solvable problems.

The above analysis can be naturally extended by considering the linear finite-difference operators containing the Jackson symbol[9]

$$Df(x) = \frac{f(x) - f(qx)}{x(1-q)}, \qquad (16)$$

where q is some number, instead of the ordinary derivative. All the above statements with minor modifications hold if, instead of the generators (6), the generators[7]

$$\begin{aligned} T^+ &= x^2 D - \{n\}x \ , \\ T^0 &= xD - \hat{n} \ , \\ T^- &= D \ , \end{aligned} \qquad (17)$$

are used, where $\{n\} = \frac{1-q^n}{1-q}$, $\hat{n} \equiv \frac{\{n\}\{n+1\}}{\{2n+2\}}$, obeying the commutation relations for a quantum algebra $g = sl_2(\mathbf{R})_q$ corresponding to the so called second Witten's deformation in a classification of C. Zachos[8] (for details, see 7). If n is a non-negative integer, the algebra (17) has (7) as a finite-dimensional representation. To this extent, all difference equations leading to q-deformed classical polynomials as eigenfunctions are reproduced as spectral problems for a certain second order polynomial element of universal enveloping algebra $U_{sl_2(\mathbf{R})_q}$. In this case the number of free parameters of quasi-exactly- and exactly- solvable operators of the second order is equal 10 and 6, respectively.

2 DIFFERENTIAL OPERATORS IN ONE REAL AND ONE GRASSMANN VARIABLE

Consider the space of all polynomials with a basis

$$\mathcal{R}_n = \langle x^0, x^1, \ldots, x^n, x^0\theta, x^1\theta, \ldots, x^{n-1}\theta \rangle , \qquad (18)$$

where $x \in R$ and θ is a Grassmann variable. The dimension of (18) is equal to $(2n+1)$. Let us consider the following set of operators

$$\begin{aligned} J^+ &= x^2 \partial_x - nx + x\theta\partial_\theta \ , \\ J^0 &= x\partial_x - \tfrac{1}{2}n + \tfrac{1}{2}\theta\partial_\theta \ , \\ J^- &= \partial_x \ , \\ T &= -\tfrac{1}{2}n - \tfrac{1}{2}\theta\partial_\theta \end{aligned} \qquad (19)$$

and

$$Q = \begin{bmatrix} Q_1 \\ Q_2 \end{bmatrix} = \begin{bmatrix} \partial_\theta \\ x\partial_\theta \end{bmatrix} , \quad \bar{Q} = \begin{bmatrix} \bar{Q}_1 \\ \bar{Q}_2 \end{bmatrix} = \begin{bmatrix} x\theta\partial_x - n\theta \\ -\theta\partial_x \end{bmatrix} . \qquad (20)$$

One can easily show that (19)-(20) obey $osp(2,2)$ (anti-) commutation relations and (19) and (20) play the role of bosonic (even) and fermionic (odd) generators, respectively[4]. If the parameter n in (19)-(20) is a non-negative integer, a finite-dimensional representation (of the form (18)) occurs.

One can ask the same questions as in previous section, but about the classification of linear operators mapping (18) to itself. In similar way we can introduce quasi-exactly-solvable $T_2(x, \theta)$ and exactly-solvable $E_2(x, \theta)$ operators.

301

Lemma 2.1 (i) Suppose $n > 1$. Any quasi-exactly-solvable operator $T_2(x, \theta)$, can be represented by a 2-nd degree polynomial of the generators (19),(20) of the algebra $osp(2, 2)$. If $n \leq 1$, the part of the quasi-exactly-solvable operator $T_2(x, \theta)$ containing derivatives up to the order n can be represented by a n-th degree polynomial in the generators (19),(20).

(ii) Conversely, any polynomial in (19),(20) is quasi-exactly solvable.

(iii) Among quasi-exactly-solvable operators there exist exactly-solvable operators $E_2(x, \theta) \subset T_2(x, \theta)$.

The proof is based on irreducibility of the space (18) with respect to the action of the operators (19),(20) and the Burnside theorem.

Analogously to the previous section, we can introduce a grading of the generators (19),(20) and formulate the necessary and sufficient conditions for the description of an exactly-solvable operator. Lemma 2.1 implies that a quasi-exactly-solvable operator is an element of the universal enveloping algebra $U_{osp(2,2)}$, one can calculate the number of free parameters: $par(T_2(x, \theta)) = 25$ and $par(E_2(x, \theta)) = 17$. In order to do so, it is necessary to take into account the quadratic Casimir operators and also some specific relations between the generators, which occur as a consequence of the realisation $osp(2, 2)$ in terms of the differential operators (19),(20). This effectively reduces the number of free parameters. The analogue of Theorem 1.1 holds as well.

3 GENERAL CONSIDERATIONS

The main conclusion of the present consideration and the papers 1, 3 can be formulated as the following theorem:

(Main) Theorem. Take a Lie algebra g, realised in terms of differential operators of first order, which possesses a finite-dimensional irreducible representation $\mathcal{P}$ (a finite module of smooth functions). Any linear differential operator, having the invariant subspace $\mathcal{P}$, which coincides with the above finite module of smooth functions, can be represented by a polynomial in generators of the algebra g plus an operator annihilating $\mathcal{P}$.

Proof. Use the Burnside theorem.

As a probable extension of this theorem, we assume that the following conjecture holds

Conjecture. If a linear differential operator T acting on functions in $\mathbf{R}^k$ does possess a finite-dimensional invariant subspace with polynomial basis, then either
(i) this finite-dimensional space coincides to a certain finite-dimensional representation of some Lie algebra of differential operators,
or
(ii) T also possesses another invariant subspace coinciding to a certain finite-dimensional representation of some Lie algebra of differential operators,
or
(iii) T preserves an infinite flag of spaces of polynomials.

An interesting question is how to describe finite modules of smooth functions in $\mathbf{R}^k$, which can serve as invariant sub-spaces of linear operators. Evidently, if we could find some realisation of a Lie algebra g in differential operators, possessing an irreducible finite module of smooth functions, we can immediately consider the direct sum of several species of g acting on the functions given as the corresponding direct products of spaces, where the original algebra g acts in each of them. For example, the direct sum of two $sl_2(\mathbf{R})$ in the representation (6) leads to operators acting on the functions on $\mathbf{R}^2$ possessing a finite-dimensional invariant subspace given by the rectangular Newton diagram. Taking the direct sum of k species of $sl_2(\mathbf{R})$ acting on $\mathbf{R}^k$, we arrive at a k-dimensional parallelopiped as the invariant sub-space. In this case the algebra $sl_2(\mathbf{R})$ plays the role of a *primary* algebra giving rise to a multidimensional geometrical figure (Newton diagram) as the invariant sub-space of some linear differential operators. In the space $\mathbf{R}^1$ there is only one such "irreducible" Newton diagram – a finite interval, in other words, a space of all polynomials in x of degree not higher than a certain integer. Thus, in $\mathbf{R}^2$ the rectangular Newton diagram is "reducible" stemming from the product of two intervals as one-dimensional Newton diagrams. In $\mathbf{R}^2$ the irreducible convex Newton diagrams are exhausted by different triangles, connected to the algebras $sl_3(\mathbf{R})$ and $\{gl_2(\mathbf{R}) \ltimes \mathbf{R}^{r+1}\}$. We could not find any other convex Newton diagrams in $\mathbf{R}^2$.

Before presenting our present knowledge about Newton diagrams for the $\mathbf{R}^k$ case, let us recall the regular representation (in terms of first order differential operators) of the Lie algebra $sl_N(\mathbf{R})$ given on the flag manifold, which acts on the smooth functions of $N(N-1)/2$ real variables $z_{i,i+q}$, $i = 1, 2, \ldots, (N-1)$, $q = 1, 2, \ldots, (N-i)$. The explicit formulas for the generators are given by (see ref. 10)

$$D(e_i) = \frac{\partial}{\partial z_{i,i+1}} - \sum_{q=i+2}^{N} z_{i+1,q} \frac{\partial}{\partial z_{i,q}},$$

$$D(f_i) = \sum_{q=1}^{i} z_{q,i+1} z_{i,i+1} \frac{\partial}{\partial z_{q,i+1}} + \sum_{q=1}^{i-1} (z_{q,i+1} - z_{q,i} z_{i,i+1}) \frac{\partial}{\partial z_{q,i}}$$
$$+ \sum_{q=i+2}^{N} z_{i,q} \frac{\partial}{\partial z_{i+1,q}} - n_i z_{i,i+1},$$

$$D(\tilde{h}_i) = -\sum_{q=i+1}^{N} z_{i,q} \frac{\partial}{\partial z_{i,q}} + \sum_{q=1}^{i-1} z_{q,i} \frac{\partial}{\partial z_{q,i}} - \sum_{p=1}^{i} n_{i-p}, \tag{21}$$

where we use the notation e_i, f_i, $\tilde{h}_i$ for generators of positive and negative roots, and the Cartan generators, respectively. If n_i are non-negative integers, the finite-dimensional irreducible representation of $sl_N(\mathbf{R})$ will occur in the form of inhomogeneous polynomials in variables $z_{i,i+q}$. The highest weight vector is characterised by integer numbers n_i, $i = 1, 2, \ldots, (N-1)$.

Connected to the space $\mathbf{R}^3$ two, at least, irreducible convex Newton diagrams appear: a tetrahedron, connected to a degenerate finite-dimensional representation of $sl_4(\mathbf{R})$ (the highest weight vector has two vanishing labels and one non-vanishing), and a certain geometrical figure, corresponding to the regular representation of $sl_3(\mathbf{R})$ given on the flag manifold. We do not know if these exhaust all irreducible Newton diagrams or none. For the general case, $\mathbf{R}^k$ there also exist polytopes connected to $sl_{k+1}(\mathbf{R})$ (except one, all labels, characterising the highest weight vector are vanishing) and geometrical figures related to some finite-dimensional regular representations of $sl_{k+1-i}(\mathbf{R})$, $i > 0$. It is easy to show, since the algebra $so_{k+1}(\mathbf{R}) \subset sl_{k+1}(\mathbf{R})$ has the polyhedron as the invariant sub-space,[§] any quadratic polynomial in the generators of the algebra $so_{k+1}(\mathbf{R})$ with real symmetric coefficients can be reduced to the form of the

---
[§]The corresponding finite-dimensional representation is unitary and reducible.

Laplace-Beltrami operator plus a scalar function by means of a gauge transformation (see the proof and discussion in ref. 6).

The above procedure can be extended to the case of Lie super-algebras of first order differential operators and also quantum algebras of first order differential operators. For the former, for linear operators acting on functions of real and Grassmann variables finite-dimensional invariant subspaces other than finite-dimensional representations can appear (see ref. 3). However, these subspaces are connected with the representations of polynomial elements of the universal enveloping algebra and, correspondingly, they can be constructed through finite-dimensional representations of the original super-algebra. For the latter, we could not find any irreducible (non-trivial) Newton diagrams in $\mathbf{R}^k$ for $k > 1$; they will occur, if a quantum space, like the quantum plane where $yx = q\,xy$, is considered instead of the ordinary one.¶ Also one can construct quasi-exactly-solvable and exactly-solvable operators considering "mixed" algebras: a direct sum of Lie algebras, Lie superalgebras and quantum algebras, realised as first order differential and/or difference operators, possessing a finite-dimensional invariant subspace.

We have described above linear differential operators which possess a finite-dimensional invariant sub-space with a polynomial basis. It is certainly possible, using changes of variables and gauge transformations, to obtain operators with an invariant subspace with a non-polynomial basis (but emerging from polynomial one). The open question is: is it possible to find operators possessing a finite-dimensional invariant sub-space with some explicit basis, which can *not* be reduced to a polynomial one using a change of variables and gauge transformation? Up to now no examples of this type have been found (see the Conjecture above). Since the original aim of this whole investigation was the construction of quasi-exactly-solvable Schrödinger operators [1,4-6], a resolution of this problem plays a crucial role in leading to a complete classification of the quasi-exactly-solvable Schrödinger operators.

REFERENCES

[1] A.V. Turbiner, Spectral Riemannian surfaces of the Sturm-Liouville operators and quasi-exactly-solvable problem, *Sov. Math.–Funk. Analysis i ego Prilogenia*, **22** (1988) 92; A.V. Turbiner, Quasi-exactly-solvable problems and $sl(2, R)$ algebra, *Comm. Math. Phys.* **118** (1988) 467.

[2] C.M. Bender and A.V. Turbiner, Analytic continuation of eigenvalue problems, Preprint WUHEP 92-13 (March 1992) *Phys. Lett. A* **173** (1993) 442;
A.V. Turbiner, A new phenomenon of non-analyticity and spontaneous supersymmetry breaking, *Phys. Lett. B* **276** (1992) 95, (E) **291** (1992) 519.

[3] A.V. Turbiner, Lie-algebraic approach to the theory of polynomial solutions. I. Ordinary differential equations and finite-difference equations in one variable, Preprint CPT-92/P.2679 (1992);
A.V. Turbiner, Lie-algebraic approach to the theory of polynomial solutions. II. Differential equations in one real and one Grassmann variable, and 2×2 matrix differential equations in one real variable, Preprint CPT-92/P.2679 and ETH-TH/92/21 (1992);
A.V. Turbiner, Lie-algebraic approach to the theory of polynomial solutions. III. Differential equations in two real variables and general outlook, Preprint ETH-TH/92-34 (1992).

[4] M.A. Shifman and A.V. Turbiner, Quantal problems with partial algebraisation of the spectrum, *Comm. Math. Phys.* **126** (1989) 347.

[5] A. Gonzales-Lopez, N. Kamran and P.J. Olver, Quasi-exactly-solvable Lie algebras of differential operators in two complex variables, Preprint of the School of Math., Univ. Minnesota 2/23/91 (1991).

[6] A.Yu. Morozov, A.M. Perelomov, A.A. Rosly, M.A. Shifman and A.V. Turbiner, Quasi-exactly-solvable problems: one-dimensional analogue of rational conformal field theories, *Int. Journ. Mod. Phys.* **A5** (1990) 803.

¶I am grateful to O.V. Ogievetski for discussion of this topic.

[7] O. Ogievetski and A. Turbiner, $sl(2, \mathbf{R})_q$ and quasi-exactly-solvable problems, Preprint CERN-TH: 6212/91 (1991).

[8] C. Zachos, Elementary paradigms of quantum algebras, *in* Proceedings of the Conference on Deformation Theory of Algebras and Quantization with Applications to Physics, Contemporary Mathematics, J. Stasheff and M.Gerstenhaber (eds.), AMS, in press (1991).

[9] H. Exton, "q-Hypergeometric Functions and Applications", Horwood Publishers, Chichester, 1983.

[10] W. Ruhl, $SL(N)$ Kac-Moody Algebras and Wess-Zumino-Witten Models, *Ann. Phys.* **206** (1991) 368.

TORUS ACTIONS, MOMENT MAPS, AND THE SYMPLECTIC GEOMETRY OF THE MODULI SPACE OF FLAT CONNECTIONS ON A TWO-MANIFOLD

Lisa C. Jeffrey[*][1] and Jonathan Weitsman[††][2]

[1]Downing College,
Cambridge CB2 1DQ, UK

[2]Isaac Newton Institute for Mathematical Sciences
20 Clarkson Road
Cambridge CB2 0EH, UK

ABSTRACT

We summarize recent work ([W],[JW91a], [JW92]) on the symplectic geometry of the moduli space of flat connections on a two-manifold. This work is based on the existence in these moduli spaces of Hamiltonian torus actions. Using these torus actions and the images of the corresponding moment maps we find a simple description of the moduli spaces, and show how it can be used to compute symplectic volumes and other quantities arising in the geometry and topology of the moduli space.

1   INTRODUCTION

In this talk we will review some recent work on the structure of some moduli spaces associated to two-manifolds. We will be particularly interested in the structure of these spaces as symplectic manifolds. The moduli spaces we consider can also be viewed as Kähler varieties once a Riemann surface structure is put on the underlying two-manifold, and the study of the Kähler geometry of the resulting projective varieties

---

[*]Address after Sep.1, 1993: Mathematics Dept., Princeton University, Princeton, NJ 08544, USA
[†]Address after Jan.1, 1993: Mathematics Dept., Columbia University, New York, NY 10027, USA
[‡]Supported in part by NSF Mathematical Sciences Postdoctoral Research Fellowship DMS 88-07291

has been a topic of much interest since the 1960's (see for example [AB]). In our work we show that the study of these spaces as *symplectic* varieties reveals a good deal of structure not obviously present in the Kähler setting.

Let us describe more explicitly the spaces in question. We consider a compact, connected, oriented two-manifold Σ^g of genus g, and associate to it the space of conjugacy classes of representations $\bar{S}_g = Hom(\pi_1(\Sigma^g), G)/G$, where G is a compact simple Lie group which in this talk will be $G = SU(2)$. The space $\bar{S}_g$ contains an open dense set S_g which is a symplectic manifold; we denote the symplectic form on S_g by ω.

The space $\bar{S}_g$ makes its appearance in various areas of mathematics. In *gauge theory*, $\bar{S}_g$ appears as the moduli space of flat connections on the trivial principal G-bundle on the two-manifold Σ^g. In *topology* $\bar{S}_g$ appears in relation with the Casson invariant of homology three-spheres; this invariant is given roughly by the intersection number of two Lagrangian subvarieties in $\bar{S}_g$. In *algebraic geometry* $\bar{S}_g$ appears once a conformal structure is chosen on the underlying two-manifold Σ^g; the space $\bar{S}_g$ is then the moduli space of semistable holomorphic vector bundles on the corresponding Riemann surface, with rank 2, degree zero, and fixed determinant. The final role of $\bar{S}_g$ and the one most closely related to the topic of this conference is its appearance in connection with *topological field theory*. For $\bar{S}_g$ is the classical phase space of Chern-Simons gauge theory, and as such is conjectured to be related to the Witten-Reshetikhin-Turaev (WRT) invariants of three manifolds. To relate our work to that of the other speakers, we will phrase our results about the moduli space in connection with this topic.

Recall that a $2 + 1$ dimensional topological field theory would assign to every two-manifold Σ^g a Hilbert space $\mathcal{H}(\Sigma^g)$, and to every three-manifold M bounding Σ^g an element $v(M) \in \mathcal{H}(\Sigma^g)$. The WRT topological invariants arise from a family of topological field theories, one for each positive integer level k. In these theories the Hilbert space $\mathcal{H}(\Sigma^g)$ is naturally isomorphic to a vector space with a basis naturally identified with a family of marked trivalent graphs; these graphs are obtained from a *pants decomposition* Γ of Σ^g as follows. We decompose Σ^g into $(2g-2)$ pairs of pants P_γ, $\gamma = 1, \ldots, 2g-2$; this gives $(3g-3)$ boundary circles C_i, $i = 1, \ldots, 3g-3$. We can associate to this decomposition a trivalent graph given by assigning a vertex to each pair of pants and an edge connecting two vertices to every boundary circle shared by the two corresponding pairs of pants (see figure 1). A *marked trivalent graph* corresponding to the pants decomposition Γ is a labeling of each edge of the graph by a real number (see figure 2); we denote the real number assigned to the edge associated to the circle C_i by x_i. A marked trivalent graph will be called *admissible* if for every $i = 1, \ldots, 3g-3$ we have

$$0 \leq x_i \leq 1 \quad (A1)$$

and if in addition, for each pair of pants P_γ with boundary $C_{i_1(\gamma)} \cup C_{i_2(\gamma)} \cup C_{i_3(\gamma)}$ we have

$$0 \leq x_{i_1(\gamma)} + x_{i_2(\gamma)} + x_{i_3(\gamma)} \leq 2, \quad (A2)$$

$$|x_{i_1(\gamma)} - x_{i_2(\gamma)}| \leq x_{i_3(\gamma)} \leq x_{i_1(\gamma)} + x_{i_2(\gamma)}. \quad (A3)$$

Let $k \in \mathbb{Z}_+$. A marked trivalent graph will be called *k-integral* if for every $i = 1, \ldots, 3g-3$ and every $\gamma = 1, \ldots, 2g-2$ we have

$$x_i \in \frac{1}{k}\mathbb{Z}, \quad (I1)$$

$$x_{i_1(\gamma)} + x_{i_2(\gamma)} + x_{i_3(\gamma)} \in \frac{1}{k}2\mathbb{Z}. \quad (I2)$$

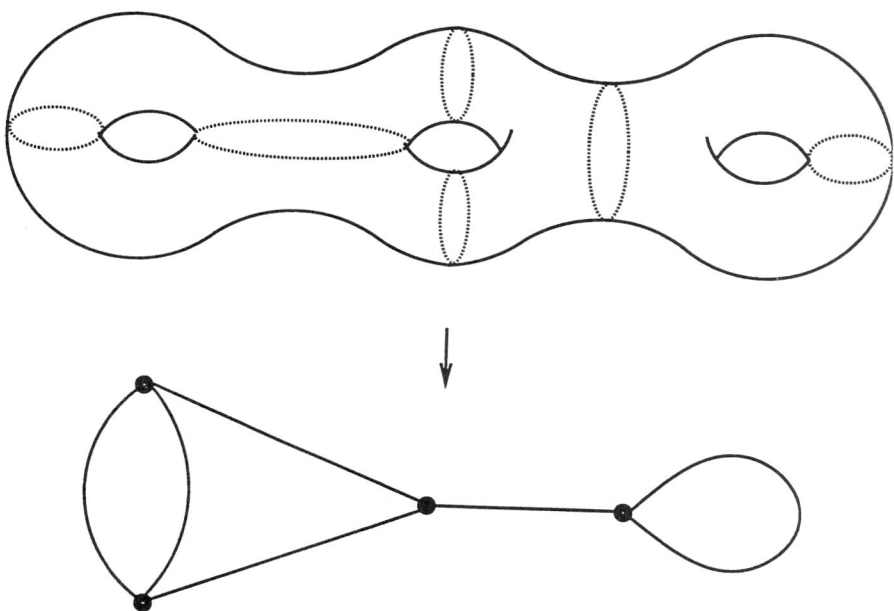

Figure 1: A pants decomposition of a two-manifold and the corresponding trivalent graph.

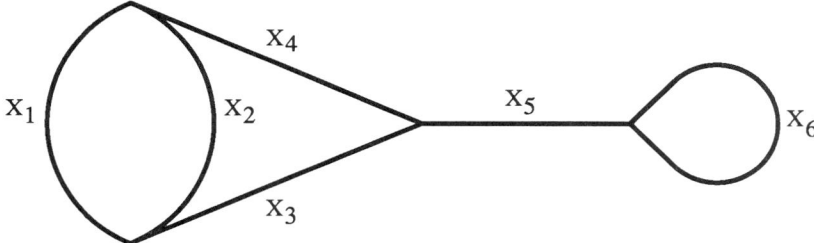

Figure 2: The trivalent graph of figure 1, labelled by real numbers $X_1, \ldots X_6$.

Then the Hilbert space assigned by the level k WRT topological field theory to a surface Σ^g of genus g is naturally isomorphic to the vector space with basis given by generators corresponding to the set $\mathcal{D}(g, k, \Gamma)$ of k-integral admissible markings of the trivalent graph associated to *any* pants decomposition Γ of the surface Σ^g. In particular the dimension of the Hilbert space is equal to the number $D(g, k)$ of such markings (which is in fact independent of the pants decomposition; see [MS]).

The moduli space $\bar{\mathcal{S}}_g$ enters this picture in the following way. On $\bar{\mathcal{S}}_g$ there exists a line bundle $\mathcal{L}$ with connection of curvature ω (see *e.g.* [RSW]). If we choose a Riemann surface structure on Σ^g and work with the corresponding Kähler structure on $\bar{\mathcal{S}}_g$, this line bundle acquires the structure of a hermitian holomorphic line bundle. We may then form a vector space by *geometric quantization* from the space $\bar{\mathcal{S}}_g$ and this Kähler

particular the dimension of the Hilbert space is equal to the number $D(g,k)$ of such markings (which is in fact independent of the pants decomposition; see [MS]).

The moduli space $\bar{\mathcal{S}}_g$ enters this picture in the following way. On $\bar{\mathcal{S}}_g$ there exists a line bundle $\mathcal{L}$ with connection of curvature ω (see *e.g.* [RSW]). If we choose a Riemann surface structure on Σ^g and work with the corresponding Kähler structure on $\bar{\mathcal{S}}_g$, this line bundle acquires the structure of a hermitian holomorphic line bundle. We may then form a vector space by *geometric quantization* from the space $\bar{\mathcal{S}}_g$ and this Kähler polarization; this quantization is none other than the space $H^0(\bar{\mathcal{S}}_g, \mathcal{L}^k)$ of holomorphic sections of the line bundle $\mathcal{L}^k$ for any $k \in \mathbb{Z}$. Then the following theorem holds:

Theorem (Verlinde Dimension Formula): ([V];[Don],[K],[NR],[Sz],[BSz],[Th])

$$\dim H^0(\bar{\mathcal{S}}_g, \mathcal{L}^k) = D(g,k).$$

The purpose of this lecture is to show how the geometry of $\bar{\mathcal{S}}_g$ is mirrored in this remarkable combinatorial formula. In brief, we will show that the symplectic variety $\bar{\mathcal{S}}_g$ is "well approximated" by the toric variety constructed from the convex polyhedron described by the inequalities $(A1) - (A3)$. Morally, then, we would expect the holomorphic sections of a line bundle on this toric variety to be associated to characters of the torus, and hence to the integral points of the polyhedron—that is, to points satisfying $(I1) - (I2)$. The rest of this talk will be devoted to explaining the terms used above, to constructing the torus actions which give the toric variety structure, and to a precise delineation of the words "well approximated." We will then prove the following result:

Theorem 1.1: ([W],[JW91a],[JW92]) Fix a pants decomposition Γ of Σ^g. There exists a dense open set $U \subset \bar{\mathcal{S}}_g$ which is symplectically diffeomorphic to a noncompact toric variety. The convex body associated to this toric variety (via the image of the moment map of the torus action) has as its closure the convex polyhedron described by the inequalities $(A1) - (A3)$.

It is important to note what this theorem does *not* say. We do not claim that the natural complex structure on the toric variety has anything to do with any Kähler structure coming from a Riemann surface; in other words we do not show that the torus actions preserve the complex structures on $U \subset \bar{\mathcal{S}}_g$ coming from any conformal structure on Σ^g. Thus we cannot directly compute the dimension of the space of holomorphic sections by counting integral points. Instead we must content ourselves with computing quantities such as the symplectic volume of the space $\bar{\mathcal{S}}_g$ which can be deduced from the toric structure on U. In one of our papers [JW92], however, we explain how knowledge of such volumes can be combined with techniques in the literature to compute the dimension of this space of holomorphic sections.

A second remark is directed at physicists, who are accustomed to different terminology for the toric structure we describe. In physics language our result is that the symplectic manifold U the phase space of a classical integrable system. Now a classical integrable system may be quantized using the Bohr-Sommerfeld rules; the allowable quantized momenta (expected for a compact phase space) are just given by the integral points of the polyhedron $(A1) - (A3)$, and are therefore in correspondence with the integral admissible markings of the trivalent graph corresponding to the pants decomposition Γ. In [JW91a] we show that this Bohr-Sommerfeld procedure can indeed be justified despite the fact that the integrable system structure does not extend from U to all of $\bar{\mathcal{S}}_g$; and our result shows that the quantization of this space in our real polar-

for an overview of this beautiful theory. We then proceed in section 3 to show how these ideas can be applied to the moduli space $\bar{S}_g$, and to construct the S^1 actions on the toric variety of Theorem 1.1. We follow this in section 4 with a description of the polyhedron, which completes the proof of Theorem 1.1. In section 5 we show how our result can be applied to calculate the symplectic volume of S_g. An application of our methods to the construction of three-manifold invariants is described in Lisa Jeffrey's talk at this conference (see also [JW91b]).

2 SYMPLECTIC MANIFOLDS, TORUS ACTIONS, AND TORIC VARIETIES

Let (M^{2m}, ω) be a $2m$-dimensional, compact, connected symplectic manifold. Suppose we are given a smooth action of the circle group S^1 on M. This action is called *Hamiltonian* if the vector field X which generates this action is symplectically dual to an exact one-form; that is, if there exists a function $\mu : M \to \mathbb{R}$ such that

$$d\mu|_p(v) = \omega|_p(v, X|_p) \qquad (2.1)$$

for any tangent vector v to M at any point $p \in M$.

The function $\mu : M \to \mathbb{R}$ associated to a Hamiltonian circle action in this way is called the *moment map* of this circle action. More generally, we may consider several (say j) commuting circle actions on M. Then the corresponding moment maps may be combined to form a moment map $\mu : M \to \mathbb{R}^j$. The following result characterizes the image of the moment map associated to a Hamiltonian torus action on a compact symplectic manifold.

Convexity Theorem:([A], [GS]) The image $\mu(M)$ of the moment map is a convex polyhedron in $\mathbb{R}^j$.

A particularly interesting case of this situation is where the torus acts effectively and has half the dimension of the manifold; that is where $j = m$. In this case the fibres of the moment map of the torus action on M are the orbits of this action. In other words every point in M is uniquely specified by giving the values of the moment map along with the values of a collection of parameters describing the location of the point along the torus orbit. In classical mechanics the components of the moment map are known as *action variables*, while the parameters describing the orbits can be chosen to be *angle variables*. Since generically the orbits will be diffeomorphic to $(S^1)^m$, these angle variables will be given on an open dense set $V \subset M$ by functions $\phi_i : V \to S^1, i = 1, \ldots, m$. In terms of these functions the symplectic form ω may be written

$$\omega|_V = \sum_i d\mu_i \wedge d\phi_i. \qquad (2.2)$$

Thus a symplectic manifold of dimension $2m$ endowed with a Hamiltonian $(S^1)^m$ action is completely determined by the image of the moment map μ; the manifold is given by the fibering above $\mu(M)$, while the symplectic form is determined by the "globalized" Darboux formula (2.2). A manifold of this type is symplectomorphic to a *toric variety*; that is, to the closure of the orbit of a linear $(\mathbb{C}^*)^m$ action in some complex projective space $\mathbb{C}P^N$, equipped with the symplectic form inherited from $\mathbb{C}P^N$ (see for example [Del]).

We pause to give some simple examples of this situation. The first is just the two sphere S^2, endowed with the usual coordinates θ, ϕ and the symplectic form $\omega = d\cos\theta \wedge d\phi$. The circle action given by rotation about the z-axis has as its generating vector field $\frac{\partial}{\partial \phi}$. Hence the moment map $\mu : S^2 \to \mathbb{R}$ must satisfy

$$d\mu(v) = \omega\left(v, \frac{\partial}{\partial \phi}\right) = d\cos\theta(v);$$

thus $\mu(\theta, \phi) = \cos\theta$ and the moment map is just projection onto the axis of rotation. The image of S^2 under this map is the interval, a convex polyhedron!

Exercise: Show that the convex polyhedron given by the image of the moment map of the linear $(S^1)^m$ action on $\mathbb{C}P^m$ is the m-simplex.

In light of the results summarized in this section, we can rephrase our task in proving the main theorem of this lecture. It is to construct on an open dense set in $\bar{S}_g$ the action of a $3g-3$ dimensional torus, and to compute the image of the corresponding moment map. As we shall see, this moment map will extend naturally to a continuous (but not differentiable!) function defined on all of $\bar{S}_g$, whose image will be the convex polyhedron described by $(A1) - (A3)$.

3 TORUS ACTIONS ON THE MODULI SPACE

We now come to the main focus of this work, which is the construction of the torus action on $\bar{S}_g$. We wish to construct $3g - 3$ commuting Hamiltonian circle actions on this space. Equivalently, we may give the moment maps of those circle actions; these will be functions $\mu_i : \bar{S}_g \to \mathbb{R}$, $i = 1, \ldots, 3g-3$. The functions μ_i are defined as follows.

Recall that a *pair of pants* is a copy of the two holed disc, or alternatively, the three holed sphere. Let us fix a decomposition Γ of the surface Σ^g into pants P_γ, $\gamma = 1, \ldots, 2g-2$. Each of the boundary circles C_i, $i = 1, \ldots, 3g-3$ of the pairs of pants will then give rise to a function $\mu_i : \bar{S}_g \to \mathbb{R}^{3g-3}$. To give a formula for μ_i, we recall that $\bar{S}_g$ is the quotient of $Hom(\pi_1(\Sigma^g), G)$ by the conjugation action of G. So we may define a function on $\bar{S}_g$ by giving a conjugation invariant function $\tilde{\mu}_i : Hom(\pi_1(\Sigma^g), G) \to \mathbb{R}$.

To define such a function we note that given a choice of basepoints, arcs, and orientations, each circle C_i gives a homotopy class $[C_i] \in \pi_1(\Sigma^g)$. Given an element $\rho \in Hom(\pi_1(\Sigma^g), G)$, that is, a representation of $\pi_1(\Sigma^g)$ into G, we may evaluate ρ on $[C_i]$, to give an element of $G = SU(2)$. The trace of $\rho([C_i])$ is conjugation invariant; and being the trace of an $SU(2)$ matrix, it is equal to $2\cos\theta_i$ for some angle $0 \leq \theta_i \leq \pi$. Thus we may define

$$\tilde{\mu}_i : Hom(\pi_1(\Sigma^g), G) \to \mathbb{R}$$

by

$$\tilde{\mu}_i(\rho) = \frac{1}{\pi} \cos^{-1} \tfrac{1}{2} \mathrm{tr}\, \rho([C_i]). \tag{3.1}$$

We have the following lemma.

Lemma 3.1: The function $\tilde{\mu}_i$ is independent of the choices of basepoints, arcs, and orientations, and descends to a function $\mu_i : \bar{S}_g \to \mathbb{R}$.

We claim that the function μ_i is the moment map for a densely defined circle action on $\bar{S}_g$. The formula (3.1) for $\tilde{\mu}_i$ shows why this cannot hold everywhere in $\bar{S}_g$; for where $\rho([C_i]) = \pm 1$, the function $\tilde{\mu}_i = \frac{1}{\pi} \cos^{-1} \tfrac{1}{2} \mathrm{tr}\, \rho([C_i])$ will have infinite derivative, so that the vector field X generating the putative S^1 action (according to formula (2.1)) would be ill-defined.

Away from these bad points, however, the following theorem of W. Goldman applies:

Theorem 3.2:([G]) Let $U_i = \mu_i^{-1}((0,1))$. The function $\mu_i|_{U_i}$ is the moment map for a Hamiltonian circle action of period 1 on U_i.

We thus obtain a circle action in $\bar{S}_g$ for each boundary circle in the pants decomposition Γ. In order to obtain a torus action in $\bar{S}_g$, we must show that these circle actions commute. This is in fact true, and is essentially due to the fact that the corresponding boundary circles are *disjoint*. We summarize this as follows.

Theorem 3.3:([G]) Let $U = \cap_{i=1}^{3g-3} U_i \subset \bar{S}_g$. The functions $\mu_i|_U$ combine to give a function $\mu : U \to \mathbb{R}^{3g-3}$ which is the moment map for an $(S^1)^{(3g-3)}$ action on U.

Note that Theorem 3.3 proves the existence of the toric variety structure on the dense open set $U \subset \bar{S}_g$. This is the main part of the proof of Theorem 1.1. The image of the moment map will be computed in the next section.

Note that circle actions will commute if the corresponding generating vector fields commute. In view of the formula (2.1), this amounts to saying that the functions μ_i form a Poisson commuting family of functions on U. Geometrically, the orbit of any point $x \in U$ under the torus action spans a torus T_x with $\omega|_{T_x} = 0$; this torus is generically of dimension $\dim T_x = 3g - 3$. Hence these tori generically foliate $U \subset \bar{S}_g$ by *Lagrangian* submanifolds. Such a foliation of a symplectic manifold is called a *real polarization* [W].

4 THE IMAGE OF THE MOMENT MAP

We have seen that the function $\mu : U \subset \bar{S}_g \to \mathbb{R}$ gives a moment map for a $(S^1)^{(3g-3)}$ action on U. Since $\dim U = 6g - 6$, this shows that U is fibred by tori over the image $\mu(U)$; in order to understand the structure of U as a symplectic manifold, we have to compute the image $\mu(U) \subset \mathbb{R}^{(3g-3)}$, or, equivalently, its closure $\mu(\bar{S}_g)$. This is the purpose of this section.

To do this we must find which $(3g-3)$-tuples of real numbers $(x_1, \ldots, x_{(3g-3)})$ occur as

$$x_i = \frac{1}{\pi} \cos^{-1} \tfrac{1}{2}\text{tr } \rho([C_i]) \tag{4.1}$$

where C_i is a curve in the fixed pants decomposition Γ of Σ^g, and ρ is a representation of $\pi_1(\Sigma^g)$ in $G = SU(2)$. In order to answer this question, it is useful to observe that it is enough to check whether the real numbers $x_{i_1(\gamma)}, x_{i_2(\gamma)}, x_{i_3(\gamma)}$ associated to each pair of pants come via equation (4.1) from a representation of the fundamental group $\pi_1(P_\gamma)$ of the pair of pants—and to do so for each pair of pants. The easiest way to see this is via gauge theory, and we refer the reader to [JW91a] for details. We state the result as a lemma.

Lemma 4.1: Let there be given a decomposition of the surface Σ^g into pairs of pants $P_\gamma, \gamma = 1, \ldots, 2g - 2$. Label the boundary circles of the pants decomposition by $C_i, i = 1, \ldots, 3g - 3$. Let $(x_1, \ldots, x_{(3g-3)}) \in \mathbb{R}^{(3g-3)}$. Suppose that each pair of pants P_γ has as its boundary $C_{i_1(\gamma)} \cup C_{i_2(\gamma)} \cup C_{i_3(\gamma)}$. Suppose that for each $\gamma = 1, \ldots, 2g - 2$, there exist $g_1(\gamma), g_2(\gamma), g_3(\gamma) \in G$ with

$$x_{i_j(\gamma)} = \frac{1}{\pi} \cos^{-1} \tfrac{1}{2} \operatorname{tr} g_i(\gamma), \quad i = 1, 2, 3$$

and with

$$g_1(\gamma) g_2(\gamma) g_3(\gamma) = 1.^*$$

Then there exists a representation $\rho \in \operatorname{Hom}(\pi_1(\Sigma^g), G)$ such that (4.1) holds.

Remark: This representation is *not* unique, and not even unique up to conjugation; in fact, if it corresponds to a generic point in $\bar{S}_g$, the circle actions will give a $(3g - 3)$ dimensional torus' worth of such representations.

The lemma reduces the task of finding the image of μ to a computation in $SU(2)$; we must find which real numbers x_1, x_2, x_3 occur via equation (4.1) as traces of matrices g_1, g_2, g_3 which multiply to give the identity. The answer to this is the following lemma.

Lemma 4.2: Let $g_1, g_2, g_3 \in SU(2)$ and suppose that $g_1 g_2 g_3 = 1$. Let $x_i = \frac{1}{\pi} \cos^{-1} \tfrac{1}{2} \operatorname{tr} g_i$, $i = 1, 2, 3$. Then

$$0 \leq x_i \leq 1, \tag{4.2a}$$

$$|x_1 - x_2| \leq x_3 \leq x_1 + x_2, \tag{4.2b}$$

$$x_1 + x_2 + x_3 \leq 2. \tag{4.2c}$$

Conversely, every triple of real numbers satisfying (4.2a-c) will occur in this way.

Note that the conditions (4.2a-c) require that x_1, x_2, x_3 form a spherical triangle.

We can now read off the image of the moment map $\mu(\bar{S}_g)$. Combining Lemmas 4.1 and 4.2 we see that

Theorem 4.3: The image $\mu(\bar{S}_g)$ of the map μ is the polyhedron given by the inequalities (A1),(A2),(A3) of Section 1.

Combining Theorems 3.3 and 4.3 we see that we have proved Theorem 1.1 of the Introduction.

5 APPLICATION: THE SYMPLECTIC VOLUME

We now show how Theorem 1.1 can be used to calculate symplectic invariants of the moduli space $\bar{S}_g$. As we noted in the introduction, the torus action on U does not extend to all of $\bar{S}_g$; furthermore, this action does not preserve the complex structure coming from a Riemann surface. Thus we cannot use the standard theorems on holomorphic sections of line bundles on toric varieties (see [Oda]) to prove the Verlinde formula. Instead we content ourselves with the computation of the volume of the moduli space.

Let us return for a moment to the general case of a symplectic manifold (M^{2m}, ω). Its symplectic volume is the integral

$$\operatorname{vol}(M) = \frac{1}{m!} \int_M \omega^m. \tag{5.1}$$

Suppose now that M is symplectomorphic to a toric variety with an effective torus action corresponding to a moment map $\mu : M \to \mathbb{R}^m$. Then by equation (2.2), there exists a dense open set $V \subset M$ with

---

*This is of course just the condition that $g_1(\gamma), g_2(\gamma), g_3(\gamma)$ give a representation of $\pi_1(P_\gamma)$ in G.

$$\omega|_V = \sum_i d\mu_i \wedge d\phi_i$$

for some functions $\phi_i : V \to (S^1)^m$, which parametrize the orbits of the $(S^1)^m$ action. Thus (5.1) becomes

$$\text{vol}(M) = \int_V \bigwedge_{i=1}^m d\mu_i \wedge \bigwedge_{i=1}^m d\phi_i.$$

We may however perform the integrals over the orbit; we have normalized the volume of the orbits to 1, so that

$$\text{vol}(M) = \int_{\mu(V)} \bigwedge_{i=1}^m d\mu_i;$$

in other words,

$$\text{vol}(M) = \text{vol}(\mu(M)). \tag{5.2}$$

Equation (5.2) may be looked upon as a simple case of the Duistermaat-Heckman theorem [DH].

Now $\mu(M)$ is a convex polyhedron in Euclidean space, and we may compute its volume by counting integral points. More precisely suppose that we are given some lattice $\Lambda \subset \mathbb{R}^m$. This lattice will be called *volume-approximating* if a fundamental domain for the lattice action on $\mathbb{R}^m$ has volume 1. Then the volume of any polyhedron P is given by

$$\text{vol}(P) = \lim_{k \to \infty} \frac{1}{k^m} \#(\Lambda \cap kP) \tag{5.3}$$

where the symbol $\#$ is used to denote the number of points in a finite set, and where kP is the polyhedron given by expanding the polyhedron $P \in \mathbb{R}^m$ by a factor of k (that is, the image of P under the self map of $\mathbb{R}^m$ given by componentwise multiplication by k.)

We now combine equations (5.2) and (5.3); if we are given a volume-approximating lattice $\Lambda \subset \mathbb{R}^m$, we have

$$\text{vol}(M) = \lim_{k \to \infty} \frac{1}{k^m} \#(\Lambda \cap k\mu(M)). \tag{5.4}$$

Let us apply this relation to the toric variety $U \subset \bar{\mathcal{S}}_g$. For our integral lattice, we choose the lattice Λ given by conditions $(I1), (I2)$ of the introduction.[†]

$$\text{vol}(\mathcal{S}_g) = \lim_{k \to \infty} \frac{1}{k^m} \#(\Lambda \cap k\mu(U)) = \lim_{k \to \infty} \frac{1}{k^m} D(g, k). \tag{5.5}$$

Equation (5.5) appears in the literature in various guises; see for example [Wit]. In our approach the appearance of the volume as the leading term in the large-k asymptotics of the Verlinde dimension $D(g, k)$ is a simple consequence of Theorem 1.1, which is an expression of the symplectic geometry behind this formula.

[†]There is a slight difference between our setting and that of equation (5.4) in that the torus action corresponding to the moment map μ on U is not effective. There does exist, however, an effective Hamiltonian action on U of a quotient of this torus by a finite group. The same methods then apply, except that the lattice Λ must be substituted for the volume-approximating lattice given by the sole condition $(I1)$; see [JW91a].

REFERENCES

[A] M.F. Atiyah. Convexity and commuting Hamiltonians. *Bull. Lond. Math. Soc.* **14**, (1981) 1.

[AB] M.F. Atiyah, R. Bott. The Yang Mills equations over Riemann surfaces. *Phil. Trans. Roy. Soc. London* **A 308** (1982) 523.

[BSz] A. Bertram, A. Szenes. Hilbert polynomials of moduli spaces of rank 2 vector bundles II. Harvard preprint (1991).

[Del] T. Delzant. Hamiltoniens periodiques et images convexes de l'application moment. *Bull. Soc. Math. France* **116**, (1988) 315.

[Don] S.K. Donaldson. Gluing techniques in the cohomology of moduli spaces. Oxford preprint (1992).

[DH] J.J. Duistermaat, G. Heckman. On the variation in the cohomology of the symplectic form of the reduced phase-space. *Inv. Math.* **69**, (1982) 259.

[G] W. Goldman. Invariant functions on Lie groups and Hamiltonian flows of surface group representations. *Inv. Math.* **85**, (1986) 263.

[GS] V. Guillemin, S. Sternberg. Convexity properties of the moment mapping. *Inv. Math.* **67**, (1982) 491.

[GS84] V. Guillemin, S. Sternberg. "Symplectic techniques in physics", Cambridge University Press, 1984.

[JW91a] L.C. Jeffrey, J. Weitsman. Bohr-Sommerfeld orbits in the moduli space of flat connections and the Verlinde dimension formula. *Commun. Math Phys.* **152**, (1992) 593.

[JW91b] L.C. Jeffrey, J. Weitsman. Half density quantization of the moduli space of flat connections and Witten's semiclassical manifold invariants. IAS preprint IASSNS-HEP-91/94; *Topology*, to appear.

[JW92] L.C. Jeffrey, J. Weitsman. Toric structures on the moduli space of flat connections on a Riemann surface: volumes and the moment map. Institute for Advanced Study Preprint IASSNS-HEP-92/25; *Adv. Math.*, to appear.

[K] F. Kirwan. The cohomology rings of moduli spaces of bundles over Riemann Surfaces. *J. Amer. Math. Soc.*, **5**, 853 (1992).

[MS] G. Moore, N. Seiberg. Classical and quantum conformal field Theory. *Commun. Math. Phys.* **123**, (1989) 77.

[NR] M.S. Narasimhan, T. R. Ramadas. Factorization of generalized theta functions. Tata Institute Preprint, 1991.

[Oda] T. Oda, "Convex Bodies and Algebraic Geometry", Springer Verlag, New York, 1988.

[RSW] T.R. Ramadas, I.M. Singer, J. Weitsman. Some comments on Chern-Simons gauge theory. *Commun. Math. Phys.* **126**, (1989) 409.

[Sz] A. Szenes. Hilbert polynomials of moduli spaces of rank 2 vector bundles I. Harvard preprint (1991).

[Th] M. Thaddeus. Conformal field theory and the cohomology of the moduli space of stable bundles. *J. Diff. Geom.*, **35**, 131 (1992).

[V] E. Verlinde. Fusion rules and modular transformations in 2d conformal field theory. *Nucl. Phys.* **B300**, (1988) 351.

[W] J. Weitsman. Real polarization of the moduli space of flat connections on a Riemann Surface. *Commun. Math. Phys.* **145**, (1992) 425.

[Wit] E. Witten. On quantum gauge theories in two dimensions, *Commun. Math. Phys.* **140**, (1991) 153.

GEOMETRIC QUANTIZATION AND WITTEN'S SEMICLASSICAL MANIFOLD INVARIANTS

Lisa C. Jeffrey[*,1] and Jonathan Weitsman[†‡,2]

[1]Downing College,
Cambridge CB2 1DQ, UK

[2]Isaac Newton Institute for Mathematical Sciences
20 Clarkson Road
Cambridge CB2 0EH, UK

1 INTRODUCTION

The Witten-Reshetikhin-Turaev invariants of three-manifolds may be defined using the axioms of topological quantum field theory. These axioms assign a complex vector space $\mathcal{H}(\Sigma^g)$ (equipped with an inner product) to an oriented surface Σ^g of genus g, and an element $v(N) \in \mathcal{H}(\Sigma^g)$ to a three-manifold N with boundary Σ^g. The vector space $\mathcal{H}(\Sigma^g)$ may be defined combinatorially, but the work of Witten suggests that alternative definition ought to be obtainable using geometric quantization (with respect to a Kähler polarization) of the Kähler variety $\bar{S}_g$ of gauge equivalence classes of flat $SU(2)$ connections. In this framework it is difficult to understand the assignment of $v(N)$ to the three-manifold N.

We shall describe an alternative approach to constructing three-manifold invariants by associating a vector space to a surface Σ^g. In contrast to the difficulty of assigning a holomorphic section of a line bundle over $\bar{S}_g$ to a three-manifold N with boundary Σ^g, there is a natural assignment of a Lagrangian submanifold $L_N \subset \bar{S}_g$ to such a three-manifold. Hence we are led to consider quantization of $\bar{S}_g$ in a *real* polarization, where elements of the quantization are naturally associated to Lagrangian subvarieties. We shall consider *closed* three-manifolds N with a Heegaard decomposition $N = H^g \cup_f H^g$, where H^g is the genus g handlebody and f is a diffeomorphism of $\Sigma^g = \partial H^g$. We

---
[*]Address after Sept.1, 1993: Mathematics Dept., Princeton University, Princeton, NJ 08544, USA
[†]Address after 1 Jan. 1993: Mathematics Department, Columbia University, New York, NY 10027, USA

construct a real polarization of $\bar{S}_g$, one fibre of which is the Lagrangian submanifold L_H associated to $H = H^g$; using this polarization, we define a quantization $\mathcal{H}_r(\Sigma^g)$ of $\bar{S}_g$, in which the subvariety L_H gives rise to a natural choice for $v(H^g)$. We then define an inner product on $\mathcal{H}_r(\bar{S}_g)$, and more generally a pairing between the vector spaces $\mathcal{H}_r(\Sigma^g)$, $\mathcal{H}'_r(\Sigma^g)$ corresponding to two different real polarizations, one of which contains the fibre L_H and the other its image f_*L_H under the diffeomorphism f. Using this pairing, we are able to construct a three-manifold invariant $Z(N)$.

One feature of our three-manifold invariant is that it is closely related to the conjectured large k limit of the Witten-Reshetikhin-Turaev invariant $Z(N,k)$. For $k \in \mathbb{Z}$, Witten used the stationary phase approximation to the heuristic path integral to give the prediction that as $k \to \infty$, the WRT invariant should satisfy an asymptotic expansion in powers of $1/k$, the leading order term of which is of the form (if the gauge equivalence classes of flat connections $\mathcal{A}_F(N)/\mathcal{G}(N)$ on N form a finite set)

$$Z_0(N,k) = \sum_{[A] \in \mathcal{A}_F(N)/\mathcal{G}(N)} e^{2\pi i k CS(A)} \tau^{1/2}(N,A). \tag{1}$$

Where this assumption holds, our three-manifold invariant turns out to be equal to $Z_0(N,k)$. Our result shows that the semiclassical invariant $Z_0(N,k)$ (which in Witten's approach has no obvious relation to the Kähler quantization of $\bar{S}_g$) can be derived from geometric quantization. In addition, this quantization provides a natural assignment for $v(H^g)$.

In the case when the gauge equivalent class of flat connections $\mathcal{A}_F(N)/\mathcal{G}(N)$ form a finite set, our construction of three-manifold invariants is essentially due to D. Johnson (ref. 4); Johnson constructed invariants similar to ours using the Reidemeister torsion $\tau(N,A)$ alone.

A detailed account of our work is given in ref. 2.

2 QUANTIZATION OF THE MODULI SPACE OF FLAT CONNECTIONS VIA A REAL POLARIZATION

(a) The moduli space

Let $G = SU(2)$, $\mathfrak{g} = \mathrm{Lie}(SU(2))$, and let Σ^g be a closed oriented surface of genus g. We define $\bar{S}_g = \mathrm{Hom}(\pi_1(\Sigma^g), G)/G = \mathcal{A}_F(\Sigma^g)/\mathcal{G}(\Sigma^g)$, where $\mathcal{A}_F(\Sigma^g) = \{A \in \Omega^1(\Sigma^g) \otimes \mathfrak{g} \mid F_A = dA + \frac{1}{2}[A,A] = 0\}$ is the space of flat connections on Σ^g, and $\mathcal{G}(\Sigma^g) = \mathrm{Maps}(\Sigma^g, G)$ is the gauge group. The space $\bar{S}_g$ has dimension $6g - 6$, and on the dense open set S_g where $\bar{S}_g$ is smooth, there is a natural symplectic form ω, normalized so that $[\omega]$ is the generator of $H^2(S_g, \mathbb{Z})$.

(b) Real polarizations

Let M be a symplectic manifold of dimension $2n$ with symplectic form ω. A *real polarization* of M is a foliation of M by Lagrangian submanifolds, i.e., a surjective map $\pi : M \to B$ onto a manifold B of dimension n, such that the fibre $\pi^{-1}(b) = L_b$ is Lagrangian (at least for almost all $b \in B$).

In the case of $\bar{S}_g$, we obtain a polarization as follows (see ref. 1 and ref. 6). A simple closed curve C determines a function $\mu_C : \bar{S}_g \to [0,1]$ by $\mu_C([A]) = \frac{1}{\pi} \cos^{-1}(\frac{1}{2} \mathrm{Tr}\, \mathrm{Hol}_C A)$. If $(C_1, \ldots, C_{3g-3})$ are the boundary circles of a decomposition of Σ^g into trinions or "pairs of pants", then the map $\pi = (\mu_{C_1}, \ldots, \mu_{C_{3g-3}}) : \bar{S}_g \to B = \pi(\bar{S}_g) \subset [0,1]^{3g-3}$ gives a real polarization of $\bar{S}_g$.

(c) The Bohr-Sommerfeld set

Suppose in addition that M is equipped with a complex line bundle $\mathcal{L}$ and a *prequantum connection* ∇ on $\mathcal{L}$, i.e., a connection ∇ on $\mathcal{L}$ whose curvature equals the symplectic form. We may then consider the *Bohr-Sommerfeld set* $\Lambda = \{b \in B \mid \mathcal{L}|_{L_b}$ has a section s_b satisfying $\nabla s_b|_{L_b} = 0\}$. In the case of $\bar{\mathcal{S}}_g$ (equipped with a line bundle $\mathcal{L}$ with connection ∇ for which the curvature is equal to the symplectic form ω, so that $\mathcal{L}^k$ corresponds to the symplectic form $k\omega$ if $k \in \mathbb{Z}$) we have that $\Lambda = \Xi \cap B$, where Ξ is a certain sublattice of $((1/k)\mathbb{Z})^{3g-3}$. We define a vector space $\mathcal{H}_r(\Sigma^g)$ with a basis given by the sections s_b lying above the points $b \in \Lambda$; this is the vector space arising from quantization in the real polarization. (See section 3 for the construction of the covariant constant section over one fibre L_H in the real polarization of $\bar{\mathcal{S}}_g$.)

(d) Half density quantization

If V is an n-dimensional vector space, a *half density* on V is a map $\rho : V^n \to \mathbb{C}$ such that if E is a linear automorphism of V, then

$$\rho(Ev_1, \ldots, Ev_n) = \sqrt{|\det E|}\, \rho(v_1, \ldots, v_n). \tag{2}$$

We now suppose in addition that each fibre L_b, $b \in \Lambda$ is equipped with a half density, i.e., a map $\rho_b(x) : (T_x L_b)^n \to \mathbb{C}$ satisfying (2) exists for every $x \in L_b$. We define the quantization $\mathcal{H}_r$ to have a basis corresponding to the Bohr-Sommerfeld set Λ:

$$\mathcal{H}_r \cong \oplus_{b \in \Lambda} \mathbb{C} s_b \otimes \rho_b.$$

Introduction of the half densities enables us to put the structure of an inner product space on $\mathcal{H}_r$. (See section 4 for the construction of the relevant half density on the fibre L_H in $\bar{\mathcal{S}}_g$.)

(e) Pairings of half densities

We shall wish to define a pairing $\mathcal{H}_r \otimes \mathcal{H}_r \to \mathbb{C}$, in other words an inner product on $\mathcal{H}_r$. This is obtained as

$$\langle\langle s_b \otimes \rho_b, s_{b'} \otimes \rho_{b'} \rangle\rangle = \begin{cases} 0 & \text{if } b \neq b'; \\ \int_{\pi^{-1}(b)} \langle s_b, s_{b'} \rangle \rho_b \rho_{b'} & \text{if } b = b'. \end{cases}$$

More generally, suppose that $\pi : M \to B$ and $\pi' : M \to B'$ are two different polarizations of M, with corresponding quantizations $\mathcal{H}_r$ and $\mathcal{H}_r'$. We shall then define a pairing $\mathcal{H}_r \otimes \mathcal{H}_r' \to \mathbb{C}$, which will be used to construct our three-manifold invariants. See section 5 for the general construction, and section 6 for its application to $\bar{\mathcal{S}}_g$ which yields three-manifold invariants.

3 COVARIANT CONSTANT SECTION OF THE LINE BUNDLE OVER ONE LEAF

(a) Within $\bar{\mathcal{S}}_g$ we define the submanifold

$$L_H = \{A \in \mathcal{A}_F(\Sigma^g) \mid A \text{ extends to a flat connection } \tilde{A} \text{ on } H^g = \partial \Sigma^g\,\}/\mathcal{G}(\Sigma^g).$$

This has a natural identification with the space of representations into G of the handlebody fundamental group, which (since H^g is homotopy equivalent to a wedge of g circles) is just the free group on g generators:

$$L_H \cong \operatorname{Hom}(\pi_1(H^g), G)/G = G^g/G.$$

(b) To construct a covariant constant section of the line bundle $\mathcal{L}^k$ over L_H, we must recall the construction of the line bundle given in ref. 5, using the *Chern-Simons cocycle*. For connections $\tilde{A}$ on *closed* three-manifolds N, the Chern-Simons functional

$\operatorname{CS}(\tilde{A}) \in \mathbb{R}/\mathbb{Z}$ is gauge invariant. This is however no longer true if $\partial N = \Sigma^g \neq \emptyset$: in this case we have for $\tilde{A} \in \Omega^1(N) \otimes \mathfrak{g}$ and $\tilde{g} \in \mathcal{G}(N) = \operatorname{Maps}(N, G)$,

$$e^{i\operatorname{CS}(\tilde{A}^{\tilde{g}})} = e^{i\operatorname{CS}(\tilde{A})}\Theta(A, g), \qquad (3)$$

where the *Chern-Simons cocycle* $\Theta(A, g) \in U(1)$ depends only on $A = \tilde{A}|_{\Sigma^g}$ and $g = \tilde{g}|_{\Sigma^g}$. One sees from this (see ref. 5) that $e^{i\operatorname{CS}(\tilde{A})}$ is not naturally a complex number but rather an element of the fibre $\mathcal{L}_{[A]}$ of the line bundle $\mathcal{L}$ above $[A] \in \bar{\mathcal{S}}_g$. One may define a section of $\mathcal{L}^k$ over L_H given by

$$s([A]) = e^{ik\operatorname{CS}(\tilde{A})},$$

where $\tilde{A}$ is any extension of A to a flat connection over H (which exists by definition of L_H). The section s turns out to be covariant constant with respect to the natural prequantum connection ∇ on $\mathcal{L}^k$. Now L_H is a leaf $\pi^{-1}(b)$ of the real polarization of $\bar{\mathcal{S}}_g$ introduced above; hence we see that it is indeed a Bohr-Sommerfeld leaf, i.e., $\pi(L_H) = b_H \in \Lambda$.

4 CONSTRUCTION OF THE HALF DENSITY

Recall the following:

Lemma 1 *Suppose we have an exact sequence*

$$0 \longrightarrow V_1 \longrightarrow V_2 \longrightarrow V_3 \longrightarrow 0$$

of vector spaces, and suppose that each V_i is equipped with a half density. Then a choice of half densities on any two of the V_i determines a half density on the third.

Now we observe that there is an exact sequence

$$0 \to H^0(H^g, \mathfrak{g}_{\tilde{x}}) \to \mathfrak{g} \xrightarrow{d} \oplus_{j=1}^g \mathfrak{g} \to H^1(H^g, \mathfrak{g}_{\tilde{x}}) \to 0 \qquad (4)$$

for any $x \in L_H$. Here, $H^i(H^g, \mathfrak{g}_{\tilde{x}})$ denotes the i'th cohomology group of the handlebody H^g with respect to the differential given by a representation $\tilde{x} : \pi_1(H^g) \to G$ in the equivalence class x. Using this sequence, one sees that the natural half density on $\mathfrak{g}$ (obtained from the Killing form) determines a half density ρ_{b_H} on L_H.

We have thus established a natural assignment

$$v(H^g) = s_{b_H} \otimes \rho_{b_H} \in \mathcal{H}_r \qquad (5)$$

for the element of $\mathcal{H}_r$ corresponding to the handlebody H^g under the axioms of topological field theory.

5 PAIRINGS OF HALF DENSITIES

We now explain how to construct a pairing $< v(H^g), f_*v(H^g) >$ of the elements $v(H^g) = s_{b_H} \otimes \rho_{b_H} \in \mathcal{H}_r$ (corresponding to L_H) and $f_*v(H^g) \in \mathcal{H}'_r$ (corresponding to the image of L_H under the action of f on $\bar{\mathcal{S}}_g$). This pairing, which must exist according to the axioms of topological field theory, will enable us to construct three-manifold invariants.

(a) Suppose we have two Lagrangian subspaces V_1, V_2 of a symplectic vector space (X, ω). We then obtain an exact sequence

$$0 \longrightarrow V_1 \cap V_2 \xrightarrow{i} V_1 \oplus V_2 \xrightarrow{j} X \longrightarrow Y = \operatorname{Cok} j \longrightarrow 0, \qquad (6)$$

where $i(v) = (v,v)$ and $j(v_1, v_2) = v_1 - v_2$. The symplectic form induces a pairing $Y \cong (V_1 \cap V_2)'$ of Y with the dual of $V_1 \cap V_2$.

(b) Suppose in addition that V_1, V_2 are equipped with half densities ρ_1, ρ_2. There is a natural half density Ω on X associated to the symplectic form: thus we get a natural half density on $(V_1 \cap V_2) \oplus Y'$, i.e. a *density* $\sigma(\rho_1, \rho_2, \Omega)$ on $V_1 \cap V_2$.

(c) Suppose L_1, L_2 are two Lagrangian submanifolds in a symplectic manifold M, equipped with half densities ρ_1, ρ_2 and covariant constant sections s_1, s_2 of the prequantum line bundle $(\mathcal{L}, \nabla)$. The sequence (6) is then defined for any $V_i = T_x L_i$, if $x \in L_1 \cap L_2$, and $X = T_x M$. Under appropriate hypotheses, we may then define a density on $L_1 \cap L_2$:

$$(s_1 \otimes \rho_1, s_2 \otimes \rho_2) = \langle s_1, s_2 \rangle \sigma(\rho_1, \rho_2, \Omega). \tag{7}$$

6 THREE-MANIFOLD INVARIANTS

Consider a closed three-manifold N given as $N = H^g \cup_f H^g$, where $f : \Sigma^g \to \Sigma^g$ is a diffeomorphism. We shall investigate the considerations of the previous section for the case where $L_1 = L_H$ and $L_2 = f_* L_H$ is the image of L_H under the corresponding element of the mapping class group, and see that the quantity (7) has the form (1).

We now recall several properties of the *Reidemeister-Ray-Singer torsion*, which will enable us to show that our pairing gives rise to the square root of the torsion. Associated to a manifold N equipped with a flat G connection A there is a chain complex $(\Omega^*(N) \otimes \mathfrak{g}, d_A)$ and hence cohomology groups $H^i(N, d_A)$, where if $\alpha \in \Omega^p(N) \otimes \mathfrak{g}$, $d_A \alpha = d\alpha + A\alpha - (-1)^p \alpha A$. The Reidemeister-Ray-Singer torsion $\tau(N, A)$ takes values in *$\otimes_{i \text{ even}} |\Lambda| H^i(N, d_A) \otimes \otimes_{i \text{ odd}} |\Lambda| H^i(N, d_A)'$. It has the following properties:

1. Suppose a manifold N is formed by gluing together N_1 and N_2 along a common boundary Σ. Suppose also that A is a flat connection on N. Then the cohomology groups $H^i(N, d_A)$, $H^i(N_\alpha, d_A)$, $H^i(\Sigma, d_A)$ are related by the Mayer-Vietoris sequence. Under this identification, the torsion satisfies

$$\tau(N, A) = \tau(N_1, A) \tau(N_2, A) / \tau(\Sigma, A). \tag{8}$$

2. Suppose A is an (irreducible) flat connection on the closed surface Σ^g. Then the tangent space $T_{[A]} \tilde{S}_g$ is identified with $H^1(\Sigma^g, d_A)$, which is the only one of the groups $H^i(\Sigma^g, d_A)$ which does not vanish. The torsion $\tau(\Sigma^g, A)$ then identifies naturally with the density coming from the symplectic form.

3. Suppose $\tilde{A}$ is an (irreducible) flat connection on H^g, which restricts on Σ^g to the flat connection A. Then $H^1(H^g, \tilde{A}) = T_{[A]} L_H$ is the only nonvanishing cohomology group, and the square root of the torsion is equal to the half density defined above.

In this situation, the exact sequence (6) is just the Mayer-Vietoris sequence. The three properties listed above suffice to show that if $\tilde{A}$ is a flat connection on $N = H^g \cup_f H^g$, then

$$\sigma(\rho_1, \rho_2, \Omega)([\tilde{A}]) = \tau^{\frac{1}{2}}(N, \tilde{A}), \tag{9}$$

which is a density on $H^1(N, d_{\tilde{A}})$.

---

*For an n-dimensional vector space V, we denote by $|\Lambda|(V)$ the space of densities on V, which is defined (cf. the definition of half densities above) as the space of functions $\phi : V^n \to \mathbb{C}$ satisfying $\phi(Ev_1, \ldots, Ev_n) = |\det E| \phi(v_1, \ldots, v_n)$ for all automorphisms E of V.

When the set of conjugacy classes of representations of $\pi_1(N)$ in G is finite, the spaces $H^1(N, d_{\tilde{A}}) = 0$ for all flat connections $\tilde{A}$ on N, and then $\tau^{1/2}(N, \tilde{A})$ and $\sigma(\rho_1, \rho_2, \Omega)([\tilde{A}])$ are just real numbers. In this case the identification (9) of $\tau^{1/2}(N, \tilde{A})$ with $\sigma(\rho_1, \rho_2, \Omega)([\tilde{A}])$ is due to D. Johnson (ref. 4, section XI, Theorem 1).

Under suitable hypotheses, we thus identify the symplectic invariant (7):

Theorem 2 *The density (7) on $\mathcal{A}_F(N)/\mathcal{G}(N)$ is given for $[\tilde{A}] \in \mathcal{A}_F(N)/\mathcal{G}(N)$ by*

$$(s_1 \otimes \rho_1, s_2 \otimes \rho_2)([\tilde{A}]) = \langle s_1, s_2 \rangle \, \sigma(\rho_1, \rho_2, \Omega)([\tilde{A}]) = e^{i\text{CS}(\tilde{A})} \tau^{\frac{1}{2}}(N, \tilde{A}). \tag{10}$$

This density is an invariant of the symplectic manifold N since the torsion and Chern-Simons invariants are known to be independent of the Heegaard splitting. Under some additional hypotheses the density (10) may be integrated over $\mathcal{A}_F(N)/\mathcal{G}(N)$ to yield a $\mathbb{C}$-valued three-manifold invariant which has the form (1).

REFERENCES

[1] Jeffrey, L.C., and J. Weitsman, Bohr-Sommerfeld Orbits in the Moduli Space of Flat Connections and the Verlinde Dimension Formula. *Commun. Math. Phys.*, **152**, (1992) 593.

[2] Jeffrey, L.C., and J. Weitsman. Half Density Quantization of the Moduli Space of Flat Connections and Witten's Semiclassical Manifold Invariants. *Topology*, in press.

[3] Jeffrey, L.C., and J. Weitsman. Toric Structures in the Moduli Space of Flat Connections on a Riemann Surface: Volumes and the Moment Map. Institute for Advanced Study Preprint IASSNS-HEP-92/25; *Advances in Math.*, to appear.

[4] Johnson, D. A Geometric Form of Casson's Invariant and its Relation to Reidemeister Torsion. Unpublished.

[5] Ramadas, T.R., Singer, I.M., and J. Weitsman. Some Comments on Chern-Simons Gauge Theory. *Commun. Math. Phys.* **126**, (1989) 409.

[6] Weitsman, J. Real Polarization of the Moduli Space of Flat Connections on a Riemann Surface. *Commun. Math. Phys.* **145**, (1992) 425.

Index

6j-symbols 12, 15, 31, 47-50, 55, 57-58, 67
Action variables 311
Alexander moves 36, 45-46
Alexander polynomial 25
Alexander theorem 35
Beltrami-differential 170-171
Beraha number 203, 208-210
Bernoulli polynomials 101
Bethe ansatz 95, 99, 188, 206, 211
Birman pair 28
Birman-Wenzl algebra 191, 195
Black holes 159, 160, 162-165, 177-181
Boulatov's model 51, 54-55, 57
BPZ equations 132-133, 137-138
Bratteli diagrams 186-187, 195-197
BRST cohomology 114, 117-118
BRST operator 160, 162-163, 165-166
BRST symmetry 109, 113, 159, 272-273, 275
Calabi-Yau manifolds 143-155
Casimir effect 95
Casimir operator 180, 254, 270, 283, 300, 302
Cell decomposition 36, 37, 38, 39, 75
Central charge 95-97, 101-106, 110, 113, 119, 132-133, 147, 155, 160, 169-170, 173-175
Chern class 75, 144-147, 151-152, 155
Chern-Simons theory 19-20, 27, 110, 269-277, 280, 308
Chromatic polynomial 203-204
Clebsch-Gordan series 31
Clifford algebra 222, 224, 229
Coadjoint orbit 232, 240, 241
Conformal field theory 20, 87, 96, 100, 104, 110, 159, 165, 178, 187, 214, 257-258, 261, 292, 307
Conway group 289, 291
Coproduct 47-48, 215-216
Coset models 109-110, 118, 120, 159-160, 166, 169, 177
Coulomb gas method 133
Coxeter number 23
Dehn twist 24
Dijkgraaf-Witten invariant 31
Dijkgraaf-Witten model 47, 49
Dilaton 165
Dilogarithm 95, 97, 99-100, 101, 103-104, 106
Dirac equation 221-222, 225-226

Donaldson invariants 110
Doplicher-Haag-Roberts 213-214, 248, 249
Drinfeld-Sokolov reduction 123-125, 131
Duistermaat-Heckman theorem 76, 315
Euler characteristic 37
Euler number 52-54, 62, 66, 69, 152-153
Four Colour Theorem 1-2, 5, 11, 90, 204
Fredholm determinant 231-234, 237
Gauss decomposition 128
Gaussian unitary ensemble 231-232, 234
Gepner conjecture 143
GSO projection 147-148
Griess algebra 290
Ground ring 159, 163, 169
Haag-Kastler axioms 249, 256
Haag duality 214-215, 219, 247-249, 257-258
Handlebodies 59, 66, 69
Hard hexagon problem 209
Hartle-Hawking state 61, 63, 67
Hecke algebra 183-184, 189-190, 195
Hodge diamond 144
Hodge number 144-148, 153
HOMFLY polynomial 19-20, 25, 28-29
Hopf algebra 47-50, 193, 195, 213-214, 216-217, 221
IRF invariant 19-20, 24, 26, 28-29
Ising model 213
Jones polynomial 1, 12, 19-20, 25, 27-28, 270, 280
Jones Wenzl projectors 1, 10, 12-13
Kac-Moody algebra 96, 109, 112, 117, 123-124, 290
Kac table 131-132, 135, 139
Kazama-Suzuki action 111
KdV hierarchy 73-75, 110
Klein group 90
Knots 12, 20
Knizhnik Zamolodchikov equation 23, 131-132, 134, 136, 138, 184
Konsevich integral 110
Kontsevich model 73, 75
Kryuchkov conjecture 5
Landau-Ginzburg theory 144-149, 152, 154, 155
Laplace operator 304
Leech lattice 289, 292-293
Lefschetz number 201
Lens space 67, 279-281, 287

323

Lie algebra 21, 101, 123-126, 129, 133, 194-196, 239, 269, 297, 299, 302-304
Liouville theory 110, 159-163, 169, 178-181
Matveev moves 36
Matrix models 53-60, 62, 70, 73-76, 85-87, 91
Minimal models 96
Miura transformation 127
Mirror manifolds 151-152
Modular group 220, 291, 293-294
Moduli space 73, 75, 124, 307-308, 311-312, 314
Moment maps 243, 307, 310-311
Monodromy 19-22, 27, 242, 266, 273-274
Moonshine module 289-290
Newton diagrams 303
Nonlinear σ-model 164
Orthogonal polynomials 59, 74, 86-88, 300
Partition function 1, 15, 19-21, 31, 35-37, 39, 44, 52, 54, 57, 59-60, 67, 73, 75, 80, 83, 87, 89-90, 96-97, 118, 120, 140, 270-271, 280, 292-294
Poincaré duality 153, 282
Potts model 203, 206, 209, 211
Primary fields 96-97, 102, 132-134, 169-174
Quantum gravity 2, 20, 51-54, 73-74, 143
Quantum groups 11, 31, 47, 54-55, 58, 214, 221-222, 226, 229, 261
Quaternions 3-4
R-matrix 216, 229, 232, 239
Random matrices 73, 87-88, 231-233
Ray-Singer torsion 279-281, 321
Reidemeister moves 5
Reidemeister torsion 279, 317-318, 321
Relative cohomology 116
RSOS models 103-104
Schensted construction 203, 211
Schrödinger equation 297-299
Schur polynomials 76-80, 82
Schwinger-Dyson equation 51, 55-57, 59-70
Shadow world 1, 14-15
Simplicial manifolds 54-55, 59, 61
Simplicial quantum gravity 51-54

Singular vectors 124, 132, 137-139
Skein relation 19, 20, 22-28, 207, 270
Spin models 213-215, 218-220
Spin networks 2
String equation 74, 87-88
Sugawara construction 160
Tachyons 159, 162
Temperley-Lieb algebra 1, 5-11, 13-14, 185, 203, 205, 207, 211
Toda theory 123-124, 127,
Topological quantum field theories 35, 47, 109-110, 269, 279-280, 308, 320
Triangulation 31-32, 35-36
Turaev-Viro axioms 31
Turaev-Viro invariant 35, 51, 55, 59, 61
Verlinde formulae 310, 314
Vertex models 85, 87-89, 193, 203, 206
Vertex operator 133, 178, 180, 195-196, 199, 289
Verma modules 132, 137, 139
Virasoro algebra 96, 118-119, 123, 133, 169-170, 172-174
Virasoro anomaly 118-119
Virasoro central charge 113
Virasoro minimal models 134
von Neumann algebra 19, 256
W algebra 96, 123-125, 127, 129, 139, 164
W_n minimal models 119-120
W_∞ algebra 159, 162-163, 169
Wakimoto bosonisation 132, 134
Whitney polynomial 205
Wilson line 20-21, 24-26, 28-29
Wilson loops 270, 273, 280
Witten mapping 76, 81-83
Witten-Reshetikhin-Turaev invariants 1, 11, 14, 59, 308, 317-318
WKB approximation 59
WZW models 101, 131, 133, 135, 140, 165, 177-178, 261-266, 270
XXZ model 94-95, 187-189
Yang-Baxter equation 19, 26-27, 223